D1247671

The New Map of Empire

Territories Settled by Commissaries after the Treaty of Utrecht
The Course of all these Rivers is not discover'd yet
NEW FRANCE
CANADA
Abitibis
Temiscamings
ALGONQUINS or ADIRONDAKS
LAKE SUPERIOR
Messesagues
SIOUX
Eastern Sioux
OUTAOWAS
LAKE HURON
L. MICHIGAN or ILINOIS
IROQUOIS
L. ONTARIO or CATARAQUI
LAKE ERIE or OKSWEGO
Mascoutens
Myamis
Great Meadows
ILINOIS or
TWIGHTWEES
CHIKTAGHEEKS
PENSILVANIA
Endless Mountains
NEW JERSEY
Philadelphia
MARYLAND
Delaware Bay
Shawnoe
VIRGINIA
THE COUNTRY BY THE
Chesapeak Bay
C. Charles
NORTH CAROLINA
Edinton
C. Hatteras
C. Lookout
C. Fear
C. Carteret
CHICASAWS
CATAWBAS
SOUTH CAROLINA
Charles Town
Port Royal
AKANSAS
Cadodaquious
LOUISIANA
Natsitoches
CHACTAWS or CHATTAS
Upper Creeks
Middle Creeks
Lower Creeks
GEORGIA
APALACHES and TIMOOKAS
Extirpated by the Carolinians
Mouths of the Misisipi or St. Louis
Apalaches Bay
Southern Bounds by Charter 1665
Here Ferdinand Soto Landed 1539
FLORIDA
BAHAMA
The Hole in the Rock
GULF OF

THE NEW MAP OF EMPIRE

How Britain Imagined America before Independence

S. Max Edelson

Harvard University Press
Cambridge, Massachusetts
London, England
2017

Copyright © 2017 by the President and Fellows of Harvard College
All rights reserved
Printed in the United States of America

First Printing

Library of Congress Cataloging-in-Publication Data

Names: Edelson, S. Max, author.
Title: The new map of empire : how Britain imagined America before independence / S. Max Edelson.
Description: Cambridge, Massachusetts : Harvard University Press, 2017. |
Includes bibliographical references and index.
Identifiers: LCCN 2016043748 | ISBN 9780674972117 (alk. paper)
Subjects: LCSH: Cartography—America—History—18th century. |
Surveying—America—History—18th century. |
Great Britain—Colonies—America—Administration.
Classification: LCC GA401 .E36 2017 | DDC 526.097/09033—dc23
LC record available at https://lccn.loc.gov/2016043748

For

Jennifer, Benjamin, Leo, and William

Contents

Maps

A Note on the Maps

The maps for this book are available online at http://mapscholar.org/empire.

In 2007–2008, as the Kislak Fellow in American Studies, I examined the Library of Congress's remarkable collection of American manuscript maps. I found few that depicted the long-settled colonies, and many more that focused on the new territories acquired by Britain in 1763. I began the research behind this book to explain the intensive mapping of places such as Nova Scotia, Pensacola, and Dominica in favor of places like Massachusetts, Philadelphia, and Virginia. Out of all proportion to their immediate commercial value and population, imperial officials as well as metropolitan merchants focused their attention on these frontiers in order to gauge the perils and promise of British America. Beyond the sheer number of images of remote rivers, islands, forts, coasts, boundaries, and sea-lanes in North America and the Caribbean, I was also struck by their shared graphic qualities. This common spatial sensibility, reflected in the style, scale, and detail of so many maps of such diverse, far-flung places, originated in the Board of Trade's vision for imperial expansion and reform, articulated in 1763. This body of images emerged from a long-standing view of America that took form in the halls of British power over the course of the eighteenth century and culminated in an unprecedented project of state-controlled colonization.

As I studied the maps laid out for my inspection on the tables of the Geography and Map Division's reading room, I learned to see them as particular, regional expressions of a unified imperial vision of space and power. I also began experimenting with digital tools to share my visual experiences in the archives with a larger audience of readers. Map history has always been limited by the challenges of reproducing dense images printed or drawn on fragile paper. Working within the constraints of print technologies, scholars have privileged a few landmark works of distinction as the true subjects of the history of cartography, leaving aside the messy, scattered, and much larger manuscript record that occupies the pages of this book. The massive push by libraries to preserve collections through digitization has opened access to the rarest old maps, a development that, along with new interest in spatial humanities, is drawing innovation to a traditional field. By making new use of this growing digital map archive, I attempt to leave behind this artificial economy of image scarcity in map history publishing and embrace the abundance of maps now available online.

Mapmakers in the service of the British state produced thousands of images of America in the eighteenth century. I have curated a collection drawn from this enormous output to illustrate shifts in geographic ideas about America before 1763, to create a broad sample of the manuscript maps of the new territories drafted between the Seven Years' War and the American Revolution, and to show how some noteworthy published maps derived from these original images. The text references these maps by number within brackets, and the enumerated Map Bibliography at the back of the book provides full citations, including URL links to online resources maintained by the libraries and archives that hold them. Instead of attempting to publish reproductions of these maps in the glossy pages of a large-trim-size edition, I have posted all 257 enumerated maps in this collection online. The best way to read this book is with a computer screen close at hand so that you can view the maps, plans, and charts mentioned in its pages. The War of Independence relegated thousands of manuscript images from working documents of imperial development to the status of artifacts that have since been deposited at libraries and archives in the United Kingdom and the United States. A primary task of this book is to reassemble a representative sample of this cartographic corpus before your eyes so that you can see how Britain attempted to take command of America and how comprehensive, provocative, and serious this effort was.

I have created seven digital atlases, one for each of the book's chapters, that present these maps in sequence on MapScholar, a web-based HTML5 platform purpose-built for map history research and display. This open source tool makes use of the increasing availability of high-resolution digital copies of historic maps and the capacities of modern web browsers to draw together distributed text, images, and geospatial data into dynamic visualizations. Research Professor Bill Ferster and I created MapScholar at the University of Virginia's Sciences, Humanities, and Arts Network of Technological Initiatives (SHANTI) to enable a new mode of writing and reading map history. These digital atlases display historic maps overlaid onto a global basemap, a visual format that helps us see them less as independent expressions whose meanings are contained, like works of art, by the frames inscribed around their edges and more as parts of a larger effort to generate a body of geographic information through surveying and mapmaking. Such visualizations provide a wider frame of reference for individual images, show multiple maps at the same time, and encourage comparisons between historic and modern representations of space. Presenting maps in this way also draws attention to differences in their scales of representation. Printed reproductions size maps to the space available on the page, a practice that diminishes our appreciation of the sometimes vast distance between maps that depict towns, tracts, fields, and structures at "larger" scales and those maps that depict the broad sweep of colonies, regions, continents, hemispheres, and the globe itself at "smaller" scales. The lexicon of British mapmaking during this generation focused on both ends of this spectrum of geographic representation. This pattern becomes visible by witnessing the jarring transitions between higher-resolution maps of small spaces and lower-resolution maps of large spaces, an experience simulated by following the navigation prompts programmed into MapScholar's digital viewer. The importance of scale in the spatial representation of American territories is thus a central concern of this book, and viewing the maps as objects that bear a defined relation to geographic space, as they appear on the website, underscores in visual terms the importance of scale to Britain's attempts to recolonize America.

By putting these maps online, organized and annotated to draw the reader's attention to details and contexts in high resolution, I hope to encourage an eighteenth-century mode of map viewing in which understanding (as well as pleasure) was to be gained by becoming immersed in mapped space, locating places described in texts, and forming views of unfamiliar parts of the world in the mind's eye. As you read this book, I encourage you to open these digital atlases and flip through their pages in sequence. Clicking on hyperlinked text and icons on any page will zoom in on details mentioned in the text and open supplemental material.

In anticipation that this resource will, inevitably, degrade, I have posted the basic map images in more durable, if less vibrant, online repositories and appended a list that will allow future readers to locate the original maps, many more of which will be published online in the future by the archives that hold them. This book relied on a number of important manuscript charts housed at the UK Admiralty Library in Portsmouth, England. Because these were unavailable for reproduction, most of these images were excluded from the Map Bibliography as well as the digital atlases created to illustrate this book. They are cited, where appropriate, in the notes.

The New Map of Empire

Introduction

During negotiations to end the Seven Years' War, Great Britain's diplomats used the leverage that came with conquests in Canada, India, Africa, and the West Indies to gain large territorial cessions from France and Spain. After the terms of the Peace of Paris went into effect on February 10, 1763, colonial British America extended from Hudson Bay to the Florida Keys, from the Atlantic coast to the Mississippi River, and across the Caribbean—at least on paper. After the ink had dried on the treaty, the king and his Privy Council charged the Lords Commissioners of Trade and Plantations (commonly known as the Board of Trade) to propose a system for managing Britain's new dominions. In response, the Board of Trade put forward a detailed plan for the new territories' occupation, development, and defense. It called for settlement across the mainland's coastal plain and in the islands, command over coastal and Caribbean navigation, and a limit to colonization in the North American interior. More ambitious yet, the Board's plan described how settling these acquired lands in improved ways would reshape British America as a whole to better secure its dependency. To support this initiative, it called for the comprehensive mapping of British America, beginning with new surveys of the edges of this enlarged empire to affirm possession of long-contested frontiers, accumulate strategic knowledge in anticipation of future French and Spanish intrigues, and implant self-sustaining settler societies.

From 1763 to 1775, agencies of the imperial state worked within the parameters of the Board's plan to secure these scattered places, dispatching surveyors to mark new boundaries, lay out forts, chart coastlines, and divide the land for settlement. Commanded to map these new territories to rigorous scientific standards, the surveyors brought the resources of a rising global empire to bear on the task of better understanding American lands and waters. They documented their discoveries with ink and paper, composing maps in the field and dispatching them to London to report on the progress of this mission to prepare American spaces for intensive, regulated colonization. In *The New Map of Empire,* I interpret a visual archive of hundreds of manuscript maps produced by these surveyors in the 1760s and 1770s. These images, considered as a cartographic corpus, describe British America transformed. This book explains the vision behind this vast project of improvement, the expeditions it set in motion, and the meanings of the images these geographic surveys produced in the generation before the American Revolution.

Across five zones—the Gulf Coast, the maritime northeast, the trans-Appalachian interior, peninsular Florida, and the southeastern Caribbean—surveyors examined places that, with the notable exception of the colonial frontier with Native North America, historians have seen as marginal to the history of early America. Their mission was to fill a prospective cartographic archive in the imperial metropole that could describe America comprehensively. The Commission for the Sale of Lands in the Ceded Islands surveyed Dominica, St. Vincent, Grenada, and Tobago to accelerate the division and sale of plantation tracts as part of a larger scheme to expand slavery, revenue, and security in the British West Indies. For more than a decade following the Royal Proclamation of 1763—George III's order to establish new colonies and regulate land grants in the acquired territories—Britain's superintendents of Indian affairs convened congresses with indigenous spokesmen to determine the specific location of an Indian boundary line, which promised to regulate commerce and settlement in the interior. The General Survey of North America, along with Admiralty and army surveying expeditions, worked to promote colonization; preserve resources; and open strategic areas to scrutiny in Quebec, the Gulf of St. Lawrence, Nova Scotia, Maine, East Florida, West Florida, and along North America's great rivers.

The thousands of maps, plans, and charts these surveyors drafted document how Britain attempted to take command of eastern North America and the West Indies after the Seven Years' War. At the heart of this new vision was a working philosophy of empire informed by ideas about the nature of interest, government, and social development that aimed to connect far-flung sites of occupation through the integrative power of commerce. This political economy of empire held that overseas territories could be secured in the short term by settling British people on the land and giving them the means to develop it through transatlantic trade and, in the long term, by using the imperial state's control over the distribution of land to shape these territories into model colonial dependencies.

The Board of Trade had long sought greater control over haphazard colonial expansion and hoped to put a stop to the violence and instability that it caused. From the time of the Board's inception in 1696, its commissioners, secretaries, advisers, and clerks took up the task of thinking broadly about the empire as more than just a collection of individual colonies. Long before the Peace of Paris, they tabulated British America's rising value to the nation's economy and tallied the many threats that could lead, after a history marred by smuggling, encroachment, speculation, and insubordination, to the loss of this valuable western empire. By the second decade of the eighteenth century, they cultivated a vision of American colonies joined together as parts of larger regions defined by latitude and economy; noted surging populations and increasing integration across and beyond the continental seaboard; and drew sharp contrasts between these growing mainland settler societies and the lucrative but vulnerable British West Indies. They envisioned Britain's western empire as an interdependent commercial system nestled within a much larger space that was controlled by rival empires and populated by Indians. Such proximity posed serious strategic threats to British America, but it also created prospects for a new era of dominance in the American trades. In fits and starts but with persistence across a century disrupted by transatlantic warfare, the Board of Trade pursued a program of reform that followed from these assumptions. The Crown's charge to the Board to design a plan for dividing and developing new territories provided a golden opportunity to impose order on a disordered colonial world.

Wartime mapmakers produced an array of images of North America and the West Indies that described previously obscure places and structured attempts by the British to establish authority over them. All of the senior surveyors appointed to map the new territories honed their craft during the conflict and brought a distinctive military eye to the images they produced from 1763 on. Their maps envisioned an empire that, shattered by the War of Independence, never came to be. The fruits of this effort were in fact used to break the empire apart. When war came to North America and the Caribbean, military hostilities disrupted these surveys and ended this great imperial mapping project, as well as the vision of expansive empire that was behind it. Some surveyors in the field were captured and imprisoned by Continental forces, while others were reassigned to map strategic locations in the settled colonies as British commanders prepared their battle plans. Copies of a pocket-sized military atlas derived from these surveys were distributed to officers before they sailed west to invade rebellious colonies. After the 1783 Treaty of Paris, Britain was forced to abandon the idea of a settler empire that featured a contiguous band of territory spread across the Atlantic Ocean's northwestern edge. The maps endured, however, as a new foundation for geographic knowledge about America. Even as the vision behind them fell apart, they were transformed into magnificent works of published cartography. In the voluminous printed sheets of *The West-India atlas, The Atlantic Neptune, The American Atlas,* and other works, images drawn from these surveys presented breathtaking new views of a lost Atlantic world. Imperial surveyors created more images of North America and the West Indies in this single generation than all of those who had taken up compasses, chains, and plane tables in the previous two centuries of British New World exploration and colonization. Repurposing old maps to describe a new nation, Americans reimagined the same images to see an expansive continental empire in which the restrained energies of settler colonialism would be unleashed and redirected toward the west.

The most important recent insight regarding the history of cartography is that maps, far from being dispassionate pictures, are created by people with agendas for how to represent the world. Mapmakers—then and now—shape their images with ideas about who has the right to command geographic space, the uses to which it should be put, and the qualities of the people who inhabit it. The maps discussed in this book are full of such meanings, and their value as sources rests on their richness as cultural as well as empirical objects. It is clear, however, that these maps of diverse and scattered places are like pieces of a vast puzzle whose overarching significance can only be revealed by bringing them together and seeing the body of knowledge they form. Once united, they reveal a coherent British vision of western empire that sought to displace settlers as meaningful agents in favor of a regulated system of colonization orchestrated from London.

A century of engagement with France, Spain, and indigenous peoples in North America and the West Indies convinced British thinkers that their productive and populous colonies faced ruin because they lacked a secure geographic position from which to expand. Although the Seven Years' War was triggered by contention over which power possessed the Ohio River valley, Britain broadened the meaning of the peace that concluded this conflict by conceiving of a plan to secure the postwar empire along its many unstable frontiers. The prospect that unregulated territories might be assimilated by rival powers convinced imperial officials that they should govern these scattered colonies—from the fishing stations at Newfoundland to the sugar plantations of the Caribbean—as parts of one system. The prewar fear of French encroachment was explicitly about space and who could possess it, yet imperial thinkers also focused on the problem of time and the challenges that stood in the way of reproducing colonial societies, economies, and populations into the future. The violence that surged across every contested imperial frontier between the Glorious Revolution of 1688 and the Peace of Paris underscored the fundamental spatial disorder that threatened the integrity of Britain's American empire.

Imperial reformers created settlement systems in each of the new territories to improve on a prevailing mode of colonization. On the theory that frontiers should be secured before waves of settlers advanced inland from the coast, the army opened corridors of settlement along West Florida's rivers, stretching between interior forts and Gulf of Mexico port towns. The Board of Trade's commissioners accused New England merchants of plundering the maritime northeast in search of short-term profits, leaving it unsettled and undefended. As the centerpiece of their plan to take possession of the Gulf of St. Lawrence for Great Britain, they apportioned the entire island of St. John into sixty-seven 20,000-acre townships. The Privy Council formally granted these early modern fiefdoms to metropolitan insiders on the condition that they draw settlers to their lands to inhabit new farming and fishing communities. The planters of the British Caribbean became a magnificently wealthy elite as they reaped the profits from their sugar estates, monopolized the best lands, and worked slaves to death in brutal conditions. An impartial auction distributed planting land to investors in the Ceded Islands who paid a premium to the Treasury for the opportunity to claim fertile new acres for sugar. Lowcountry planters in South Carolina and Georgia had established powerful family dynasties at the price of overpopulating the southeast with restive African slaves and acquiring far more land than they could cultivate. In East Florida, the Board encouraged township proprietors to experiment with indentured servant labor as well as new commodities, and surveyors laid out tracts along the full length of the St. Johns River to organize rapid plantation settlement. To secure the volatile North American Indian frontier, the Board persuaded the king to prohibit unregulated western expansion behind the settled colonies and recognize the territories of indigenous nations. In each of these theaters for reformed colonization, geographic surveys and mapmaking enabled Britain to begin settling new colonies without having to delegate the task to a cadre of autonomous founding colonists who, if history was any guide, would shape the trajectory of development to suit their own interests regardless of the consequences.

What these maps mean depends on what we think they are. Among other things, these maps were the source materials from which commercial mapmakers engraved images for publication and broad public consumption; they were a form of discourse composed to persuade viewers of an ideological message, especially about claims of sovereignty; they established routes and relationships across space as they recorded the findings of organized journeys of conquest, exploration, and survey; they marked abstract boundaries across visible landscapes; they terrified and tantalized, making fears and aspirations palpable by giving them geographic form; they were technical documents produced by the state to gather instrumental knowledge of places in which it had a material interest, particularly regarding defense and landownership; and they served as patronage performances drafted by their makers to demonstrate their expert knowledge of place in hopes of preferment. Above all, these maps represented knowledge to demonstrate authority over American places. This book is not, strictly speaking, about the making, dissemination, and consumption of maps; it is, rather, a spatial history of empire that focuses on maps because they enabled British officials to see distant lands in high resolution after the Seven Years' War, a capacity that emboldened them to take command of new colonial territory directly from London in new ways—and with new purpose.

The Board of Trade was a comparatively small, undernourished, but influentially placed committee, which oversaw British America by generating and controlling information. This challenge of exerting power outward, across vast reaches of space, preoccupied Britain's army and navy as well, both of which developed a deep administrative infrastructure in the second half of the eighteenth century that mobilized people, resources, expertise, and technology to project force into the Atlantic Ocean and around the world. Historians have failed to appreciate the scope of the Board of Trade's plan, in part because few scholars have seen the voluminous body of maps it produced gathered in one place, which by its sheer volume and breadth testifies to the seriousness of Britain's intentions. This tendency to dismiss imperial reform also reflects a number of assumptions about American colonists and British officials that this book challenges. Historians of the imperial school have generally regarded new colonial regulations as inoffensive and necessary for the proper administration of a stable overseas empire. Those who have argued that colonial protest emanated from an unsubstantiated fear of centralized power, steeped in an eccentric strand of radical English republicanism, have likewise emphasized the benign intentions of Britain's American policies. This view underestimates not only how provocative these policies were to colonial sensibilities but also how unprecedented they were as attempts to extend the power of a central government into the economic lives of its subjects. Although Americans were famously undertaxed compared with Britons at home, they had generally enjoyed the freedom to claim land where they chose and develop it as they wanted, and this landed liberty was at the center of their identities as settlers and subjects. Britain stepped between colonists and American land in the years before the American Revolution by prohibiting the legitimate ownership of interior tracts, imposing new costs and restrictions on obtaining grants in the new territories, favoring metropolitan insiders over American elites in the distribution of new lands, and threatening to extend such oversight to the settled colonies. Moreover, the explanations behind the royal proclamations, acts of Parliament, and modes of enforcement that colonists objected to, voiced by senior officials as well as writers who published their commentaries in widely circulating pamphlets, stigmatized colonists as selfish, reckless, and obstinate, and dismissed what the Americans regarded as their greatest historical accomplishment—the occupation and improvement of so much land with so many people—as a looming disaster for Britain. The imperial state's intrusion into the process of colonization and the language used to justify it denigrated colonists' stature within the British Atlantic world and threatened their future as meaningful agents of empire.[1]

In their attempts to chart the complex struggles for power among shifting factions in eighteenth-century Britain, historians have downplayed the Board of Trade's role in imperial reform. They have focused on the Earl of Halifax's robust leadership of the Board in the 1740s, capped off by the settlement of Nova Scotia, as the apogee of its influence. The revocation of its power to appoint colonial officials, as well as the failure of Halifax's campaign to secure a seat on the Privy Council for its president, seemed to indicate its decline into irrelevance. Using appointments to high offices to track the winners and losers in this roiling pageant of competing patronage networks and shifting political alliances ignores the practical power that came with collecting information, determining its meaning, and articulating policies based on expert knowledge. Although many of its initiatives failed to bear fruit, the Board exercised such power both through its constant communication with colonial officials over mundane as well as extraordinary affairs and through its many reports and proposals to the secretary of state for the Southern Department and the king-in-council, many of which gained approval with minimal revision. By the turn of the eighteenth century, the Board of Trade was codifying information that flowed from the outer reaches of a vast, discontinuous maritime realm on which most believed Great Britain's fortunes depended. As it calculated the stunning rise in the value of overseas colonies to national wealth, it assumed a place at the center of a nexus of data about empire that gave its recommendations enormous weight in councils of state. The Board of Trade's ability to use this command of vital information about the Atlantic world to effect change oscillated over the course of the eighteenth century with the outbreak and negotiated conclusions of three major wars. It was the "peacetime consigliere" of Britain's Atlantic empire, declining in influence during wartime but rising in stature after the conclusion of conflicts, when it came time to organize territorial gains and address the vulnerabilities the fighting exposed. Few historians have questioned the view that the Board generated many words but few meaningful actions when it came to the colonies. I contend that not only did the Board of Trade matter, but its new conception of America in 1763 changed the course of empire.[2]

Because British ministries changed hands so frequently between 1762 and 1783, the resulting political turmoil generated inconsistent legislative policies toward America, which lacked any coherence. Political leaders rose and fell in brisk succession during this tumultuous period. The different heads of short-lived ruling coalitions—Bute, Grenville, Rockingham, Chatham, Grafton, North, and Shelburne—along with their allies and followers, expressed, in word and deed, different views of taxation, regulation, and the proper response to increasingly intense and coordinated colonial protests. Despite this range of stances—some sympathetic to American concerns, others hostile to them—all agreed on the necessity of establishing new colonies with greater metropolitan oversight, securing America's economic and political dependence on Great Britain, and forestalling any moves toward independence. A number of prescriptions featured in the Board's plan in 1763—including the jurisdiction of the trans-Appalachian northwest, the administration of commerce with Native Americans, and the deployment of soldiers in remote forts and garrisons—changed over this period. Much of the Board's work was transferred in 1768 to the new secretaries of state for the colonies, the Earl of Hillsborough (1768–1772) and the Earl of Dartmouth (1772–1775), but these were the same men who had headed the Board for much of the 1760s and who continued to see America through the lens of the Board's long-standing mission for reform. Even when the king named George Germain colonial secretary in 1775 to manage Britain's war against America from the Colonial Office, surveyors continued to draft maps and charts, and new colonists obtained land in the new territories on restrictive terms. Although disrupted by the war, the core vision first articulated by the Board of Trade under the Earl of Shelburne in 1763 persisted until Shelburne, as prime minister, oversaw the negotiations that ended the War of Independence twenty years later.[3]

Mapmaking created images of American places that represented what those places were, how they came to be, and what they should become. Although this book cannot do justice to the full historical experiences of European, African, and indigenous peoples who inhabited the territories that were formed into colonies at the frontiers of this empire, understanding Britain's surveys of them helps us view these places less as fractured, underpopulated frontiers and more the way the Board's commissioners saw them when they first sketched British America's new boundaries. By joining these disparate places together on a new map of empire, they imagined a way to define and secure them as growing dependencies. These maps were designed to make distant places visible and bring the process of their settlement and development directly within the purview of officials in London. These schemes for new colonies, which could bind the whole of the British Atlantic system into a new order, rested on the idea that this space could be revealed, and thereby administered, by mapping it.

Modern states have used technologies of information gathering to implement broad reformist schemes, from harvesting forests to building cities. To represent the nation in ways that prepared it for regulation, taxation, and social control, technocrats reduced its complexity to salient elements that could be managed by sweeping policies. The "crowning artifact of this mighty simplification," James C. Scott has argued, "is the cadastral map," a document that lists landowning inhabitants and displays the bounds of their individual properties. To make such maps of an entire nation legible required the elimination of extraneous detail as well as significant abstraction. This resulted in a restricted view of land and people that saw them solely in terms of the state's interests. Socio-ecological systems became board feet of lumber, bodies to be housed, and acres to be taxed. Broad official consensus about the soundness of the aims of rational control and human improvement gave these visionaries access to considerable resources and the power to use them. Their lofty schemes, however, failed to account for the importance of varied and particular conditions, and their techniques of information gathering and analysis ignored the local knowledge of place that had been built up at the scale of human experience.[4]

Great Britain's attempt to map the new territories of its American empire—perhaps the first effort to grasp the potential of using geographic information to enact a new kind of colonial rule—anticipated the totalizing schemes in which modern states put such information to use. In the eighteenth century, however, Britain lacked the resources to gather data comprehensively as well as the administrative capacities to store, analyze, disseminate, and access the information it did accumulate. And it proved unable to reconcile ideas of place that drove policy at the broadest scale with the particular information its surveyors returned, especially when these particularities conflicted with the general vision. Britain enlarged its command of fiscal resources, forged a political consensus about the need to improve colonial administration, and demonstrated the ability to project military force to the ends of the earth, but even this most powerful of early modern states lacked the capabilities of its modern successors to transform whole regions to exacting new standards. As a result, the attempt by the Board of Trade to control colonization from London through the rigorous representation of land and space differed from these twentieth-century social experiments in a number of ways. The Board's commissioners shared the precept of early modern political economy that the state lacked the power to alter fundamental interests, which drove behavior from a deep structural level of society that was, to a large degree, beyond its reach. Instead of forcing an idealized way of work and life on British subjects abroad, the Board imposed conditions on the granting of the king's conquered and ceded lands, guiding colonists by these restrictions to act in ways that it believed would benefit Britain. In this way, its colonization schemes sought to create a new relationship between the settler and the state, which was an extension of the negotiated governing practices that had shaped colonial America from its inception. In the end, it depended on colonists to act in ways approved by the state. Britain's effort to map America was also profoundly empirical, as concerned about revealing the distinctive qualities of places as it was in creating generalized schemas. Such practical empiricism was part of a long tradition of English inquiry that respected tried-and-true methods, particularly in agricultural science. As applied to the challenge of mapping America, this openness to vernacular knowledge also reflected the utter lack of durable information about the places under scrutiny. Instead of a confident, top-down imposition of a rigid technocratic order, Britain's scheme for the new colonies was highly prospective and rested on assumptions that the surveys were commissioned to examine. Britain launched its plan before it possessed the knowledge needed to execute it, leaving it open to new conceptions and prone to failure.[5]

The War of Independence fatally disrupted the surveys and the schemes they supported only twelve years after they began, making it difficult to assess them. In this brief time, Britain negotiated a continental boundary line with Native American nations, oversaw the settlement of nascent colonies of fishermen and farmers on St. John Island and planters and slaves in East Florida, opened routes of river and coastal navigation, established a sugar plantation economy in Grenada, and occupied a number of strategic military sites. At the same time, it failed to prevent squatters from encroaching on indigenous lands, relinquished control over Indian commerce to provincial governments, fought a bloody war of extermination to a stalemate against St. Vincent's Black Caribs, struggled to establish a viable settler society in West Florida, and withdrew soldiers from a number of recently occupied forts and garrisons in the North American interior. Judged against the Board of Trade's lofty vision, these disappointing outcomes resulted in part from the outsized ambitions of an overreaching state.

Britain succumbed to the predictable hubris of modernizing states, so confident in their technologies of governance that their expectations for order almost inevitably fracture when faced with unforeseen contingencies, especially when people resist schemes that they believe aim at their subjugation. Its struggle to impose a particular spatial order on America generated its own undoing, which went beyond a contribution to the general discontent that drove colonial protest. Like Narcissus, the British state beheld the geographic images it produced and was beguiled by them. In fact, George III became so enamored of the maps dispatched from America that he plundered Whitehall offices to stock a personal archive, now the King George III Topographic Collection at the British Library. Ambitious officials soon learned that making elegantly drafted presentation copies of surveyors' manuscripts was a good way to gain an audience before the king. But like the monarch surrounded by his maps, which seemed to open even the most remote coasts to view, Britain failed to grasp America's vastness and the enormous distance between the general scale of sovereign claims, on the one hand, and the particular scale of land as it was occupied, on the other.

The "new map of empire" is both a way to describe this plan to reform British America and a real cartographic object: the first map the Board of Trade used to visualize it. Robert Sayer's Fleet Street press produced, in all likelihood, hundreds of copies of Emanuel Bowen's *An Accurate Map of North America Describing and distinguishing the British, Spanish and French Dominions on this Great Continent According to the Definitive Treaty Concluded at Paris 10th Feb[ruar]y 1763* (1763). The Board purchased one of these, marked it up to illustrate the geographic plan it proposed, and annexed the image to its official report to the king, which it submitted on June 8, 1763. This map can be retrieved at shelfmark MR 1/26 from its sturdy metal cabinet at the National Archives of the United Kingdom at Kew. Across its surface, lines and washes of color cut across the latitudes that defined original colonial charters and prewar provincial boundaries. These annotations marked spaces along the Atlantic watershed to which mainland settler societies were to be confined, compressed the formerly expansive provinces of New France and Spanish Florida, highlighted certain places earmarked for intensive colonization, and set off others as off-limits to legal settlement. The Board of Trade's commissioners made visible a transatlantic system of political economy on this map and assumed that it could take hold within the spaces the map represented.

As geographers who have wrestled with the problem of scale have recognized, maps tend to perpetuate the false notion that scale is a natural property of space. The idea that cartographic space exists absolutely to contain territories, populations, economies, and societies made it particularly attractive to imperial reformers, who sought to establish a new standard of metropolitan command over the colonial periphery. Part of the value of cartography for administering empire was that its absolute scale of measurement contains a clear hierarchy of relationships that shows precisely how smaller places are embedded within larger ones. But this idea of space as infinitely reconcilable is an illusion. In reality, the nature of space changes at different levels of resolution. Those who neglect this characteristic of scale—like Britain's centralizing, mapmaking, senior colonial officials—and "use data at one scale to make inferences about phenomena at other scales," commit what geographers call a "cross-level fallacy." When it assumed that its macro-level judgments about economy and society for each of British America's regions would persist at the micro level of the individual tract of land, the Board of Trade committed the cross-level fallacy known as "scaling down." Benjamin Franklin ridiculed Josiah Tucker for making this mistake when the famed political economist proposed that Britain clear a strip of land a mile wide across the North American frontier to prevent Indian attacks, as if such an impossible task could be performed as simply as drawing a line across a paper map.[6]

When it called for the mapping of small spaces at high resolution and presumed that this intimate knowledge would confer a general understanding over larger regions, the Board fell into the converse error of scaling up. Before 1763, few maps captured the detailed sense of space at the scale at which inhabitants experienced American places. Their working knowledge can be glimpsed from time to time in references that reveal the boundaries they drew between places of security and danger, prosperity and poverty, and civility and savagery. Most took this lived geography for granted and did not bother to describe, much less explicitly draw, these mental maps, which they carried with them to impose a sense of place on their surroundings. The spatial pattern by which places might be characterized by these opposing values shifted from generation to generation, as settlement expanded, violence flared and subsided, economic activity intensified, and populations grew. Spatial sensibility in early America was never locked in place the way that restorations of historically preserved houses might lead us to believe. Colonists expanded the territorial reach of their provinces by developing regionally rooted systems of agriculture, trade, and settlement. The resident elites who governed these societies controlled the legal process by which their colonies grew, establishing new jurisdictions and opening the territory they contained to acquisition through warrants, surveys, grants, and sales. Because land was the ultimate source of wealth in colonial British America, this creole class of American-born leaders, who knew these geographies by experience, devoted itself to acquiring and mastering a body of firsthand knowledge about the landscapes that contained this real property. Britain adopted a plan for American empire in 1763 at the general scale of the continent. In place of an open-ended process of social development, it imposed a desired end point for growth: colonies fully organized to contribute commodities to the Atlantic economy derived from the distinctive resources of their environments. To make this general plan a reality, the Board of Trade initiated intensive and extensive new surveys of American coasts to generate firsthand knowledge of new colonial spaces at much higher resolutions. But it underestimated the difficulties of scaling down this vision to the level of settlement, where settlers took possession of land, tract by tract, by planting it.

Maps were a technology that permitted Britain to direct the settlement and development of new territories from a distance. Surveys and maps at the high-resolution scale of the individual tract simulated direct access to the land as it was seen by those who lived on it, revealing the local character of its terrain, the boundary lines by which landowners carved it up into a "private property mosaic," and the ways in which these places joined together to form a regional pattern of settlement. Particularly after 1748, the Board of Trade demanded new descriptive reports of natural resources, new statistical accounts of trade, and especially new maps that depicted the landscape at this salient scale for colonization, from one-half to two miles to the inch. With such precise accounts of British America at its disposal, the Board laid plans to supplant creole expertise. Mapmakers attempted to produce credible facsimiles of the world, and this effort produced an illusion of direct correspondence between cartographic images and the underlying spaces they represented. Sophisticated eighteenth-century map viewers scrutinized such claims of accuracy, critiquing maps they found to have errors and pointing out their discrepancies with known places. Colonizers nevertheless relied on maps to know the places that they traversed, governed, and defended. Without the representational credibility of their maps, the colonial enterprise in which they were engaged, resting as it did so fundamentally on imagining, distributing, and possessing distant land, would have seemed as uncertain as it in fact was.[7]

Not only did British officials fail to appreciate how American spaces, and the capacities of people to understand them, changed at different scales, but they also radically underestimated the immensity of the territory acquired in 1763. The dream of creating an imperial archive that mapped all of British America at the scale of the individual tract was impossible given the resources and technologies of an eighteenth-century state. In the twenty-first century, the United States Coast Survey continues to draw on data collected more than 150 years ago to produce its state-of-the-art coastal charts—there are simply not enough ships, days, and funds to resurvey all 3.4 million square nautical miles of the United States' territorial waters anew. When the Board of Trade's commissioners spread Bowen's *Accurate Map of North America* before them, it allowed them to inspect the length and breadth of the continent and mold its regions into new jurisdictions. The map's certainty gave them a seemingly clear picture of space that was in fact immeasurably and unfathomably vast. Those who made these general maps understood their limits but were drawn by the craft of cartographic representation to render uncertainties as absolutes. Once these images of frontier spaces entered into the workings of colonial policy making and governance, these general maps could make ambitious projects of spatial control seem reasonable.[8]

Britain was far from alone among Atlantic powers in initiating programs to gather data, improve agriculture, draft new laws, and map space in the hopes of reconfiguring colonial societies as part of an "[e]nlightened state agenda." Neither were its efforts limited to the Western Hemisphere. In addition to the new American territories, the Peace of Paris granted Britain exclusive control over the slaving factories along the Senegal River as well as stronger command over Bengal, the Carnatic coast, and the Deccan plateau in India. Just as the Board of Trade launched its ambitious surveys of Britain's new American territories in 1763, the East India Company began to survey its extensive territories in Bengal, acquired after the British victory at the Battle of Plassey in 1757. It appointed Hugh Cameron to the position of surveyor of the new lands in 1761, to trace the boundaries of the territory, followed by James Rennell as surveyor general, charged to "form one general chart" by which these new domains could be governed. In 1761, it commissioned sea captain Bartholomew Plaisted to chart the "Harbours, Rivers, Shores, Shoals,

[and] Soundings" between "Calcutta & Chittagong" and, in 1779, appointed Alexander Dalrymple as its chief hydrographer. Like their American counterparts, these Indian surveyors described "New Lands" acquired by conquest, assessed their value for agriculture, delineated routes of communication, and identified strategic points for defense. These episodic surveys were later organized into a project that aimed to situate rivers, roads, and coastlines with new precision within a "vast map of Bengal." By 1774, inland and coastal surveys were joined together in a body of maps that depicted the growing territorial reach of the Company at a scale of five miles to the inch. From 1767 to 1774, British surveyors mapped the southern district of Madras at two miles to the inch, the same baseline scale employed by the General Survey of North America. The movement of British armies across the subcontinent during the Second Anglo-Mysore War (1780–1784) turned these early forays into a systematic surveying effort. Rennell's *Bengal Atlas* (1779–1781) and his *Map of Hindoostan* (1782) consolidated old maps and new surveys to present published images to the world proclaiming Britain's command of the subcontinent.[9]

Although strong parallels link these American and South Asian mapping efforts, they were disconnected from each other in practice. The Offices of the Secretaries of State were split between the Southern Department (Southern England, Wales, Ireland, America, and Catholic Europe) and the Northern Department (Northern England, Scotland, and Protestant Europe), and although the same ruling ministries oversaw cabinet offices together, they operated as separate entities with distinct purviews. The unique charter of the East India Company constituted a third branch of government that was detached from this structure of administration entirely. Although India increasingly occupied the imaginations and fiscal calculations of Britain's rulers, they regarded America and its integration into a more centralized empire as the paramount concern to ensure Britain's integrity as a world power.

That such comparable mapping projects should emerge separately at this moment shows that within those British government agencies that looked outward toward the wider world, there was such a thing as the "imperial state," which cohered around a new ethos of systematic governance. Only by looking back at the history of British imperialism from the present, however, can we see a clear progression from England's collection of Atlantic trading outposts established in the seventeenth century to Britain's global territorial empire of the nineteenth century, turning on the hinge of the American Revolution's disruptions. Rather than signaling a coherent shift from one imperial mode to another, Britain's attempts to map America and India in the second half of the eighteenth century reveal this to have been a time of profound disorientation, in which imperial actors grappled with the global scale of the early modern world, to which the objectives and actions of the Seven Years' War had brought new focus. Generated from rigorous methods, with great seriousness and at great expense, these new maps of empire gave authority to the idea of extensive rule over vast spaces, serving as proxies for an imagined degree of control that did not in fact exist. By mapping contested spaces and demarcating extensive new frontiers, Britain created images that concealed imperial weakness as much as they functioned as instruments that enabled the state to exercise real power over remote spaces.

Americans understood that the Board of Trade's schemes posed a direct challenge to their identities as colonizers. Bringing civil order to spaces that lacked it was not, they believed, a task that could proceed from principles dictated by noblemen gathered around a map across a mahogany table in Whitehall. It required the brand of experiential knowledge that they possessed, expertise that adapted standards for agriculture, society, economy, and culture to the varied places that made up colonial British America. Geography played a role in making such expertise indispensable. Because the Americas stretch from north to south, the colonies that could be founded along its Atlantic shores and islands occupied distinct latitudes, each with its own climate and natural resources. As many commentators on empire recognized, this environmental variety gave Britain the ability to produce a wide range of valuable organic commodities and made specialization (in fish, timber, wheat, tobacco, naval stores, rice, indigo, coffee, cotton, and sugar) a driver of the western empire's economic importance. In each distinctive region, colonists fashioned versions of English culture and society. The most important social event in their development was the rise of an American elite, born and bred in the colonies, that subjugated servants and slaves to perform most of the labor, amassed family fortunes, and assumed the right to govern these rapidly growing societies.

Adam Smith measured the "improvement" of American colonies by their capacity to occupy territory and integrate new lands into the civil and economic order of their societies. By the 1760s, the colonies had unquestionably met this definition as an imperial success. Thus, the critique of settler colonialism at the heart of the Board of Trade's plan for empire was surprising, because it ran counter to this commonplace narrative of improvement. It described the history of American colonization as a litany of disorders caused by the self-interested pursuit of wealth and power by squatters, speculators, traders, and legislators. What the Board's vision took from the provincial leaders who governed the colonies was not their capacity to preside over their societies but their power to extend this dominance in future plans for demographic, territorial, and commercial growth. This plan left much of the negotiated authority and provincial autonomy established during the empire's creation intact, but by assuming rigorous control over the relation of individual American provinces to one another and subsuming them within an empire whose borders were regulated and policed from London, it boded ill for the future reproduction of colonies as prosperous, rising societies seeking new thresholds of civility, wealth, and cultural achievement.[10]

Geographic ideas shaped the future of postwar British America. The Board of Trade, with its bold plan for reconfiguring America, initiated a contest for spatial order that was joined on both sides of the Atlantic by colonists, military officers, imperial officials, indigenous peoples, speculators, intellectuals, merchants, and many others. Their views of space embodied homelands, defined corporate identities, represented fears of diminution and disorder, and articulated collective aspirations for security as well as stature. These understandings were expressed through as well as shaped by maps, a medium of representation that put them into circulation and held them up for scrutiny and debate. Most important, they connected human experience at the scale of the individual passing through a landscape to panoramic views of vast spaces that gave these known points broader meaning. By visualizing these spaces, maps gave form to the project of improving America.

BAY
LAKE MISTASSIN
ST LAWRENCE RIVER
LAKE HURON
L. ONTARIO OR CATARAQUI
LAKE ERIE OR OKSWEGO
ALGONQUINS OR ADIRONDACKS
Temiscamings
Abitibis
Nipissings
IROQUOIS
YORK COUNTY
NEW HAMPSHIRE
MASSACHUSETTS
CONNECTICUT
NEW YORK
NEW JERSEY
PENSILVANIA
MARYLAND
ONEYDOES
ONONDAGAES
TUSCARORAS
CAYUGAES
SENECAS
Philadelphia
Boston
Cape Anne
C. Cod
Nantucket I.
Nantucket Shoals
C. Charles
CHESAPEAK BAY
Edinton
C. Hatteras
C. Lookout
C. Fear
C. Carteret
Charles Town
Penobscot B.
WES
The Limits of His Majesty's several Provinces are here laid down as they at present exercise their Jurisdictions: But the Limits of the Massachusets Province with New York, New York with New Jersey, Connecticut with New York, and Pensilvania with Maryland, are not yet finally determined.
Nor is the Boundary of North & South Carolina yet settled. Or of South Carolina & Georgia.
XVIIth Article.
His Britannick Majesty shall cause to be demolished all the Fortifications which his Subjects shall have erected in the Bay of Honduras, & other places of the Territory of Spain in that part of the World, four Months after the Ratification of the present Treaty: & his Catholick Majesty shall not permit his Britannick Majesty's Subjects, or their Workmen, to be disturbed or molested, under any pretence whatsoever, in the said Places, in their Occupation of Cutting, Loading & carrying away Logwood: And for this purpose, they may build without hindrance, & occupy without interruption, the Houses and Magazines which are necessary for them, for their Families, & for their Effects: And his Catholick Majesty assures to them, by this Article, the full Enjoyment of those Advantages, & Powers, on the Spanish Coasts and Territories, as above stipulated, immediately after the Ratification of the present Treaty.

ONE

A Vision for American Empire

On June 8, 1763, the Lords Commissioners of Trade and Plantations drafted a bold new vision for western empire, illustrated it with a hand-colored map, and sent it to the king. The Board of Trade's "Report on Acquisitions in America"—a master plan for colonial British America after the Seven Years' War—imagined taking command of newly acquired lands, imposing limits on long-settled colonies, and forging far-flung territories into a tightly integrated commercial economy. As the commissioners annotated their copy of Emanuel Bowen's *An Accurate Map of North America* (1763), they made visible an Atlantic system by which Britain's diverse collection of New World settlements could work together for the benefit of the mother country. Above all, they sought to strengthen bonds of colonial dependence based on new characterizations of the environments, populations, and possibilities of American places. The "Report on Acquisitions in America" critiqued the history of colonization before 1763 and proposed an alternative for the future, in which carefully managed trade and centralized control of land could harmonize a cacophony of provincial interests. In this momentous year, the Board described a zone of settlement arrayed across the "Whole Coast" of North America and the islands of the Caribbean Sea.

In addition to imposing firm borders on the colonies, the commissioners narrated a reckless history of American expansion over the better part of two centuries, which featured heedless acquisition, fraud, and self-dealing at the expense of the nation's security and prosperity. By dictating how the new territories would be settled, developed, and defended, the Board imagined channeling the expansive energies of migrating settlers to occupy the Florida colonies, the Gulf of St. Lawrence, and the southern islands of the Caribbees. Dedicated reformers affiliated with the Board of Trade attempted to reinvent colonization by establishing a regulated method for planting people on these new frontiers. Their aim was to reconstitute colonial British America from its contested edges inward, for which they sought unmediated views of American lands. They initiated the General Survey of North America and charged the Indian superintendents, the Commission for the Sale of Lands in the Ceded Islands, the Admiralty, and the army to contribute to a vast cartographic effort to represent these places comprehensively. To produce a record of place necessary to command the settlement and development of colonies from a distance, British reformers made mapmaking an imperial priority.

Representing American Spaces, 1660–1762

From its formal establishment in 1696 until its dissolution in 1782, the Board of Trade kept watch over the world with which England had entangled its interests through commerce and colonization. It remained a prescriptive body, charged to find facts and give advice, but by the early eighteenth century it did so with a wider purview, sharper analytical tools, and an expanded portfolio of places and trades to oversee. In 1622, the English state appointed the first of several councils, committees, and commissions to examine its American colonies. From 1660, however, precursors to the Board began a sustained quest to understand and critique American places. From the beginning, members of the Board of Trade fought a rearguard battle against the decentralized process of overseas settlement as they attempted to bring chartered colonies under the direct supervision of the Crown. Over decades, they articulated a consistent perspective on the problems of American colonization, one that pointed to a solution: a general plan for the administration of all the colonies, which hinged on direct control over the granting of American land. Only by holding each of these distinctive places to a common standard could Britain bring the riot of jurisdictions and provincial interests that had taken root since 1607 into an overarching system under its scrutiny.[1]

The Board of Trade's history as a self-conscious body charged to bring order to the chaos of colonization began on July 4, 1660, when Charles II appointed a new Committee for Trade and Plantations, empowered to examine matters "concerning the plantations, as well in the Continent as Islands of America." Staffed with a number of "understanding able persons," including merchants, ship captains, and colonists, this early incarnation of the Board put a premium on direct experience with America, which these "knowing active men" had acquired firsthand. The idea of taking the "Severall Pieces" of England's scattered dominions in America and drawing them together into a more "certain, civill, and uniform waie of Government" was part of its formal instructions, as was the idea of establishing a "continuall correspondencie" about the resources, commerce, wants, and governments of each colony, so that "all of them" could be "collected into one viewe."[2]

From the beginning, this Committee for Trade and Plantations gathered information about America and synthesized it into new general representations on which policies for regulation could be based. New instructions issued in 1672 to a reconstituted Council for Foreign Plantations urged it to discover the colonies' boundaries, climates, lands, and jurisdictions so that they might be reshaped to better serve the kingdom's economic interests. Such a study would allow the Council's members to examine the methods of cultivation that settlers had devised to adapt agriculture and husbandry to their distinctive environments and suggest more profitable commodities so that "Custome be not nourished to the pr[e]judice of Trade." The Crown charged the Council in the late seventeenth century to reveal England's American colonies as they were, estimate the value of what they could become, and bridge the distance between the two by devising new schemes of development and regulation. To record and transmit this information, it generated texts that made use of the languages of political economy and natural history; summarized the flow of manufactures, commodities, vessels, and revenue across the Atlantic with statistics; and collected maps to create authoritative, comprehensive, and densely inscribed representations of American places. "[By] all Way[s] and means," ordered the king in 1672, the Council was to "procure exact Mapps, Platts or Charts of all and Every of our said Plantations abroad." In particular, it was to obtain detailed "Mapps and Descriptions of their respective Ports, Harbours, Forts, Bayes, Rivers with the Depth of their respective Channells coming in or going up, and the

Soundings all a long upon the said respective Coasts from place to place," opening the riversides and coastlines around which colonial settlements clustered to maritime commerce. Moreover, it was "carefully to Register and Keepe" these maps, establishing an official cartographic archive of empire.[3]

The Council undertook to promote the "well governing of [the king's] Forraine Plantations" in 1671, when it met at the Earl of Bristol's London house in a "large room furnished with atlases, mapps, charts, [and] globes." In 1675, the secretary of the reformed Committee of the Lords of Trade and Plantations, Robert Southwell, worked to gather together the Council's scattered papers and assemble a permanent library of charters, maps, books, and documents, even questioning his predecessor, John Locke (who averred that "for globes and maps he never had any"), about lost materials that may have passed into private hands. In the early 1680s, Southwell's successor, William Blathwayt, compiled fifty-four maps and charts in a leather-bound guard book fitted with paper stubs so that he could glue images in place as he acquired them. These were far from the only maps that England's seventeenth-century colonial officials used to view America from London. Indeed, the Committee owned several published atlases and wall maps (including John Seller's "large Chart in 16 sheets of the Maritime Coast of America") and outfitted its offices in Scotland Yard with a number of large cabinets for "papers, books, and Maps," suggesting a much larger collection of cartographic materials. However, because each image pasted into the guard book had to be individually selected, the forty-eight surviving sheets of Blathwayt's atlas reveal which places the Committee deemed most worthy of inclusion as it pursued the king's command to assemble a map archive.[4]

The paper label affixed to this volume's spine reads: "Islands / Plantation / Mapps." The atlas's contents bore out this title's suggestion that English America should be understood, first and foremost, as a collection of islands. An early chart of "Terra Nova," or Newfoundland, showed the cod fishery's stations inked in vivid color on vellum; Abraham Peyrounin's *Carte de lisle de Sainct Christophle* traced the boundaries by which England and France divided the rising sugar island of St. Christopher (or St. Kitts, as it was commonly called) and sited the forts they had built to defend their portions of its fertile tropical acres; and an English copy of a Dutch sea chart of the "Caribe Islands," or Lesser Antilles—colored to show the nations that claimed them—was intersected with rhumb lines to aid navigation into the Caribbean Sea. [**1–3**] Images of Bermuda, Tobago, Martinique, Long Island, Barbados, Guadeloupe, and Montserrat appeared in its sheets, along with four different maps of Jamaica, the most represented place in the atlas. As the sugar trade transformed England's economy after mid-century, the Crown struggled to defend its vulnerable and valuable West Indian islands, exposed to the predation of imperial rivals.[5]

Islands were not only the most profitable, and thus the most interesting, part of these overseas dominions; they also offered an apt geographic metaphor for English America as a whole. Seventeenth-century colonies were outposts of settlement, perched on seacoasts and connected by water routes. The early mainland colonies thus "functioned, in their relationship to one another and to the Atlantic world at large, much like islands." Following two world maps that opened the volume, Blathwayt pasted in Morden and Berry's *A New Map of the English Plantations in America, both Continent and Islands* [1673], one of the first published maps to offer a comprehensive view of English America. [**4**] It pictured a space for settlement and seafaring, bisected by the Tropic of Cancer and evenly weighted between North America and the West Indies. The individual images of English colonies that follow focus, in roughly equal numbers, on continental as well as island colonies. Those steeped in geographic learning at the English universities would have recognized Blathwayt's island-heavy volume as a later-day *isolario*—an illustrated encyclopedia of "maritime cosmography," which was a hallmark of Renaissance publishing. It likewise fractured the world into numerous articulated pieces, presenting an array of exotic places, one after another, but forming no unified picture of a larger empire.[6]

"Plantation," the second term in the volume's title, introduced the atlas as a visual record of English colonization. A handful of images acknowledged the presence of Spanish, French, and Dutch colonies within the geographic frame Morden and Berry created to display "the English Plantations in America" (as well as a few—of Brazil, Paraguay, the Straits of Magellan, Africa, and Mumbai Harbour—that fell beyond it). For the most part, however, Blathwayt brought together images of places in which English colonists and their enslaved, indentured, and free dependents had improved wild wastes and established the rudiments of English society abroad. These maps presented high-resolution images of colonial territories, opening to inspection how settlers had taken up land within them. Francis Lamb's *Jamaicae Descriptio* located the "most Eminent Settlements" that clustered in the island's southern parishes, siting each noteworthy plantation and numbering it to correspond with a named proprietor. **[5]** Richard Ford's *A New Map of the Island of Barbadoes* depicted every "Parish, Plantation, Watermill, Windmill, & Cattlemill" that vouched for this well-settled island's capacity to grow and grind sugarcane. **[6]** Roanoke Island had once occupied the center of English colonizing activity, but on James Lancaster's 1679 chart of Albemarle Sound, the island languished at the margin, apparently uninhabited. **[7]** To the west stretched a coastal landscape of river inlets interspersed with twenty-two house icons: red-roofed, two-story structures, each with a chimney that showed these remote settlements to be hearth-bearing English homes. As shown in Thomas Clarke's map, Bermuda's distinctive strips of land gave each tenant a cross-section of vital resources, from the ocean shore to scarce interior forests. **[8]** New England's centrifugal town system had pushed outward to establish some five-dozen communities scattered across coastlines and riversides on an untitled manuscript map from 1678. **[9]** A dizzying bird's-eye view of Montserrat sketched in 1673 offered chorographic profiles of its settled plains wrapped around its coastline. **[10]** An imagined grid of property lines showed how English colonists had taken up grants of arable coastal land for their slaves to clear. Stocked with these and many other images of improvements made by enterprising settlers, Blathwayt's atlas was a pictorial record of English colonization in America.[7]

When they leafed through its pages to locate the places mentioned in incoming colonial letters, the Lords of Trade saw images of overseas territories produced at an assortment of scales and in a multitude of contrasting colors and styles. Hand-inked manuscripts followed printed copies produced by copperplate presses in London, Paris, and Amsterdam. The book itself—an album displaying images made by several mapmakers at different times and for different purposes—revealed how dependent official English geographic knowledge was on the output of more sophisticated cartographic cultures in Holland and France. Like those influential early compilers of travel narratives, Richard Hakluyt and Samuel Purchas, William Blathwayt knew the merchant adventurers that the Committee interviewed to obtain firsthand accounts of Atlantic trade, and it was through their hands that his atlas acquired some of its unique manuscript images. Spain's *Casa de Contratación* oversaw official chart making from its establishment in 1503, and France's *Dépôt des cartes et plans de la Marine* organized the navy's hydrographic efforts to form a comprehensive surveying agency from 1720. Before Britain's Admiralty established its own hydrographic service in 1795, the Board of Trade took the lead in compiling maps for the use of the government, generating new surveys, and working with commercial mapmakers to engrave and print new maps.

In 1696, William III reestablished this agency "for promoting the trade of our kingdom and for inspecting and improving our plantations in America and elsewhere" as the Lords Commissioners of Trade and Plantations. This new Board of Trade took its place within a rising administrative culture—exemplified by the office of the Inspector-General of Imports and Exports, also established in 1696—which valued systematic rigor in the gathering of information to characterize England's subjects, territories, commerce, and wealth. At the turn of the eighteenth century, the Board joined other key government offices (including the Admiralty, Treasury, and Secretaries of State) in compiling increasingly large volumes of "useful knowledge" that administrators deemed indispensable to the task of identifying the kingdom's interests and governing with them in mind. The Board's charge to "inspect," "examine," "inquire," and "take account of" English America meant setting up a new plantation office in Whitehall as a repository of information for the imperial state. With a new appreciation that the "strength and riches" of England in a "great measure depend[ed]" on its overseas trade, the king specifically empowered the Board to bring to light flaws in colonial agriculture, manufacturing, and legislation, and to envision schemes to improve colonial economy and administration. In particular, the commissioners were to investigate the natural capacities of the colonies, down to the level of how the "Limits of the Soyl" indicated the commodities that each should produce.[8]

The Board's clerks rescued its archive of papers, books, and maps from the Whitehall Palace fire of 1698, which destroyed its quarters, and moved this growing archive into a series of cramped offices within England's sprawling governmental center along the Thames. In 1746, the Board and its papers found a more permanent home in the new Treasury Building. Architect William Kent designed this neo-Palladian structure, completed in 1736, to instill Roman virtue at the center of Prime Minister Robert Walpole's parliamentary state. Its purist conception of classical forms expressed the Board's raison d'être: to promote the common good of the British nation as a whole over the narrow interests of the particular, private, and provincial. Its clean lines and absolute symmetry proclaimed the "virtue of the state" and projected British "cultural competence," compared with the decadent palaces that housed the governments of European rivals.[9]

Within its chambers, four rooms (or sets of rooms) each played a role in managing the flow of information from the wider Atlantic world. In one, as many as a dozen clerks copied incoming correspondence, drafted letters, and annotated and archived the voluminous paper record of the overseas empire. Before leaving for the day, each clerk turned over his papers to the secretary, who kept the sensitive commercial and political intelligence they contained under lock and key. In the secretary's room, the chief plantation administrator opened letters, transmitted accounts, scheduled the commissioners' deliberations, drafted correspondence with colonial governors, and compiled reports. Although he was an official who lacked the stature of a commissioner, a long-tenured secretary served as the keeper of the Board's institutional memory and wielded his command over the minutia of colonial affairs as a respected expert. Higher-ranking commissioners came and went with the shifting fortunes of the political factions with which they were aligned, but secretaries tended to remain in Whitehall, steeped in the world of Atlantic affairs. The Board's first secretary, William Popple, began an administrative dynasty that spanned forty-one years, from 1696 to 1737, during which he handed down the office to his son and namesake, who, in turn, bequeathed it to his son, Alured Popple. John Pownall spent nearly a quarter-century organizing the work of the Board, holding the offices of chief clerk, joint secretary, and secretary during the period 1745–1768.

In the waiting rooms of the Board's apartments, those with firsthand knowledge of America—including ship captains, merchants, governors, and Indian dignitaries—waited to be summoned to the council chamber, where the commissioners would interview them about matters of war, peace, trade, frontiers, governance, agriculture, fraud, exploration, settlement, disorder, and money, among other topics. Seated around the chamber's great table, the commissioners reviewed these testimonies, read the abstracts prepared by the secretary from the documents that had passed through their clerks' hands, and deliberated in formal sessions. They submitted their opinions and proposals, occasionally in detailed formal representations and reports, to the king-in-council. Although the Crown's vast American dominions touched the lives of millions of people, everything that monarchs from William III to George III knew about them flowed through these four rooms, managed by a cadre of clerks, eight commissioners, and one secretary. Salaries aside, the Board's expenses—roughly £2,000 per year—consisted chiefly of the costs of transatlantic postage and the means and materials for writing, reading, and archiving: not only books and maps but also pens, paper, ink, sand, folders, and candles. By 1774, the Board had taken in so much information that the sheer volume of the "[b]ooks and papers" stockpiled over the previous eight decades threatened to overwhelm the office's human occupants.[10]

The Board of Trade's stock of informal authority plummeted during wartime, when more powerful political actors worked to advance royal and national interests in Europe, America, and the wider world. After each of the major eighteenth-century conflicts that pitted Britain against France and Spain, however, the Crown called on the Board to put its knowledge to use in reforming American colonies. In the wake of peace settlements negotiated at Utrecht (1713), Aix-la-Chapelle (1748), and Paris (1763), the Board attempted to make good on its long-standing mission to bring the colonies under tighter control by initiating new settlement schemes and ramping up its capacity to gather information about American places. Once the diplomats had signed off on treaties that redrew boundaries; traded conquests; and set out the terms by which France, Spain, and Britain might coexist in contested Atlantic spaces, the shape and character of overseas colonies was open for the Board of Trade to influence.

As the War of the Spanish Succession (1701–1714) drew to a close, Britain negotiated the Treaty of Utrecht (1713), by whose terms France granted Britain full control over the previously shared sugar island of St. Christopher and ceded its claims to Acadia, Newfoundland, and Prince Rupert's Land along Hudson Bay. The Crown called on the Board to provide detailed information about these territories that, its commissioners reported with chagrin, they simply did not have. Taking stock of an eclectic collection of seventeenth-century "maps of the Plantations," many copied by hand by the Board's clerks, the commissioners found "few here that are to be depended on." To form a "true idea of those parts" on which they must deliberate from across an ocean, they needed better maps. The agreements solemnized at Utrecht established, for the first time in formal diplomatic terms, that the world was made up of sovereign realms within international law. Regardless of how flagrantly such high-minded ideas might be violated on the ground, after Utrecht, no European state could make a credible claim to territory without first documenting how these places had been discovered, occupied, and rightfully obtained. Territories acquired by conquest and treaty presented the Board with an opportunity to take command of the process of colonization itself by granting this land on restrictive new terms, designed to improve security and promote development. As it pursued settlement schemes to populate ceded territories in St. Christopher and Nova Scotia, it called for detailed new surveys. The Treaty of Utrecht marked Britain's first bid at achieving an "English Atlantic hegemony," and the Board of Trade's attempts to implement it revealed how little it knew about the spaces it sought to secure for the nation.[11]

To remedy this lack of credible geographic knowledge, the Board requested that the minister to France search the print shops of Paris for the "best maps of America that can be had" of the "Islands, Provinces and Settlements made by any Europeans there." In 1720, the commissioners realized they could not locate the Canso Islands off the coast of Nova Scotia, a flashpoint for renewed conflict with the French. To populate an official imperial archive with dependable representations of America, it would be necessary to go beyond compilation and actively engage in surveying these territories firsthand. Until such a grand enterprise could begin, the Board worked to integrate original mapmaking into its regular communications with the colonies. It ordered the governors to locate the boundaries of their provinces, describe the most remote English settlements "on the frontiers towards the Lakes and mountains," and document French or Spanish encroachments "together with a chart or map." It adjudicated boundary disputes between neighboring colonies and appointed commissions to resolve contested borders, extend them into the west with new surveys, and produce maps that showcased the results.[12]

From its inception in 1696, the Board of Trade took up the task of thinking broadly about the empire as more than a collection of colonies. After the Treaty of Utrecht, it did so forcefully and systematically, using the spatial language of maps to describe a territorial empire in peril in North America. The Board's 1721 report on the "State of His Majesty's Colonies and Plantations on the Continent of North America"—authored by Commissioner Martin Bladen—put forward a new way of seeing eastern North America as a continental space over which France and Britain vied for supremacy. From his appointment to the Board in 1717 until his death in 1746, Bladen was Britain's most influential colonial official, determined to carry on its mission to populate, regulate, and redeem the colonies. The "State of His Majesty's Plantations" report encapsulated each of the mainland colonies—from Nova Scotia to South Carolina—in a description that began with a precise geographic account of the boundaries laid out by its charter. Bladen then asked a straightforward question of each place: How well did its population, economy, and government secure the extent of this territory for the Crown? In each province, the answer was clear: none had fully realized the potential value of its resources, and all were vulnerable to disorders of maladministration from within and attack from without. Accounts of the colonies' populations pointed out the disparities between the numbers of men capable of bearing arms and the vast domains they could not possibly be mobilized to defend. Ten statistical tables derived from the "Custom House books" for the years 1714–1717 summarized British America's enormous economic value, perhaps for the first time. Every year, the colonies consumed roughly £1,000,000 worth of goods and exported £800,000 worth of commodities, in addition to filling the Treasury's coffers with customs duties, enhancing Britain's independence from foreign markets, and bolstering its maritime capacity.[13]

After enumerating what Britain stood to lose by neglecting the colonies, Bladen looked at a French map of eastern North America and saw a threatening network of encircling and intersecting waterways. Since the voyages of Louis Hennepin and Robert de La Salle established the navigability of the Mississippi River in the 1680s, France had exploited its command of lakes and rivers to outflank Britain's populous settlements on the Atlantic seaboard. Bladen viewed Britain's colonies on "Henn[e]p[i]ns map" (published in 1698 in the English edition of *A New Discovery of a Vast Country in America*) and saw them pinned to the coast, exposed to advances from the interior. **[11]** The French image of the Great Lakes, dilated to the size of imposing seas, occupied this imagined continent's center. Along the massive trunks of the Mississippi and St. Lawrence Rivers, tributaries spread like overgrown tendrils across its length and breadth. At the heads of these lakes and rivers, the French had built forts and settlements and established routes of "communication between Canada & Mississippi"—made visible by dotted lines "prick'd down" on the map—thus stitching together an interior empire across the "vast tract of land" within their watersheds. By "one view of the Map of North America," Bladen wrote, "Your Majesty will see the danger your subjects are in, surrounded by the French." Bladen's interpretation of the French threat as depicted on Hennepin's map set the agenda for "securing, improving & enlarging so valuable a possession" as North America for the next half century. This meant taking steps to command the space between the British mainland and the contested interior. His report urged the Crown to "settle" the boundaries of the mainland colonies, fortify the frontier, and establish stronger diplomatic ties with Indian nations. These policy prescriptions initiated a spate of fort building and elevated the reform of the Indian trade from a nagging concern to a central imperial objective.[14]

In addition to defending the settled mainland colonies against the geographic machinations of the French, Britain would also have to integrate them into a tighter imperial administrative unit in order to secure their benefits. Bladen's report codified long-standing charges of colonial misconduct into a focused critique. Destructive private interests had destabilized British America, whose governance was disrupted by distance, precedent, and the interfering prerogatives of corporate charters.

Colonists carried on "illicit trade with foreigners," gave "shelter to pirates and outlaws," refused to "contribute to the defence of the Neighbouring Colonies," and destroyed "timber fit for the service of the Royal Navy." The leverage the Crown had to design a more compliant and defensible colonial world—its ownership of American land—had been squandered, as "whole provinces ha[d] been granted" to irresponsible proprietors, whose charters should be revoked, because "no power in the plantations should be independent of England." Left to their own devices and disconnected from British authority, colonists flouted trade laws, debased the value of coinage, and failed to provide troops in times of war. Even governors appointed by the king made "exorbitant grants to private persons" after they became co-opted by the local elites over whom they were sent to govern. In addition to defining and securing the colonies' external borders, the Board shaped a vision of internal control that focused on schemes to monitor land grants, collect quitrents, preempt speculation, and shape the social profiles of newly colonized communities by regulating the granting of land. The "State of His Majesty's Plantations" report urged an ambitious program for "improving and enlarging" British North America so that it might fully occupy the territories outlined in its charters. It reflected on prospects for colonial fragmentation and countered with a view of a single "Continent of America, from Nova Scotia to South Carolina."[15]

Commercial mapmakers produced images of eastern North America after the Treaty of Utrecht that reflected many of the Board's strategic assumptions about the geography of American empire. These maps embraced Martin Bladen's idea that the mainland colonies should be seen as one contiguous territorial empire, from Nova Scotia to Georgia, impressive in its extent, population, and commercial value. They redrew the continent in this way to counter maps that pictured New France and Louisiana joined together to demonstrate French sovereignty across the center of the continent. Guillaume Delisle's 1703 *Carte du Canada ou de la Nouvelle France,* for example, showed vast territories connected by waterways by which French fur traders traversed a landscape dense with Native towns. **[12]** By "casting an Eye upon the map" derived from Delisle's *Carte de la Louisiane* (1718), readers of one alarmist pamphlet could see the stranglehold of French rivers that surrounded "all the Provinces on the Main." [13] Such maps of the continent visualized an aggressive French gambit for "almost all North America." British cartographer Herman Moll countered this view with his 1715 *A new and exact map of the dominions of the King of Great Britain on ye Continent of North America,* a map that pictured a British America enlarged by new acquisitions in Nova Scotia and Newfoundland. [14] Overriding provincial boundaries, Moll fused the colonies together into an Atlantic-facing subcontinent south of the St. Lawrence River, which he labeled the "British Emp[ire]." Insets of a rising plantation society in South Carolina's Lowcountry demonstrated an ongoing process of populating territory that the French, despite their capacity to traverse great distances, could never match. Just as thousands of settlers had mobilized to inhabit and improve the Chesapeake Bay, the Delaware and Hudson River valleys, and New England, they would, Moll predicted, extend British society and empire into the lands to the north and south.[16]

Because Britain lacked an official cartographic agency, commercial mapmakers synthesized new knowledge about America for the use of the state as they answered a popular demand for patriotic artifacts. Henry Popple's grandfather, father, and brother had served as secretaries to the Board of Trade (and he, briefly, as one of its clerks), and he drew on these connections to gain the Board's "Approbation" for his massive, twenty-sheet *Map of the British Empire in America* (1733). [15] He filled in Moll's subcontinent with precise boundaries, finely articulated rivers, named Indian nations, and frontier forts, and placed this image of an integrated British mainland within an Atlantic and circum-Caribbean frame. The Board's commissioners allowed him access to their "Authentic Records & Actual Surveys" and purchased a copy for their map library and for each of the colonial governors but had had limited influence over the content of the image. Although they identified a need for reliable maps of the colonies, the commissioners did little to make good on this priority; commercial mapmakers like Moll and Popple filled this gap with images that celebrated the idea of empire at the scale of the continent.[17]

Two colonial wars linked to the larger European War of the Austrian Succession—the War of Jenkins's Ear (1739–1744) and King George's War (1744–1748)—contested this idea of an expansive British American empire. These conflicts threatened the security of the mainland precisely where Martin Bladen had feared it was most vulnerable: in the contested borderlands between New England and New France and across the southern frontier that separated South Carolina and Georgia from Spanish Florida. Naval warfare and island conquests paralyzed the thriving trades in plantation commodities, slaves, and manufactures centered in the West Indies. At the peace, the ministry gave the Board of Trade a brief to tighten British control over colonial defense, administration, and settlement. The Treaty of Aix-la-Chapelle (1748) has become a byword for the apparent futility of eighteenth-century colonial wars, in which France, Britain, and Spain wasted treasure and spilled blood to conquer American territories, only to return them to their original owners at the peace talks. To resolve lingering disputes over Nova Scotia and the Windward Islands, which could not be settled definitively by treaty, British commissioners, assisted by the Board of Trade, negotiated with their French counterparts to establish boundaries authenticated by histories of discovery and occupation and illustrated by maps. This chimerical quest to solve the zero-sum game for control over the continent by appealing to the authority of cartography devolved into a pointless "map war" in which each side pictured its own imperial interests in America. The failure of this British-French commission to define a boundary between Nova Scotia and Acadia gave the lie to the premise of Aix-la-Chapelle: that France and Britain could each stake out a claim to American territory that was not mutually exclusive of the other.[18]

In 1748, as these negotiators began wrangling over where French America ended and British America began, George Montagu-Dunk, Earl of Halifax, assumed the presidency of the Board of Trade intent on securing British sovereignty through the direct occupation of contested territory. His boldest initiative, "establishing a civil government in the province of Nova Scotia, and settling three thousand protestant subjects within the same," broke new ground in the Board of Trade's quest to shape British America. For the first time in its history, the state took direct charge of the planting of colonists in a New World province. Between 1748 and 1750, no detail of this enterprise was too small that it evaded the scrutiny of the commissioners, who discussed everything from the ventilation of transport ships to the ploughs best suited to maritime soils. They summoned prospective migrants to be personally interviewed in

Whitehall before issuing a certificate of permission to embark for Nova Scotia. Within two years, the Board had transported some six thousand settlers to live in planned townships surrounding a fortified port town on the Atlantic coast, a place the Mi'kmaq Indians called Chebucto and the British renamed Halifax, after its powerful patron. Britain ordered the creation of new maps of Nova Scotia to open its contested lands to view. Army captain Charles Morris surveyed the province to "see what room there was for interspersing British settlements . . . in a commodious and defensible manner among those of the French" and to make visible the geography of rivers and coasts to the Board so that it could locate forts to secure this new overseas province. As a military outpost and naval station, Halifax mirrored the French fortress at Louisbourg, roughly 150 miles to the northeast on Cape Breton Island; as a place of sizable Protestant population on the Atlantic coast, it countered the threatening presence of some nine thousand French Acadians and as many as two thousand Mi'kmaq Indians across the peninsula. In the Board of Trade's most ambitious intervention in America to date, mapmaking went hand in hand with colonization.[19]

The Board of Trade's first attempt to map the British mainland as a whole yielded a cartographic landmark, John Mitchell's *A map of the British and French dominions in North America* (1755). **[16]** This eight-sheet wall map, more than six feet wide and four feet tall, set a new standard for geographic knowledge of the continent. Although composed by Mitchell—an American physician and polymath—the map followed Halifax's "design of uniting the colonies under one general direction" in the face of French "encirclement." It projected bold British claims to sovereignty over virtually all of eastern North America. Halifax personally recruited Mitchell to undertake the map, giving him access to the Board's archive of manuscript images; when these sources proved wanting, he commanded colonial governors to draft new maps of their provinces, from which Mitchell could devise a state-of-the-art synthesis. Mitchell's map was a fantasy of dominion, which anticipated that the rapid reproduction of American settler populations would soon extend jurisdictions westward. Within the open boundaries of colonies whose charters extended to the Pacific Ocean, Mitchell identified a number of specific British settlements that had taken root across the mountains and along the Ohio and St. Lawrence Rivers. Bolstered by the credibility that came with compiling the manuscript record of settlement, the map pictured a mainland empire reaching across eastern North America.[20]

Settling Halifax was the most prominent among many initiatives between 1748 and 1754 with which Britain sought to make manifest a new "policy of strict supervision and control" over its American territories. Aware of the colonies' impressive demographic and economic growth and the importance of its trade, senior officials observed the erosion of their authority over truculent colonial assemblies during a mid-century moment in which the threat of French attack loomed. They noted how the emission of provincial paper currency diluted the value of debts owed to British merchants and logged numerous violations of the Molasses Act of 1733. Under Halifax, the Board of Trade censured governors who seemed too much in the interest of local elites and scrutinized colonial laws with new vigor, recommending that the Privy Council disallow many it deemed "repugnant" to British interests and a usurpation of the Crown's prerogative. It proposed legislation to prevent the printing of colonial paper money, resulting in the Currency Act of 1751, and submitted detailed reports on the troubled governments of New York and New Jersey, along with a plan for their renovation. Those who watched the Board's energetic efforts on these separate fronts believed that it was preparing to implement a comprehensive program of reform to "settle a general plan for establishing the King[']s Authority in all the plantations." In preparation for just such a move, it pressed the governors for updated information about their colonies—a long-standing practice of issuing official questionnaires that took on new urgency under Halifax. The Board ordered that they report their geographic situations and the locations of their boundaries; provide a history of their charters and early constitutions; and describe their natural capacities and agricultural output, the value and volume of their exports and imports, and their progress in mining and manufacturing. They were to profile their populations, enumerate acres of land granted and quitrents due the Crown, describe frontier fortifications, assess their militias, characterize the threat posed by the proximity of French and Spanish settlements, report on the state of Indian affairs, and account for the provision of funds used to pay the salaries of royal officials. The Board established the first transatlantic packet-boat service, in part to speed up the communication of this information across the ocean. In 1752, the Board gained the short-lived authority to appoint colonial officials directly, a power that further enabled its dramatic bid to take command of the colonies. The ministry created a captain general as well as two superintendents of Indian affairs—one commissioned to represent the Crown to the nations of the Native north and another for the south—taking military decisions and Indian diplomacy out of the hands of colonial governors. Although the Board of Trade reached the pinnacle of its administrative power at this moment, its efforts to reform the colonies, thwarted at every turn by colonists, left activist governors embittered and the commissioners of the Board embarrassed that so little of its decades-long mission to reform and regulate the colonies had been accomplished in the respite provided by the peace of 1748 and under the robust leadership of the powerful Halifax. After British and French forces clashed along the Ohio River in 1754, initiating the Seven Years' War, the commissioners of the Board of Trade stepped aside as ministers, generals, and admirals waged a global war. The impact of Halifax's initiatives, however, endured this diminution of explicit power. The cohort of ambitious, intellectually engaged young men he had groomed to administer the colonies shared his commitment to bringing order to a disordered colonial world. This reforming generation rose to positions of influence at the war's end, sustaining Halifax's vision after he resigned the Board's leadership. Inheritors of the Board's "searching reassessment of colonial policies and concepts of empire," they continued its decades-long mission as they framed a plan for the peace.[21]

The Political Economy of the Peace of Paris

Britain's Seven Years' War victories—by 1762 its forces commanded Quebec, Louisbourg, Havana, coastal Senegal, Grenada, Madras, Guadeloupe, Martinique, and Fort Pitt, among other places—opened the question of the form the new empire should take to sharp public debate. Against the expectation that some conquests would be traded back at the peace negotiations, advocates of a more focused commercial empire proposed returning continental territories to France and retaining West Indian islands of proven value. Those who imagined British settlement expanding into the west countered with a vision of a grand territorial empire populated by millions of subjects occupying millions of acres. As negotiators hammered out the terms of the Treaty of Paris, this clamor over how to shape British America was a moment rich with geographic possibilities. In the words of one of its conquering heroes, Admiral George Brydges Rodney, the "empire of Great Britain [was] at stake," and any number of new configurations of overseas territories might be fashioned into a new colonial world.[22]

Under the leadership of John Stuart, Earl of Bute, the ministry leveraged military conquests into a peace that attempted to square the circle of these competing visions of empire. Its terms defended North America from foreign intrigue and encroachment. The security of the "colonies upon the continent" had been the war's "original object," and it became the defining principle of the peace. A century of conflict across contentious borderlands had taught the lesson that "while France possesses any single place in America, from whence she may molest our settlements, they can never enjoy any repose." The treaty removed Britain's great rival "from our neighbourhood in America." As it restored Havana, Guadeloupe, and Martinique, it also added four new Caribbean islands to the British West Indies. The new empire would be composed of new tropical plantations as well as new mainland settlements, integrated into a more tightly regulated Atlantic commercial system.[23]

Before he assumed the leadership of the Board of Trade, twenty-five-year-old William Petty, Earl of Shelburne, stood before the House of Lords on December 9, 1762, in support of a motion to approve these "preliminaries of peace," negotiated to settle the Seven Years' War. He began with a reflection on treaty diplomacy in a simpler age, when each state sought the most "Possession of Territory" it could secure by force to obtain the largest portion of a fixed pool of wealth. Now Europe formed an "extensive Union" knitted together by the "Indissoluble System" of commerce. National power derived not from the sum total of acres capable of being farmed and taxed within a fixed domain but from the "number of Industrious people" it contained, whose work and wants enlarged the capacities of the nation to make, trade, and consume—and, in the process, amass a growing and renewable source of wealth. To "reap the Fruits of Victory" in such a system, Shelburne argued, Britain must "restore some of [its] conquests." This calculating view ran counter to the triumphal nationalism of other Britons, who reveled in the image of Great Britain standing astride an enlarging portion of the globe in dominating majesty. A young James Boswell, a like-minded intellectual who favored reason over emotion in determining matters of national policy, was disappointed that he could not get a seat in the gallery to hear the debate over the preliminaries in the Commons, but he gleefully skewered those he observed among the masses who relished the idea of conquests but failed to think of the costs of ongoing conflict. Should Britain seek to keep all the French and Spanish colonies it had taken, Shelburne cautioned, neither power would have any reason to allow it to "enjoy [its] conquest in Peace," opening a prospect of "Eternal War." The composition of the new empire, he argued, should be determined by a reckoning of which new territories seemed likely to "best serve to Increase our Wealth[,] the number of our People . . . and consequently our Future power." Securing North America for Britain, he asserted, was well worth the price of returning Havana, Guadeloupe, and other additions to the realm, despite the public's sense that these territories enhanced the nation's grandeur on the world stage.[24]

The Peace of Paris achieved what the Board of Trade had long desired: unbroken sovereignty across the western shores of the North Atlantic. In Shelburne's ringing phrase, this "universal empire of that extended Coast" seemed to establish an enduring foundation for national wealth. Removing Britain's rivals not only protected the colonies but also promised to "open a new Field of Commerce, with many Indian Nations"; secure northeastern forests for the supply of "resources for the Increase of our Naval Power"; and, most important for reaping the benefits of sustained transatlantic commerce, inaugurate a "Solid & lasting Peace." By his reckoning, exports to the mainland colonies were worth £2.5 million per year, a full quarter of the value of all the goods Britain sent abroad; and imports ranging from fish to indigo, largely re-exported to the continent, strengthened the nation's balance of trade. Without supplies of North American shipping and lumber secured by this treaty, Shelburne observed, Britain's "W[est] Ind[ian] Islands cannot subsist." The king's new lands and waters in Canada and Nova Scotia were projected to contribute £250,000 worth of fish and £60,000 of furs to this balance; French inhabitants of this expanded mainland empire, long feared as internal enemies, became assets instead of liabilities by this accounting, projected to consume "some 200,000£ value of British manufacture" every year. Claiming Florida from Spain opened new avenues for the deerskin trade and secured ports from which British ships commanded the routes of the "Homeward bound Flotas from Vera Cruz." By any rational measure, such gains easily exceeded the value of acquiring a few French sugar islands. Nullifying the threat of attack would set the war-weary minds of mainland colonists at ease, encouraging them to cultivate new lands and populate them with more people, whose demand for manufactures created employment at home in a virtuous cycle of production and consumption that guaranteed the "Wealth[,] safety & Independ[e]nce of these Kingdoms . . . for ages to come."[25]

Shelburne expressed a vision of empire that was profoundly shaped by eighteenth-century political economy. In framing his ideas about what the colonies were worth and how the state could capture their value, he consulted with the preeminent social and economic theorists of the age, including Josiah Tucker and Adam Smith, copies of whose *Theory of Moral Sentiments* he distributed as gifts to his friends. A year before his 1762 speech to the Lords, Smith had accompanied the young nobleman on a journey from Edinburgh to London. Shelburne recalled their lengthy discussions across the countryside as a transformative moment that clarified the "difference between light and darkness" in his appreciation for the way the world worked. The "principles [he] first imbibed" from Smith and other noteworthy political economists laid the intellectual foundation for an influential political career that centered on American affairs. During his five-month tenure as the first Lord of Trade, followed by a brief stint on the Privy Council, Shelburne presided over the drafting of the Board's "Report on Acquisitions in America" of June 8, 1763, the document that set in motion Britain's postwar plan for imperial reform. When William Pitt returned to power in 1768, Shelburne served for two years as the most senior official in charge of colonial affairs, the secretary of state for the Southern Department, overseeing the implementation of policies he had recommended as the Board's president. Seeking to allay American objections to direct taxation, he proposed, unsuccessfully, that the ministry retire its debt and fund its overseas administration with a more rigorous collection of the quitrent payments owed by American landowners to the Crown and by drawing more revenue from the East India Company.[26]

Shelburne joined a generation of postwar officials and commentators in imagining how the state should balance the interests of those spread across Britain's diverse Atlantic dominions. In many respects, their attempts to shape society in newly acquired territories pursued the objectives of doctrinaire mercantilism. Although Shelburne was a free-trade liberal who renounced restraints and monopolies, his vision of the new American empire was guided by the same aims as seventeenth-century regulators: to improve the resource balance by adding new lands to the national domain, to secure vital commodities, to open markets for home manufactures, to increase maritime capacities, and to support a vibrant community of traders. The imperial political economy Shelburne advocated likewise described a system of generating wealth that placed the interests of the home country at the center and viewed colonies as dependencies that must serve those interests. The means by which the end of rising national wealth could be achieved, however, had changed since Parliament passed the Navigation Act of 1651, because new thinkers had so dramatically revised understandings about the nature of interest and the proper role of the state in influencing economic behavior. To explain why some civilizations advanced while others declined, early modern theorists probed the intersections between economy, society, and the state, opening a window onto the deep structural forces on which everything seemed to depend but which lay beyond the easy manipulation of political rulers. They traced historical patterns of social progress, from primitive nomadism to polite urbanity; they investigated the psychology of human sociability; and they promoted commerce as a way to extend fellow feeling on a global scale, urging policies that would make trade, rather than war, the means by which nations as well as individuals might pursue divergent interests to their mutual benefit.[27]

Eighteenth-century political economy diagnosed the ills of the early modern world and proposed the cure. Convinced of an underlying harmony between the natural order of the world and human nature, these philosophers were reformers at heart. They searched for asymmetries—from bad laws to cultural prejudices—that kept the natural pursuit of particular interests from contributing to a general common good, and advocated changes in law and policy that would restore the universe of contending interests into a state of self-sustaining equilibrium. When early modern political economists trained their analysis on the problem of the colonies, they saw the history of colonization itself as a pathological disruption in this natural system, requiring wholesale renovation. They viewed humanity locked in an internal battle between the passions and the will, the former leading people toward immediate, selfish ends, and the latter promoting deferred goals and true interest. Individuals saw their influence and their responsibilities within an "immediate circle of concern" that was likely to produce rival groups within society, each striving against the others. The history of "grand unwieldy empires" suggested that as inhabitants spread across space, their interests would inevitably diverge, encouraging central states to subjugate their populations—a process of territorial expansion that generated, in the end, "one great source of human misery." Colonies and the violence and venality they fostered were monuments to unreason, unjustifiable monopolies of power that served as a sop to a populace eager to celebrate the exploits of military action in defense of the "imagination that they possessed a great empire on the west side of the Atlantic." Such an empire was the mirror image of a true nation that bound itself together and perpetuated its virtues through deepening commitments to free trade with the wider world.[28]

To reform colonization, they argued, the state must unleash the power of commercial exchange across space. As articulated by Francis Hutcheson and Adam Smith, unfettered commerce operated as a dynamic force for creating harmony where disorder had prevailed, reconciling a multitude of competing interests by linking them together in a web of interactions far too complex and dynamic for any state to strictly monitor, much less attempt to control. In this new political economy of empire, the state was not to attempt to secure the benefits of colonies through legal restrictions but to identify the inherent interests of places and impose policies that encouraged them to align with one another, culminating in a general imperial common good. The Peace of Paris erased contentious imperial frontiers and created the possibility that new colonies might be shaped from the outset to form a new imperial order, opening America as a great laboratory to put these ideas into practice. After 1763, the Board of Trade launched schemes to restructure colonial territories, trades, and populations into a natural state of economic dependence on Great Britain. Imperial reformers imagined that they could build commercial subordination into the very structure of these emerging societies by controlling who had access to the king's land and setting new terms on which it must be developed.[29]

"Is there any thing in the world, that should be more thought a matter of state than trade, especially in an island?" asked Charles D'Avenant in 1697. Compared with the vast extent of continental France, Britain was a mere island, constrained by its geography. American colonies enlarged the British Isles well beyond their natural bounds, solving the problem of how to challenge French power from a homeland that would always possess fewer people cultivating fewer acres. Overseas colonization added distant tracts of land to its economy, extending its productive capacities beyond what its "Native Commodities can afford." "If we keep it," promised Benjamin Franklin of Canada, "all the Country from St. Laurence to Missis[s]ip[p]i, will in another Century be fill'd with British People; Britain itself will become vastly more populous by the immense Increase of its Commerce; the Atlantic Sea will be cover'd with your Trading Ships; and your naval Power thence continually increasing, will extend your Influence round the whole Globe, and awe the World!" Franklin's optimism ran counter to the view that saw imperial extensions of the nation as artificial and thus doomed to fatal imbalances and corruptions. Earlier economic commentators had cautioned that "[e]xtended dominion" through conquest of foreign territory and "riches flowing in by trade" constituted unnatural economic nourishment, especially if distance and disorder undermined a nation's hold on remote colonies. By 1763, however, Britain depended on its transmarine provinces and could not risk the loss of the numerous productive acres that settlers and slaves had added to the national domain.[30]

The century that preceded the Peace of Paris had featured intense meditations on the nature of American empire. Colonists had created productive and populous extensions of British society in the New World, but the energies devoted to transatlantic colonization flowed away from its center, a process that spun the colonial world into an array of distinctive regional societies, from New England to Barbados. Once begun, this process of demographic diffusion seemed impossible to restrain. North American populations, largely confined to coastal enclaves and navigable rivers in the seventeenth century, surged beyond the fall line of the Atlantic coastal plain by the middle decades of the eighteenth century. Behind them, full-fledged cities rose on the sites of what had been modest ports, testifying to their development as societies in the growing stature of their architecture, the sophistication of their cultures, and—especially after 1750—the variety of imported goods available for consumption. Originating in disorder and poised to grow beyond all control, Britain's Atlantic empire appeared trapped in a tragic moment in time, just before its unstable foundations crumbled.[31]

Once it became clear that the "Well-being of the Universality" of the nation depended on its colonies, it became intolerable to consider that Britain should not command what was "essentially necessary, for its Preservation." This newly acknowledged dependence sharpened a sense of the risks to this artificial domain, distended across thousands of miles of maritime space—especially as colonies grew into more prosperous, autonomous, and developed societies. In his 1752 treatise on the legitimacy of parliamentary trade regulation, James Abercromby insisted that ways needed to be found to put the general interests of the whole above any "particular Interest." The metropolitan view was that empire should resemble a family united under a patriarchal head, creating a "uniformity of interest" structured around a just and natural authority. With these vulnerabilities laid bare, urged Thomas Pownall, Britain must take advantage of the opportunity presented by the Peace of Paris to shape its American territories so that the nation could become a single "grand marine dominion, consisting of our possessions in the Atlantic and in America united into one interest."[32]

On February 10, 1763, the Treaty of Paris expressed the new continental dimensions of British North America. To "remove forever all subject of dispute with regard to the limits of the British and French territories on the continent"—a flashpoint for violence across the maritime northeast and mainland interior—it negotiated France's departure from North America. Spain had worked to destabilize Britain's southern frontier during the first half of the eighteenth century, checking settler expansion by offering safe haven to runaway slaves, raiding the Carolina coast, and launching counterattacks that depopulated Georgia's sea islands. Now all of "Florida, with Fort St. Augustin[e], and the Bay of Pensacola," and "all that Spain possesses" from the Gulf Coast to the Atlantic Ocean became British. To write British North America into the law of nations as a continent apart, the treaty imagined a "line drawn along the middle of the River Mississippi" and asserted the right of "free navigation" along the "whole breadth and length" of its waters. Declaring this great river to possess the legal definition of a *mare liberum*—a "free sea"—aimed to open British communication between the interior and the Gulf of Mexico so that it could not legally be cut off by the interposition of New Orleans, which France ceded to Spain along with the rest of Louisiana as part of this negotiated settlement. It also "fixed irrevocably" mainland British America's western boundary at a river that, when viewed through a lawyer's lens, bore the trappings of an ocean. By defining a new mainland empire out of the interlocking waters of the St. Lawrence, the Great Lakes, the Mississippi, and the Atlantic and Gulf coasts, the Treaty of Paris established a basis in nature as well as law for a separate "Continent of America" under the exclusive sovereignty of Great Britain. In the West Indies, the treaty divided long-contested islands with new clarity and gave Britain a commanding territorial and maritime advantage across the archipelago of the Lesser Antilles. The Peace of Paris revoked the 1748 "neutral islands" compromise, which set aside the less cultivated Windward Islands as places where no European settlement was permitted. France reclaimed its conquered sugar islands and took possession of formerly neutral St. Lucia, but at the price of ceding to Great Britain long-settled Grenada, along with three formerly neutral islands it had once claimed—an act that gave the Ceded Islands its collective name.[33]

The New Map of Empire

Although Britain had battled and negotiated with France and Spain on behalf of its American colonies, it lacked a clear plan for the peace. The king charged the Board of Trade with the task of designing a coherent imperial order in America on May 5, 1763. The Board responded with its report of June 8, which divided the colonies into circumscribed jurisdictions, identified the commercial value of new lands, put forward plans for settling them, and outlined a program of conciliation with the Indian nations. After decades struggling against the intransigence of colonists, proprietors, and entrenched interests, the king handed the Board the chance to act on its mission to regulate America. Compared with past opportunities to resettle part of the island of St. Christopher, populate frontier townships in Carolina, and settle parts of Atlantic Nova Scotia, this mandate opened four islands and the better part of a continent to new schemes of colonization.[34]

This royal command came with a small library of materials—Southern Secretary of State Charles Wyndham, Earl of Egremont, dispatched thirty-nine separate enclosures to the Board of Trade—designed to inform its discussions and frame its conclusions. These included a printed copy of the Treaty of Paris, population estimates, export statistics, troop returns, articles of capitulation, a plan for deploying military forces in the North American interior, and reports by occupying generals who described the newly acquired territories they now commanded, illustrated with new maps drafted by military engineers on the spot. One of these enclosures, "Hints Relative to the Division and Government of the Conquered and Newly Acquired Countries in America," attributed to Henry Ellis, loomed above the rest as a foundational text for a reformed British America. Other documents supplied discrete information, but Ellis's presented a comprehensive scheme that set the agenda for the Board's deliberations. A second influential text, "Hints Respecting the Settlement for Our American Provinces," attributed to William Knox, expressed Ellis's ideas in the language of contemporary political economy, the de rigueur discourse of eighteenth-century policy making. The stacks of overflowing papers archived in the presses of the Plantation Office included many such short, expository essays, submitted by authors who often called them "hints." Under the cover of these humbly titled position papers, ambitious thinkers proposed daring expeditions, hatched schemes for new commodities, and imagined sweeping administrative reforms, such as Benjamin Franklin's 1754 "Short Hints toward a Scheme for Uniting the Northern Colonies." A hint was offered almost casually, as a disinterested suggestion to be adopted or discarded as the powerful officials who considered it saw fit. Beneath its nonchalant title, however, was a hard claim of credibility (as well as an implicit plea for appointments to undertake the improvements, industries, voyages, and settlements these tracts proposed). Those who wrote them proffered the kind of indispensable knowledge that could make general principles work in practice in remote places, far from the centers of power, where nature and society followed unfamiliar rules. Discard such advice, those who dropped these hints implied, at your peril. For decades before the changes set in motion by the Peace of Paris, the Board of Trade had primed the pump of solicited and unsolicited advice, which brought a flood of hints from the world's peripheries into its chambers in Whitehall. In 1763, eager projectors—as those who proposed innovative schemes of improvement were called—sent them in by the score to demonstrate their expertise in American affairs and claim the rewards of office, land, or commercial advantage in Britain's enlarged empire. The texts written by Ellis and Knox share space with sixty-one other hints, thoughts, letters, journals, plans, descriptions, drafts, remarks, estimates, extracts, accounts, petitions, propositions, catalogs, reasons, and opinions in a folio volume of "Papers and Proposals relative to North America" in Shelburne's papers, a collection that captures something of the range and volume of the Board's information-gathering and policy-making practices.[35]

Regarded as the "oracle of the ministry for all America," Henry Ellis was born into a family of English colonists in Ulster and left Ireland for the sea at the age of nineteen. At twenty-five, he entered the first rank of Britain's overseas projectors, serving as hydrographer on Arthur Dobbs's mission to discover a northwest passage to Asia. Although the expedition failed to find this elusive water route, his account of the voyage, published with detailed navigational charts, gained him a prestigious appointment as a fellow of the Royal Society. Such acclaim brought him to the attention of Halifax, who gave him command of his namesake vessel, the *Earl of Halifax,* and a mission to explore the African coast and conduct experiments in its tropical climate in the early 1750s. Halifax later appointed Ellis governor of Georgia, a "large but weak Province" that had long figured in the Board's deliberations as the key to defending the vulnerable southern mainland frontier. When Ellis arrived in Savannah in 1757, in the midst of war with France, he settled the tumult of a politically fractious colony. He conciliated the Creek Indians by defining a clear boundary to settlement and led efforts to fortify the province's frontiers and ports. Taking command over the land system, he established new parish jurisdictions, pushed initiatives to escheat vacant tracts to the Crown, and placed restrictions on new grants to steer social development away from plantation slavery and toward middling European farming. Appointed governor of Nova Scotia in 1761, he remained in London while a lieutenant presided over his patron's colony. After building a reputation as an imperial improver who had gained his expertise about American affairs firsthand in the North Atlantic, West Africa, the West Indies, and the North American southeast, Ellis returned to the metropolis. Although he formally resigned his governorship in 1763, he remained a high-level policy adviser behind the scenes of Britain's colonial administration. Ellis's influence extended so far that he was granted the opportunity to review the Board's "Report on Acquisitions in America" before its submission to the king in June. A London merchant reported seeing a copy of the king's proclamation of October 7, 1763—the document that turned the report's recommendations for America into law—"in Governor Ellis[']s handwriting" before its publication.[36]

Henry Ellis's "Hints Relative to the Division and Government of the Conquered and Newly Acquired Countries in America" proposed that the colonies, old as well as new, be bounded and contained. To govern Canada's "vast Extent," it must be divided into settled "Provinces." To finally resolve a century of warfare over the isthmuses, coastlines, and islands of the maritime northeast, these once-contested places should be absorbed into the governments of Newfoundland and Nova Scotia on the grounds that "every part of the British Dominions however circumstanced should be under some Jurisdiction or other." Georgia, once confined by "too narrow Limits" on the southeastern frontier, should now "be extended Southward" until it gained a comparable size to its neighbors on the mainland. Just as expansive New France was to be compressed within restricted boundaries, the ambiguous extent of Spanish Florida should now take concrete form as two bounded jurisdictions, peninsular East Florida and the Gulf Coast colony of West Florida. To curb the territorial ambitions of the "ancient provinces" along the Atlantic seaboard, Britain should "fix upon some Line for a Western

Boundary . . . beyond which our People should not at present be permitted to settle." Ellis imagined this line deflecting surging populations from the "Heart of America," where their distance from the coast would put them "out of reach of Government, and where, from the great Difficulty of procuring European Commodities, they would be compelled to commence Manufactur[e]s to the infinite prejudice of Britain." Setting a universal western terminus to colonial domains would encourage would-be frontiersmen to migrate to "Nova Scotia, or to the provinces on the Southern Frontier, where they would be usefull to their Mother Country" by occupying territories that needing defending. Ellis's "Hints" defined the ideal colony as a clearly delineated territory that was compact, well governed, and connected to Britain through Atlantic trade.[37]

Another Halifax protégé, William Knox, left northern Ireland for America at twenty-four in 1757 to serve as Georgia's provost marshal under Ellis. There, he invested in land and slaves and worked closely with the governor to transform a province notorious for corruption and disorder into a model royal colony. He left this post in 1762 and returned as an absentee planter to Britain. During a brief trip to France, Knox reflected on the utterly improved landscape he saw through his carriage window on a two-hundred-mile journey to Paris. In contrast to Georgia's rough frontier, where wilderness intruded everywhere, he did not glimpse a "single acre of uncultivated ground." This "immensity of the tillage and the neatness of the farms" was a foundation of arable wealth that Britain's rival taxed "with impunity" to wield power in the world. Such a contrast helped convince him of the necessity of taking firm command of American land. Back in London, Knox turned to his pen to seek out new political patronage. He circulated several schemes, hints, and remarks among a broad circle of influential officials; made an "appearance at the Board of Trade" bearing a letter of recommendation from Ellis; and eventually gained the attention of George Grenville, Charles Townshend, and Shelburne as these senior officials were in the midst of framing the Peace of Paris. He turned what he had learned in Georgia into a "political stock" of knowledge he hoped would earn him another influential appointment.[38]

Knox's "Hints Respecting the Settlement for Our American Provinces" seconded Ellis's vision and explained in greater detail why Britain must bind errant colonial societies into a new state of dependency. Knox focused attention on the mutually reinforcing fertility of American lands and migrating settlers. Where Benjamin Franklin had celebrated the extensive occupation of land as a boon for Britain (expressing in words what John Mitchell's map did in its image of colonies as bands of latitude that extended, seemingly without end, into the west), Knox saw imperial disorder. The "Prince that acquires new territory," Franklin had reasoned, "if he finds it vacant, or removes the natives to give his own people room," would lay the territorial foundation for the "generation of multitudes." Opening western land to young men and women without property would give them the wherewithal to support households and marry young, at the very start of women's childbearing years. By "clearing America of woods," their progeny would soon cover the face of the continent until it was so overspread with white bodies that this entire quadrant of the globe would appear to "reflect a bright light" if viewed from space. Knox agreed with Franklin's analysis, but he viewed the exponential growth of creole populations with alarm rather than optimism. The relentless "Copulation, and Prolificacy" of American colonists was no virtue but one of many vices that could be classed under the heading of the heedless pursuit of self-interest at the expense of an imperial common good. The deep forces of demography, he argued, operated far more powerfully than cultural affinities with Britain to guide colonists' actions. If Britain failed to "set bounds to the Increment of People, and to the extent of the Settlements in that country," the trajectory of population growth would—regardless of sentiment, religion, and patriotism—detach the colonies from the empire. Now that the king's "possessions in North America," Knox concluded, were "so many times more extensive than the Island of Great Britain, if they were equally well inhabited, Great Britain could no longer maintain her dominion over them."[39]

Knox advocated "confining our Settlements in America within proper Limits" by barring colonists from occupying new land beyond the "Banks of Navigable Rivers" that flowed to the Atlantic Ocean. He described an imaginary line of economic interest governed by the costs of transporting the "bulky Commodities" that benefited Great Britain. As settlers crossed the mountains to take up land in places like Virginia's Shenandoah Valley, they settled farms beyond the reach of low-cost river transportation to Atlantic ports. Hobbled by the burden of overland transportation across rugged terrain, such settlers could not sell the "rough materials" that mercantilism demanded at a profit, and they would have little choice but to begin making the goods they could no longer afford to purchase from Britain. Those who migrated across this geographic threshold dictated by the value-to-bulk ratio of agricultural produce would be "entirely lost to Great Britain."[40]

Knox's "Hints" seethes with negative characterizations of provincials. Every migrant to the colonies was a deserter, who harmed Britain once by depriving it of labor and then again by taking up its trades and competing against it. Instead of obedient "Children," colonists rebelled against reasonable commercial regulation to become "Rivals to the Mother Country." They hunted for game in the woods like Indians, losing their own civility and taking on the traits of savages. Their internecine competition over the spoils of the Indian trade, for which English traders were known more for fraud than for fair dealing, sparked destabilizing violence across the eastern continent. In the "Wilds of America," their claims for special "Privileges" undermined the authority of their governors as well as the monarch who appointed them, for whom they bore no natural, filial love. As subjects, the colonists were, in their "excessive licentiousness," like the worst drunks at the most "disorderly Houses." Knox, like other well-informed Britons, knew that American colonies were limbs connected to the body of Britain's national vitality. With this knowledge came a corollary fear that they might be severed. Britain's strength—so recently shown overmatching the military might of a much larger and more populous France—rested on the artificial extension of its realm through overseas colonization and maritime commerce. That the very territorial integrity of the nation rested on American colonies invited a day of reckoning, especially if their many social disorders could not be mitigated by the state. As provincials sought to burnish their British cultural credentials, influential metropolitans voiced a disparaging understanding of Americans as "foreigners," who possessed a shaky claim on the status of ethnic insiders. Knox, like other proponents of imperial reform, stressed the degeneracy of American-born creoles in his writings.[41]

Although such criticisms could be vicious, they lent urgency to proposals to reintegrate the colonies into a stable structural relationship with metropolitan Britain before the process of cultural, social, and economic divergence became too deep-seated to reverse. At this moment, British descriptions of America oscillated between antinomies that only a new imperial order could reconcile. Sharp views of colonial difference shared space with celebrations of a shared culture of liberty, Protestantism, and enterprise, which united Britons at home and abroad. Appreciation for the value and dynamism of transatlantic trade jostled in pamphlets with evidence of how smuggling undermined the nation's interests. And most persistently, the idea that colonies should occupy positions of prosperous and secure dependence within the empire vied with the infuriating prospect of American independence—the inevitable outcome of the unchecked growth of populations and expansion of settlements. Knox and his fellow critics castigated colonists as different and inferior but also hoped that they could be brought back into the fold as members of a unified Atlantic society. Their insistence that colonists bear a greater burden for the costs of this empire by paying new taxes was an invitation to join other British provinces as contributing members of the same polity under the common authority of a farsighted Parliament. Against the backdrop of a century of disruptive colonial wars, the system of well-regulated defense, settlement, and government Ellis and Knox proposed held out the promise of a future characterized by American inclusion and prosperity.[42]

Under the leadership of the reformist "triumvirate" of Halifax, Egremont, and Grenville, Ellis and Knox found themselves emboldened by powerful patrons to analyze the pathologies of American colonies in stark terms and offer prescriptions for change derived from their firsthand observations about American capacities, characters, and conditions. Part of a larger generation of reformers, they joined the Board's long-serving secretary, John Pownall; Shelburne's personal secretary at the Plantation Office, Maurice Morgann; and Thomas Whately and Charles Jenkinson at the Treasury as influential "subministers"—those junior officials who gave "clarity and precision to dimly formed colonial ideas" of their powerful superiors and seized this moment to shape the course of British

America in the second half of the eighteenth century. Although Ellis turned down a post in the newly formed Colonial Office in 1768, electing to draw back from politics, Knox worked at the vanguard of Britain's administration to promote the policies he helped define. He was a mouthpiece for successive ministries, sounding the alarm over the national debt and insisting on Parliament's right to tax the colonists in *The Present State of the Nation* (1768) and proclaiming the necessity of reducing them to dependency in *The Controversy between Great Britain and Her Colonies Reviewed* (1769). In 1770, he accepted the position that Ellis had refused, joining John Pownall to become joint under-secretary of state for the colonies under George Germain, Viscount Sackville, from which he defended the Coercive Acts in 1774 and helped plan and organize the war against America. Ellis and Knox's shared vision for reform percolated through the bureaucratic channels of the imperial administration to initiate a striking shift in Britain's American policies in 1763. Morgann outlined Ellis's and Knox's recommendations in his "Plan for Securing the Future Dependence of the Provinces on the Continent of America" and summarized justifications for them in "On American Commerce and Government in Newly Acquired Territories." John Pownall synthesized these texts in his "Sketch of a Report concerning the Cessions in Africa and America at the Peace of 1763," the document that served as a first draft of the Board's formal report of June 8.[43]

The Board of Trade redefined British America as a coastal, commercial empire in 1763. Its 6,692-word "Report on Acquisitions in America" laid a practical foundation for what Shelburne had imagined as a "universal empire of that extended Coast." Within the territories under the king's sovereignty in North America, it described the expansive Atlantic edge of eastern North America as a place of productive economies, already inhabited by populous settler societies, and a sparsely peopled littoral, whose bounty would be released through intensive new colonization. It aimed at nothing less than the "secure settling of the whole Coast of North America . . . from the Mouth of the Mississippi to the Boundaries of the Hudson's Bay Settlements," organized to capture the wealth that inhered in its lands, waters, and climates from the "whole Variety of Produce which is capable of being raised in that immense Tract of Sea Coast." To the commodities British America already produced, it projected adding new supplies of fish from the Gulf of St. Lawrence; furs and deerskins from the indigenous interior; white pine masts (a vital strategic commodity for outfitting the navy's warships) from Maine's forests; indigo, cotton, and silk from the subtropical southeast; and more sugar (and coffee and cotton) following the "speedy Settlement and Culture of the new acquired Islands."[44]

In 1755, London cartographer Emanuel Bowen pictured the contest for North America on the first edition of his *An accurate map of North America,* an image that exhibited a divided continent in the midst of the Seven Years' War. **[17]** Across the surface of the western Atlantic, engraver John Gibson inscribed blocks of text that recounted the failures of diplomatic commissions to settle the geographic meanings of old charters, treaties, maps, and accounts of occupation and discovery, and described the "Chicanery" and "Incroachment" by which France had swindled Britain out of its rightful territories. Following the Peace of Paris, wrote Secretary of State Egremont with relief, Britain's American interests were no longer subject to the "tedious and unsatisfactory method" of referring disputes to the "discussion of commissaries." In the wake of this momentous agreement, Bowen revised his map to capitalize on the patriotic fervor with which Britons everywhere celebrated the new extent of British America. Erasing the first map's truculent charges of French deceit, Gibson reengraved the copperplate with sober excerpts from the treaty, the document that quelled territorial contention by granting Britain exclusive sovereignty over eastern North America. East of Newfoundland, Article 4 proclaimed "Canada with all its Dependencies" to be British; Article 9 added Tobago and three formerly "neutral" islands—Grenada, Dominica, and St. Vincent—to the king's dominions; Article 20 transferred Florida as well as "all that Spain possesses on the Continent of North America" east of the Mississippi. Those wealthy enough to purchase and display this wall map, which measured almost four feet across, did so to marvel at the grandeur of Britain's American empire. Its elaborate cartouche pictured an idealized Indian couple seated in calm subordination amid exotic fauna beneath the map's title. Over its frame hung a heavy fishing net, indicating one of the many commodities that could be profitably extracted from North America's Atlantic edge.[45]

The Board of Trade purchased a copy of this 1763 version of Bowen's map and annotated it to describe a novel geography of bounded colonial zones. **[18]** In washes of color, the commissioners drew boundaries around Quebec, Nova Scotia, Georgia, East Florida, and West Florida. Citing natural landmarks and compass bearings in their written report, they described the courses of these lines in the precise geographic language of a formal treaty, but they also illustrated the new colonial order in this "annex'd Chart in which those Limits are particularly delineated" to give the king a "clearer Conception than can be conveyed by descriptive Words alone." Across this map, they color-coded the continent, revealing the new shape of the empire, and imagining new relationships among European rivals and between British settlers and Native Americas. For all but two colonies, these new boundaries restricted places that had once extended across geographic space. "Canada as possessed and claimed by the French consisted of an immense Tract of Country including . . . the whole Lands to the westward indefinitely," but under the Board's delimiting paintbrush, the formerly vast New France shrank in size to assume the "proper and natural Boundaries" of British Quebec, rendered as a rose-colored wedge encompassing only the territory within the watershed of the St. Lawrence River. Spanish Florida's "great Tract of Sea Coast from St. Augustin[e], round Cape Florida, along the Gulph of Mexico, to the Mouth of the Mississippi" seemed just as vast and ungovernable. It was thus "indispensably necessary that this Country should be divided into two distinct Governments" and formed into West Florida, a green-colored band along the Gulf Coast, and peninsular, yellow-tinted East Florida.[46]

Georgia and Nova Scotia alone would grow in size according to the Board's design. Once battle-scarred provinces that defended contested imperial frontiers, Georgia expanded south to the St. John's River, taking on the dimensions of a full-fledged colony; Nova Scotia grew beyond its peninsula to incorporate what remained of French Acadia and encompass two great islands in the Gulf of St. Lawrence, St. John and Cape Breton. Bringing more territory within the jurisdictions of these "well regulated" royal colonies—still free from the worst excesses of "Republican Mixture" that characterized the other mainland colonies—not only allowed them room to grow after years on the front lines of imperial conflict but also charted a course for their social and political development as subordinate dependencies of the Crown. The Board had helped found and frame both colonies as eighteenth-century models, and both featured a significant military presence, charters that strengthened the powers of the governor, and a high degree of accountability to the metropolis. The Board intended the new colonies, "where planting and Settlement, as well as Trade and Commerce are the immediate objects," to have similarly "regular Government." They would be administered without interfering legislative assemblies during their founding phase by appointed governors who should be "obliged to constant Residence" and enjoined, on pain of immediate removal, to furnish "in a regular and punctual manner[,] such Information" about their colonies as the Board demanded. Far to the southeast, the commissioners colored the newly ceded islands of Dominica, St. Vincent, Grenada, and Tobago on Bowen's map, highlighting them as places slated for rapid development as revenue-generating, slave-consuming extensions of the West Indian sugar economy. Instead of decentralizing power by establishing separate governments in each of these islands, the Board urged that they be joined together "into one general Government" so that the king's commands "will go thro' the Channel of one Person," a governor charged with the critical task—so vulnerable to corrupting influences—of surveying and dividing the land "into Lotts proper for Sugar Plantations" and carrying this program of development "most speedily into Execution." A "large Military force should be kept up" in each of these new colonies to secure British sovereignty over them and to ensure the "Publick Tranquillity" as immigrants began populating them.[47]

With a free hand to shape these conquered and ceded lands into model dependencies, the Board of Trade urged a form of initiating rule by direct royal fiat that would have provoked immediate resistance had it been attempted in any settled colony. Shelburne introduced a note of conciliation into the "Report on Acquisitions in America" with a plea to make methods for generating "[r]evenue least burthensome and most palatable to the Colonies," but others who contributed to it saw in this new map of empire a welcome image of the colonies bounded and subordinated. The Earl of Bute, who headed the ministry as the ideas in the report moved from proposal to proclamation, cared little for the subtleties of this new imperial political economy, or even for the economic promise of planting new colonies, so long as Britain pursued a strong program to "bring our old Colonies into order." Reports of illicit trade by American merchants still rankled after the peace, because wartime smuggling reflected such "base ingratitude to the mother country," as Admiral Rodney put it. New England traders had supplied the "Enemy with every Means of protracting the War," charged Josiah Tucker, "greatly to their own Profit . . . but to the lasting Detriment of this Country, whose Lands and Revenues are mortgaged for Ages to come, towards defraying the Expence of this ruinous, consuming War." The report gratified British resentment over these perceived betrayals with its plan to police the Atlantic with "utmost vigilance" from the naval station at Newfoundland to prevent "Contraband Trade" and bring to justice those "desperate and lawless persons" who, through "Neglect, Connivance and Fraud," drained wealth out of the empire through smuggling. The Board took aim at colonial speculators whose "scandalous ingrossing" of the land, aided and abetted by "Extravagant & injudicious Grants" by royal governors, bent provincial government to the interests of private "land Jobbers." It declaimed against illegal logging and wanton "Waste" in the white pine forests of New Hampshire and Maine, set aside as the king's woods.[48]

One such punitive recommendation in particular captured the attention of Americans by striking at the heart of their identities as a colonizing people. The commissioners drew a bold red line across the mountains, imposing a western border for each of the colonies between New York and Georgia, and tinted the mainland within a continuous shade of blue. This imagined boundary was the first visualization of what historians have named the "Proclamation Line" because it pictured a stable frontier beyond which new grants were forbidden by the king's proclamation of October 7, 1763, issued four months after the Board of Trade first drew it on Bowen's map. Before the edict that set this sudden terminus, many anticipated that a prime dividend of the peace would be a dramatic expansion of settler societies on the continent. Parliamentary proponents of the initial treaty terms believed the end of New France would mean that the "colonies on the continent" could "extend themselves without danger." The great "increase of population" on the mainland had stimulated trade with Britain, and the Peace promised an acceleration of colonial settlements, as "American planters would, by the very course of their natural propagation," form a market that could consume all the goods British workers might make. Stretching across diverse latitudes, the "great variety of climates which that country contained" brought a new bounty of natural resources into a British circle of commerce that would end its dependence on the rest of the world. Advocates of colonization into the interior held that the "real grandeur" of the British empire would be measured not only in terms of the "mere advantages of traffic" but especially by enlarging the "extent of territory" and "number of subjects" who could be called British.[49]

The Board of Trade's commissioners, by contrast, imposed this ban on western settlement as part of an "exact union of system" that would advance two complementary goals: keeping colonists within the coastal orbit of Atlantic markets and protecting Native Americans from a postwar invasion of western settlers. To redirect burgeoning populations from further provocative encroachments, the Board defined the new colonies of "Canada, Florida and the new acquired Islands in the West Indies" as places of "planting, perpetual Settlement & Cultivation," which could absorb settlers in search of land. Their restless energies could be put to good use by encouraging them to occupy the "extensive . . . Line of Sea Coast" set aside on the map "to be settled by British Subjects." Redefining the western empire as the "Whole Coast" of North America plus the British islands of the West Indies thus enlarged the space open to colonization, but instead of empires in miniature with designs on expansion, each of the mainland colonies would now be bound by a line of division with newly recognized "Indian Nations." The spaces that fell beyond the jurisdictions of these marked colonies remained uncolored on the map, as an interior "Indian Country." Defined as "hunting Grounds; where no settlement by planting is intended, immediately at least, to be attempted," these lands "should be considered as Belonging to the Indians," who, like any other subjects of the king, now fell under his "dominion and protection."[50]

As mainland settlers saw their claims on western land erased, they also observed how ethnic and racial "others" claimed new rights and stature within the empire. By the terms of the Treaty of Paris, Britain claimed sovereignty not only over vast new stretches of American territory but also over the many non-British people who lived there. It granted the King's "new Roman Catholic subjects" freedom of religious practice, invited Spanish residents of Florida to live "wherever they think proper," and assured French planters in Dominica and Grenada that they could keep their slaves growing sugar. It made clear that Indians who lived between the Mississippi and the mountains owed their allegiance directly to the king rather than to the governor of any particular colony. Those Acadians remaining in Nova Scotia, subject to deportation only a few years earlier, might now become "useful membe[r]s of Society." This polyglot sovereignty extended to include the wide range of Native peoples who lived beyond the British pale and organized these interior spaces into a preliminary civil order that stood outside the relationship between colonists and their provincial governments. Within the span of two paragraphs, William Knox wrote of affirming the "Colonys in their Dependence" and of defending the "Independency" of Indian "Allies." Creole settlers, long accustomed to regarding themselves as superior Britons, saw themselves on the new map of empire sharing the continent with people they despised as religious, ethic, and racial inferiors. The expansion of Britain's sovereignty reduced the stature of provincials to one among many "fellow subjects" under the king in Africa, America, Asia, and Europe.[51]

The act of creating new colonies provided an opportunity to define a new kind of colonial subject. Those who "served so faithfully & bravely during the late War" might make ideal colonists to "undertake such new Settlement," a suggestion that yielded the proclamation's provision of fifty acres of land for every private, and up to five thousand acres for every field officer, to be claimed in Quebec, East Florida, or West Florida "without Fee." Discharged soldiers and officers would be joined by new European migrants, whom Knox believed should be kept a "separate People . . . without any great intercourse with the Old Colonys." As model colonists inhabiting new colonies well regulated from the start, they would not be lured to imitate the swagger of the "sordid Legislators of Carolina," whom "no King can Govern." With Bowen's *Accurate Map* laid before them, the commissioners regarded colonists as individuals, who could be led by their interests with the right combination of incentives and limits, rather than as members of distinct societies, each with its own history of adaptive settlement and claims to privileges as members of chartered polities. Compared with the halting progress of seventeenth-century colonies that took years to secure their subsistence, defend their territories, and produce exports, the Board's June 8 report urged an accelerated timeline for "speedy Settlement" for planting productive and self-sustaining societies so that Britain could reap the benefits of new colonization "as quickly as possible." Although the Board affirmed an ideal state of colonial dependency, its "Report on the Acquisitions in America" was no prospectus for subjugation. Throughout, its language was attuned to the psychology of colonists as self-interested actors and geared toward fulfilling their material wants. Because the prospect of "getting Lands without purchase" was the great "temptation" of westward migration, such a vice could be mitigated without direct coercion simply by ending interior land grants. Those settlers formerly "discouraged" from settling Nova Scotia could now be lured there with promises of land and security on this open frontier for settlement. The Board sought to leverage the Crown's legal control over real property to encourage colonists to "risque their persons and Property in taking up new Lands" where and how the commissioners directed them to do so.[52]

Britain's plan for America sought to create an integrated pale of settlement that stretched across eastern North America's coast and terminated at a fixed boundary with Indian country. The North American mainland stood at the center of this new configuration, and questions of how the eastern half of the continent should be bounded, developed, and defined shaped the debates over the empire's composition. Through the analytic lens of political economy, a cohort of planners described how tropical islands should complement the temperate mainland; how rising populations should populate strategic territories; and how divergent economic stakes in planting, trade, and manufacturing around the British Atlantic world could be reconciled by long-distance commerce. They drew lines around the edges of this newly defined mainland and embedded it within a larger domain that extended across the mountains and along the contested archipelagoes of the West Indies. Britain imagined America anew by seeing its collection of mainland colonies as a single space at the center of a reformed imperial system. Shelburne headed the Board as it drafted the "Report on Acquisitions in America," his devotion to political economy shaping its language, analysis, and recommendations. In 1768, East Florida governor James Grant described his colony as a "child of Your Lordship's Creation," one guided by Shelburne's visionary expectation that he would "see a New World rise" on the new frontiers of British America.[53]

As Britain eyed its newly expanded American empire, the Board noted the "greatest Difficulties arising from the Want of exact Surveys of those Countr[ies], many parts of which have never been surveyed at all, and others so imperfectly, that the Charts and Maps thereof are not to be depended upon." With some seized and inherited maps and other "authentick Information" left behind by the French, Canada's geography was, at best, "tolerable well understood." In attempting to describe Florida, however, a search of the "materials in [the Plantation] Office" by the Board's clerks turned up no "Charts or Accounts on which [they could] depend." To take possession of the new territories, the first task was to appoint a "proper number of able and skilfull Surveyors" who could map these unknown lands "with all possible Accuracy." These new surveys were necessary to transform British America into the form sketched on Bowen's map. To make the abstraction of an Indian boundary separating North America into indigenous and colonial zones a reality, representatives from some fifty-four nations and seven colonial provinces would have to negotiate and survey its course and mark it across regional maps of the interior that had yet to be created. The proclamation's description of lakes, heights, watersheds, and confluences gave the report's proposed colonial boundaries legal authority, but no one knew the precise coordinates of these landmarks. Good maps represented a fundamental knowledge of the land, required to direct colonization in the way the Board intended. Only by knowing territory singled out for settlement—that is, understanding its economic potential, its topographic variety, the contours of its terrains, the courses of its rivers, and the proximity of agricultural land to ports and harbors—could authorities make decisions about where to grant land, what it should produce, and how it could be formed into a civil, defensible zone for settlement.[54]

To describe their plan for America, the Lords of Trade sketched rough boundaries across unfamiliar lands and envisioned the integration of new agricultural hinterlands via uncharted rivers to the unsounded harbors of prospective Atlantic ports. This plan imagined a new process of colonization, one that centralized land granting and monitored economic development in the hopes of displacing the volatile trial-and-error method by which previous settler colonies had struggled to connect their territories to the Atlantic economy in the first generation of settlement. It sought to mitigate the consequences of releasing independent settlers whose ambitions to acquire landed fortunes in America had been the driving force behind territorial expansion. The Board's focus on the regulation and settlement of the new imperial frontier reflected the reality that colonies established as outposts of empire in the seventeenth century had become full-fledged societies by the second half of the eighteenth century. Their developmental trajectories had long been set into path-dependent courses that would have been difficult to divert even if the provincial elites who governed them had been willing to give up control of their futures. The Board sought to mold an empire that concentrated populations, agriculture, and trade along an expansive Atlantic coast. Working from the edges of the new map of empire inward, its plan concentrated the imperial state's resources at the margins of settled society, where no settlers yet lived to challenge the authority of London to enact schemes of regulation.

MISSISSIPI R.
LANDS RE
N. C
S. CA
GEOR
Tombeche
Natches
Apalachie
N.° 7
W. FLORIDA
Mobile
Ibbeville
Pensacola
21.
31.
MEXICO

{ TWO }

Commanding Space after the Seven Years' War

The Seven Years' War brought military surveyors into intensive engagement with American spaces. To clarify North American geography for commanders, they mapped forts at high resolution and traced routes that stretched across regions. The maps they made at these disparate scales sacrificed the ideal of the high-resolution topographic survey, a model of representation that the exigencies of frontier warfare made impossible to follow. Especially when these maps were gathered together and published in atlases, they presented a fractured image of America as an assortment of strategic sites and paths. As the army and navy occupied territories in the closing years of the war, engineers and navigators mapped conquered lands and coasts with greater rigor and sophistication. Their efforts, particularly in British Canada, demonstrated the power of comprehensive surveys to prepare new colonies for settlement. As troops took command of forts and ports in West Florida, these defensive outposts and the rivers that connected them served as a frame of nodes and spokes designed to give security and structure to a new military mode of colonization along the Gulf Coast. The Board of Trade's vision for postwar America sought to surpass the heterogeneity and incompleteness of wartime mapmaking, setting a new standard for representing knowledge with which to assert command over geographic space.

The "war now kindled in America," wrote Samuel Johnson in 1756, "has incited us to survey and delineate the immense wastes of the western continent by stronger motives than mere science or curiosity could ever have supplied." Britons pored over images of obscure American places during and after the conflict, which "enabled the imagination to wander over the lakes and mountains" and visualize an expansive empire. Mapmakers gave form to British ambitions for North America before the Seven Years' War by projecting charter boundaries deeper into contested interiors. In the midst of its campaigns, army and Admiralty surveyors mapped places that featured in military operations. The idea of a "general survey" of America—a key postwar initiative of the Board of Trade—emerged from this productive but unsystematic process of wartime mapmaking to represent vast spaces systematically. Over the second half of the eighteenth century, British America's shape underwent surprising changes and revisions before the eyes of map readers. This protean moment in the history of American cartography encouraged a sense of openness to spatial change, undermining the idea of America as a place defined by nature and suggesting, instead, that those who wielded political authority could dictate the form of North American empire to suit Britain's geopolitical interests.[1]

Wartime Mapmaking

As General Edward Braddock marched to his doom in the backwoods of Pennsylvania in the summer of 1755, his engineers looked in vain for the places marked on maps published in London. After the costly victory at Lake George in the fall, a New York militia captain sketched out the three "passes" that stood between Fort Edward and the French strongholds of Fort Carillon (Ticonderoga) and Fort Saint-Frédéric (Crown Point). Because he had "never seen any correct Map" of this area, he created one himself and sent it directly to Secretary of War Henry Fox. Over the course of the Seven Years' War, military mapmakers drafted new images of the "Country on the other Side of the Allegany Mountains" to orient commanding officers as they passed through uncertain lands. Such moments of disorientation and vulnerability, when no good map was at hand at a time when this knowledge truly mattered, underscored the need for systematic postwar surveys of North America and the West Indies among senior officials.[2]

Reassembling Britain's official production of early American maps is difficult, especially because the British government never collected these images in one place. The dispersion of manuscript maps reflects not only the diversity of agencies that commissioned and received them from abroad—including the Admiralty, army, Treasury, Board of Trade, and secretary of state for the Southern Department's office—but also the failure of the state to bring these wide-ranging images together for the purpose of analysis. Although these images were copied and shared among offices, Britain did not lodge this incoming flow of data in a dynamic "center of calculation" by which it might be formed into an accessible repository for the purposes of imperial administration. Those images that circulated widely did so because London's commercial map publishers gained direct access to manuscript drafts and printed engraved versions of them, following the British custom of forgoing secrecy in favor of broad public dissemination. As these maps lost their cogency as current records of strategic sites, they became artifacts that were deposited into a number of modern libraries and archives in the United Kingdom and the United States.[3]

British surveyors produced few maps of America before 1700. More new images reached London in the first half of the eighteenth century, reflecting the state's more general turn toward information gathering as it oversaw the expanding geographic range of its strategic and commercial interests overseas. The vast majority of official maps arrived after 1750, part of the surging cartographic output during the Seven Years' War and its aftermath. Beginning in 1755, Britain's military buildup in the western empire brought with it trained surveyors who dramatically increased its capacity for mapmaking. These surveyors produced scores of military plans of harbors, barracks, and forts, as well as sketches of the routes of communication connecting these positions. Colonial governors, many of whom also served as military commanders, drew on the expertise of these surveyors to illustrate their reports to Whitehall.[4]

In 1753, twenty-one-year-old George Washington, under a commission from Virginia governor Robert Dinwiddie, delivered a letter to the French commander in the Ohio River valley demanding that he vacate lands claimed by Virginia. This young land surveyor and militia officer drafted a map to illustrate the "French purpose to advance . . . to the Ohio and beyond." **[19]** Washington inked a path that crossed rivers and mountains between the French forts at Lake Erie and Will's Creek, a tributary of the Potomac River and the site of British Fort Cumberland. His sketch dilated the breadth of the Ohio River well beyond its true size to punctuate this new geography of intrusion and underscore the threat it posed to "prevent our settlements" from expanding into a contested interior. It was one of several maps sent home by anxious American governors that gave new form to an exposed western frontier. Washington's subsequent defeat on July 3, 1754, at Fort Necessity, a site he occupied in an attempt to forestall the capacity for French movement along this backcountry pathway, was the spark that ignited the larger global conflict.[5]

As the war began, other military mapmakers traced strategic routes across previously unmapped spaces. William Alexander, secretary to Massachusetts governor (and wartime general) William Shirley, created a picture of mobility across the northeastern frontier, tracing the many waterways—in yellow, red, and blue—by which forces from New France might move against British positions or be attacked from New York and New England. **[20]** A 1758 sketch map documented efforts to control the Hudson and Mohawk River corridors. **[21]** A census of red-inked companies and regiments deployed across a network of forts and towns illustrated the magnitude of military force that could be mobilized within it. Such maps pictured Britain's northern backcountry as an undifferentiated space to be traversed by soldiers on the march, but they revealed little about its enormous topographic or social complexity as a place located between European and Native American societies.[6]

These low-resolution overviews of strategic routes shared space in American dispatches with high-resolution images of new and refurbished frontier forts, including Fort Pitt, constructed on the site of the vanquished Fort Duquesne from 1759 to 1761. Named for Britain's wartime leader, William Pitt, the site—over which the conflict had begun—served as an emblem of triumph for the British cause in North America. After the peace, Mary Ann Rocque (topographer to His Royal Highness the Duke of Gloucester) published *A Set of Plans and Forts in America* (1763), begun by her husband, John Rocque, before his death in 1762. This atlas compiled engraved versions of manuscript images first drafted on-site by military engineers for audiences eager to view the scenes of action in North America, from the dispiriting defeats of 1754–1757 to the stirring victories of 1759–1762. Its thirty plates concentrate on interior forts as well as the fortified cities of Albany, Halifax, Schenectady, Montreal, New

York, and Quebec. Lieutenant G. Wright had produced a manuscript plan of Fort Pitt on the spot, which arrived in London in a dispatch from General Stanwix; it then left the obscurity of its place within the official papers of the secretary of state's office when the Rocques engraved a new image based on the sketch and printed and bound it into copies of their atlas. [**22, 23**] Such plans focused directly on the layout and surrounding topography of fortified positions at such high resolutions that their scale is reported in feet—rather than miles—to the inch. For those who had purchased recently published histories of the war, *A Set of Plans and Forts* illustrated the sites of the war's key engagements; however, it did not locate any of these sites within a broader region or explain why any of them had strategic value. With the exceptions of the plans of Fort William Henry and Louisbourg, which show the positions of besieging forces, these images are empty of any sign of human activity other than the military structures themselves and the surrounding provision fields that fed the troops. Leafing through the Rocques' atlas transported the reader from one American location to another, illustrating well-rendered defensive positions in a discontinuous series.[7]

As they captured forts, towns, and ships, British forces took possession of French and Spanish manuscript maps and the closely held knowledge about strategic sites throughout the hemisphere that they contained. Upon assuming command of British forces in North America in 1758, Lieutenant General Jeffrey Amherst consulted an extensive map library, which included a number of captured French maps. In actions at sea during the mid-eighteenth-century wars, naval officers "found on board their respective prizes many curious Draughts and Surveys of the SPANISH Settlements." From 1755 to 1762 alone, the British navy took fifty-six French ships and confiscated the contents of their captains' map cupboards. The "abundance of new materials" seized from the enemy found their way into commanders' eclectic personal map libraries, and "when the victorious fleets returned to England," many of these images became part of official archives. The seventy-two maps and charts in Admiral Richard Howe's collection, dating from the 1730s through the 1780s, reflected a long naval career that spanned a tumultuous period from the War of Jenkins's Ear through the American Revolutionary War. Among Howe's collection were some two dozen charts taken from the

French and Spanish. French place names on the manuscript "Chart of the East Coast of Cape Breton," along with a fleur-de-lis to point toward north, betrayed its origins. [24] Others, including eight charts of bays and coasts of Cuba by Spanish hydrographer Francisco Mathias Celi, are direct copies made on tracing paper.[8] [25]

Howe's capture of the French ship *Alcide* in 1755 was a consolation prize taken when the navy failed in its mission to intercept the bulk of troop reinforcements destined for Canada that year. This ship's documents and charts, however, along with those seized at the capitulation of the French fortress of Louisbourg in 1758, yielded geographic information about the navigation of the St. Lawrence River that helped Britain win the war. Britain had attempted to invade New France in 1711 during Queen Anne's War, an initiative that foundered in the river's notoriously treacherous waters and was doomed in part by a lack of good charts. As Rear Admiral Walker Hovenden prepared to take Quebec, he called on two provincial ship captains, Cyprian Southack and John Bonner, each known to have drafted a manuscript chart of the river, and each of whom declined to pilot the fleet's fourteen ships of the line past the Gaspé Peninsula and into the heart of the continent. Southack admitted that he had never actually ventured "higher up that River" than the Seven Isles "lying just at the mouth." Using his "Chart of the Gulf and River St. Lawrence," he briefed a French ship captain, who then attempted to pilot the fleet and its eight thousand men into Canada. [26] Despite sending small craft ahead to sound a route through the fog, the ships of the fleet lost their way. The British left behind eight wrecked transports and 884 dead soldiers and sailors before returning to England, having accomplished nothing. With Quebec secure, France stood firm in 1713 and refused to consider ceding Canada in the Treaty of Utrecht.[9]

After the Admiralty's James Cook and the army's Samuel Holland participated in the siege of Louisbourg in 1758, they met aboard the *Pembroke* as it anchored off Cape Breton Island and studied the cache of French charts that the British had acquired over the course of the war. An English annotation on the back of "Plan de la Riviere de Canada" noted that it had been drawn by Jean Sharpenez, pilot of the *Lantreprenant* when he successfully navigated the ship upriver to Quebec along with the fleet in 1755, helping transport some five thousand troops to reinforce

France's American armies. Combined with some new coastal surveys, these captured images opened the St. Lawrence to safe navigation. Military mapmakers revealed hazards on "[t]he river St. Laurence" by noting "Land Marks for avoiding" the ledges and rocks that lurked just beneath the surface of the water at low tide. New routes were "Plan[ne]d from a Manuscript found on board the Alcide" titled "Remarques & Observations pour la Navagation," an official report from France's *Dépôt des cartes et plans de la Marine.* **[27]** By discovering this secret hydrographic knowledge among seized papers and representing it on new charts, British military surveyors marked clear channels from the Gulf of St. Lawrence to the Isle of Orleans. With this new information, Major General James Wolfe knew enough about the river to plan an elaborate surprise attack, floating his soldiers silently along the ebbing tide in the middle of the night of September 12, 1759, and massing them for the engagement on the Plains of Abraham before dawn the next day. The army likewise deployed troops along a stretch of the river beyond Montreal aided by "A Sketch of the River St. Laurence" "taken from a French Draught." **[28]** These charts were crucial pieces of military intelligence that made possible the British conquest of New France in 1759–1760.[10]

London mapmaker Thomas Jefferys was the first "geographer to the king" to make the title meaningful. He gained regular access to an intensifying volume of official American manuscript images after 1750, and specialized in engraving and printing versions of them. In 1768, Jefferys (in partnership with Robert Sayer) poured the contents of his substantial American catalog into *A General Topography of North America and the West Indies,* an atlas whose 106 sheets displayed ninety-three maps, plans, and charts. This overstuffed compendium unites these fragmented images of American space to tell the story of Britain's triumph in the Seven Years' War as the realization of imperial grandeur. Many critics of the peace plan had argued that Britain should return the frozen wastes of Canada to the French and keep Havana and the tropical islands of Martinique and Guadeloupe. By attempting to control the entirety of the eastern continent, they argued, Britain risked the fate of Rome by extending itself so far across so much space. As he invited Britons to visualize the military victories that earned their nation's vast sovereignty in America, Jefferys brushes aside the narrow mercantile vision of empire behind such views. *A General Topography* dazzles the eye with the immensity of new territory that had been acquired at a high cost of blood and treasure.[11]

Like many atlases, its first maps offer the broadest view before descending into detail. John Green's "A Chart of North and South America, Including the Atlantic and Pacific Oceans" pictured the hemisphere in six separate sheets that could be pasted together to stretch from the Atlantic edges of Europe and Africa to the fringes of Kamchatka and Australia. [**29**] Although British America occupies just one of these six charts (the "Chart of the Atlantic Ocean, with the British, French, & Spanish Settlements in North America, and the West Indies"), from this place at the center of the New World, British sea power was positioned to command the whole. A second copy of this chart described how Britain assumed this central geographic position in the Western Hemisphere. Pasted across North America was a plate that showed the "Claims of the French in 1756" as a threatening yellow expanse; this could be folded back to reveal the continent partitioned by a French peace proposal from 1761, which designated the lands along the Ohio as a green neutral zone. Folding this plate back revealed the actual surface of the engraved map: the true configuration of the continent, Jefferys's map showed, was its red-colored borders indicating absolute British sovereignty, with the French nowhere to be seen. [**30–32**] If one were feeling patriotic as well as destructive, it would be easy to tear off these printed plates, glued to the map to document French pretensions to North America, and throw them away.

From this hemispheric view, the atlas surveys North America and the West Indies from north to south. Jefferys reproduces a series of important regional maps—first printed in the 1750s to stake British claims to North American territory—with specific new information about the extent of sovereignty claims, the spread of jurisdictions, and actual settlement across the land, to which he added his own recently completed "Florida from the Latest Authorities." [**33–38**] This map, cribbed from seized French and Spanish manuscripts, presented the new British territories of West and East Florida. Unlike the others, these new colonies had boundaries but lacked counties, parishes, townships, and any other sign that, beyond the outposts at St. Augustine, Mobile, and Pensacola, Europeans lived there at all. These regional maps of occupation provided a geographic framework for locating the many maps, plans, and charts in the atlas that zoomed in at high resolution to narrate British defeats and victories in the Seven Years' War.

By compiling the fragmentary record of forts and routes between the covers of this atlas, Jefferys creates an illusion of a depth in British geographic knowledge of America for a few important, well-mapped places that loom large in the story of the war. After the Virginians' defeat at the Battle of Fort Necessity, Washington pledged Captain Robert Stobo as a hostage to ensure British compliance with the terms of his surrender. Unwisely given his freedom to roam the grounds of Fort Duquesne, Stobo sketched a detailed plan that a Native American intermediary then smuggled to Fort Cumberland, where British officials made copies to send to London. [**39, 40**] Jefferys's version of this image, the printed "Plan of Fort Le Quesne," initiated his cartographic narrative of the struggle to command the forks of the Ohio River, played out over successive maps in *A General Topography*. [**41**] The tale of Braddock's catastrophic march begins by showing, in high resolution, the arrangement of the general's "flying column" of thirteen hundred men to evict the French from Fort Duquesne in July 1755. As drawn by Robert Orme, each unit symbol sits in perfect alignment along a straight, cleared path through western Pennsylvania's forests. Next, a "Plan of the Field of Battle" zooms out to show this body of troops in relation to the fort and the rivers it commanded, frozen in time, "as they were on the March at the time of the Attack," when French and Indian fighters arrived to wreak havoc on this neatly arranged formation. A final map pictures the "Route and Encampments of the English Army" from high above the surface of the earth as a dotted line beginning from Maryland's Fort Cumberland and terminating at a pair of crossed sabers where these forces clashed. [**42**] Britain had hoped to drive France from its frontier stronghold but instead suffered a grave defeat at the Battle of Monongahela after a laborious effort to build a road for its army through the wilderness. Jefferys retold this national tragedy in repurposed military maps and plans, engraved, printed, and arranged in retrospect to immerse *A General Topography*'s readers within a vital and remote American place.[12]

As such images encouraged viewers to reflect in sober consideration of the dangers of defending American empire, Jefferys brought others together, allowing them to revel in the daring enterprise of Britain's armed forces in victory. "An Exact Chart of the River St. Laurence" exposed the navigation of the great river based on images and information seized from the French; "A Map of the Several Dispositions of the English Fleet & Army" descended in scale to picture these forces, safely transported to the center of New France, arrayed to begin their siege; at higher resolution still, "A Correct Plan of the Environs of Quebec" pictured the topography of mudflats, fields, and houses along the river as well as the three-masted ships of the line, brought to attack the city; and a detailed prewar "Plan of Quebec" depicted the lines of its fortifications and the arrangements of its streets alongside a narrative of the history of previous failed attempts to take the "Capital of New France or Canada." [**43–46**] Victory on the Plains of Abraham outside the city turned the tide of the war for Britain and gave it the negotiating clout it needed to claim all of eastern North America during the treaty negotiations. Jefferys's atlas demonstrated how Britain wielded geographic knowledge as a weapon of war as it told the story of the Seven Years' War in America.

As *A General Topography* celebrates the vast extent of Britain's new dominions on the continent, it also expresses the idea of expansive command over navigation in the Caribbean Basin. Thirty-one individual town plans, roughly a third of the sheets of the atlas, picture Spanish Caribbean ports (along with a few French ports) on the islands and the encircling mainland. These plans reveal the historic port of Veracruz, from which gold and silver had been exported to Spain since the days of Cortés, the English logwood cutting encampments at the Bay of Campeche, and the Isthmus of Panama, famously seized, plundered, and held for ransom by Captain Henry Morgan a century before. The threat of British commercial inroads into Spanish America loomed over matters of war and diplomacy during the eighteenth century. After 1763, British acquisitions of new territories within and around the Caribbean Sea opened up lucrative new opportunities for trade with imperial adversaries. Picturing the ports of the Spanish Main so comprehensively was not an attempt to revive the seventeenth-century dream of the Western Design, in which England's Commonwealth forces had attempted to take over Spanish territories for the glory of God and country. Instead, these maps illustrate the many points of entry through which British traders hoped to take advantage of their maritime dominance in the region by opening new channels of commerce to sell slaves and manufactures and collect commodities and specie. Such images helped Britons imagine a western world in which consumers—eager to trade for what British factories and slavers could provide—were as important as subjects.[13]

The North American sheets of the atlas narrated a successful war for continental territory; these plans of Spanish port towns helped Britons picture fleets of schooners flying the Union Flag in motion across the Caribbean Sea. The colorist who shaded Britain's territories red on the Library of Congress's copy of the "New Chart of the West Indies" showed how the Peace of Paris had given Britain a commanding position over navigation along these tropical archipelagoes and coastlines. [**47**] The newly ceded islands in the southeast guarded the sea-lanes by which most ships entered the Caribbean Basin; and the Florida peninsula, outlined in British red, guarded the straits by which they typically exited into the Atlantic. A similarly colored map of Cuba showed how the Florida Keys, the Bahamas, and Jamaica surrounded Spain's largest American island. [**48**] Although the *Casa de Contratación,* Spain's official hydrographic service, sought to keep its harbor charts out of enemy hands, Jefferys gained access to these seized manuscripts through his official connections. Each represents an imperial secret revealed for the benefit of British naval and merchant pilots. Eleven charts of Cuba, including one of the Bahía de Nipe—an image taken from the same manuscript that was copied onto tracing paper in Admiral Richard Howe's map collection—reveal every significant harbor on the island, opening the colony to British navigation for trade as well as attack. [**49**]

A General Topography moves backward and forward in time. It focuses in on the streetscapes of cities, and details the positions of forces in battle. Above all, it describes the lands that the Seven Years' War was fought to win and, by showcasing their magnitude, makes a case for why it was worth the effort. Its maps recorded the costs and benefits of American colonies to Great Britain, tallying the price paid in money and lives for every victory. Britons did not read atlases in isolation. Benjamin West's 1770 epic painting *The Death of General Wolfe* sang the praises of a national hero who gave his life to drive the French out of North America.

Thomas Jefferys's *A General Topography* assembled fractured wartime images of America and gave them form as tokens of national sacrifice that established Britain's rightful claim to North American empire.[14]

Mapping Conquered Lands

Military engineers and navigators undertook rigorous surveys of strategic zones in the final phase of the Seven Years' War. In the wake of British victories, commanders ordered surveyors to assemble representations of conquered American territories that could displace the disjointed collections that Jefferys anthologized in *A General Topography.* For lack of credible charts, naval operations had "suffered very much during this war." In 1759, the Admiralty ordered its captains to begin gathering information about every coast, harbor, and port they encountered, transforming its deployed warships into a wide-ranging survey expedition. From that year, each vessel that cruised American waters kept a remark book that recorded "Descriptions for sailing in and out of Ports with soundings, Marks for particular Rocks, Shoals &ca. with Lattitudes, Longitudes, Tides, and Variations of the Compass." Naval surveyors located nearby sources of "Provisions and Refreshments," observed "Fortifications and Landing Places," and took note of the trade and fishing they observed. In 1761, John Major, master of the *Norwich,* charted the Gut of Canso, a long-contested strait that separated the eastern end of British Nova Scotia from the recently conquered Île-Royale (Cape Breton Island), as well as the shoal waters of the Seven Isles at the mouth of the St. Lawrence River. As his ship approached remote coasts far from the sites of previous battles, he charted the great bays of the western Gulf—Chaleur and Gaspé—and sketched potential harbors on the islands of Cape Breton, St. John, and the Magdalenes. After a French squadron took the English fishing station at St. John's, Newfoundland, in June 1762, the British navy mobilized its ships to recapture the port. With St. John's back under British control by September, James Cook, master of the flagship *Northumberland,* took up the Admiralty's charge by surveying the harbors of Newfoundland, the St. Lawrence River, and part of the coast of Nova Scotia as Britain secured the maritime northeast in the war's final year. Like many other naval surveyors, Cook drafted a new chart of the harbor of Halifax, the well-trafficked home of the North Atlantic fleet and perhaps the best mapped place in America. Every captain who brought his ship safely home to England surrendered his remark book and new charts to the Admiralty, contributing to an emerging body of hydrographic knowledge.[15]

As Britain brought its naval might to bear against French and Spanish targets around the world, forces under the command of Lord Andrew Rollo captured the island of Dominica on June 6, 1761. As soon as the army dislodged the French from their stronghold in the port town of Roseau, surveyors began drafting maps of this new possession. Lieutenant Colonel Archibald Campbell, a trained Royal Military Academy engineer, set down everything he could see of the island's terrain from the deck of the *Dublin* in his manuscript "Sketch of the coast round the island of Dominique." **[50]** Beyond his sight line, which extended some four thousand yards from the shore, the interior of Britain's newest island remained a blank and unknown space. Aside from spots of cleared farmland around Roseau, Point La Soie, Point Michel, and Grande Bay, the visible coast revealed few signs of human habitation among the densely wooded hillsides through which rivers had carved steep ravines before terminating at the shore. On his map, Dominica's coastline bore names for every inlet, point, and anchorage, copied from an undisclosed French precursor. As Campbell and his army engineers began their survey of conquered Dominica, the Admiralty compiled information gleaned from the French to create a new coastal survey. Rear Admiral Richard Tyrell's "Plan of the Island of Dominica" described three geographic features that officials in London would need to plan the island's recolonization: the names and courses of its rivers (along which new plantations would be settled), the locations of preexisting settlements (which the British would need to secure), and the boundaries of French administrative districts (out of which they could frame a new government). **[51]** This new "Survey and Map" ("taken under the direction of Admiral Rodney," whose squadrons had seized Martinique, Grenada, and Havana in 1762) revealed a Dominica "already divided into ten parishes or districts." Although these preexisting French "quarters" were "more unequal in extent" than the Board of Trade had instructed newly appointed governors to create, these jurisdictional lines were worth preserving, as they captured the practical knowledge that the French had acquired of the places that were "most convenient for Settlement," governed by what the "nature of the Country," with its forbidding mountains, "would admit." Immediately following its capture in 1761, British surveyors mapped Dominica from two spatial perspectives: one that rendered its topographic complexity with great density to create a naturalistic image of the landscape; the other that abstracted salient features of its natural and human geography for the purposes of imperial planning.[16]

In addition to securing new land on which slaves could grow sugar, the conquests of Dominica and other former "neutral" islands held out the prospect of unprecedented control over the sea-lanes through which shipping from Europe and Africa entered the Caribbean Sea. During the 1760s and 1770s, Admiralty surveyors assembled a portfolio of images that revealed the coastlines, hazards, sight lines, and depths of a dozen or more bays and harbors from which the navy could command navigation. John Stephens Hall updated Jacques-Nicolas Bellin's respected chart of "Les Petites Antilles" with a newly surveyed "Chart of the Caribbee Islands," which fixed the location of the misplaced Grenadines and composed new "Directions for Sailing" that linked the new islands to known routes. Hall's harbor charts not only identified safe anchorages throughout the Ceded Islands but also demonstrated Britain's capacity to improve on France's sterling reputation for generating geographic knowledge by enhancing representations of maritime space with new military surveys.[17]

The army engineers who came to America to design forts; lay out roads; and draft maps, charts, and plans in the era of the Seven Years' War followed a mode of military surveying first developed on Britain's northern borderlands. From 1689 to 1815, Britain invested in mapping a military landscape across contested territory in its Board of Ordnance surveys. After the Jacobite Rebellion of 1745, the army established a network of forts and garrisons in the Scottish Highlands. Maintaining these new frontier positions demanded an explicit rendering of the topography of this mountainous region, in part to locate the roadways leading to and from these new posts. [52] Lessons learned in the Highlands sharpened a sense of the need for the comprehensive mapping of strategic areas. Since "every possible movement proper for an Army to make in the Field, entirely depends on a just and thorough knowledge of the Country," surveyors were commanded to examine "minutely the Face of that Country" and note its visible features. Re-creating the complexities of terrain on a sheet of paper called for high-resolution images at larger scales of representation. Directing armed men and their supplies through a contested landscape meant knowing "whether it [was] flat and level, or interrupted with hollows and deep vales." It required a new set of symbols that recognized the variation in the appearance of the terrain, such as whether the soil was "Clay or Sandy, Rocky and Stony or smooth, in Tillage or in Grass." Men on the march needed to know the locations of fences that might obstruct their path as well as the "nature of the Wood" between the fields, whether "thick Copse and impassable or grown Timber and open," should they venture off the road. As Britain worked to map its strategic frontiers at home before the Seven Years' War, its surveyors developed expectations for what military maps should look like and the density of the information they should contain.[18]

From the seventeenth century, military engineers developed cartographic techniques to give commanders the ability to direct troops more effectively. As early modern European fighting forces became more mobile, their maps had to become more comprehensive to support "armies in fast-changing tactical situations which called for large-scale delineation of the terrain, its cover and communications, in sufficient detail to permit commanders to plan and execute mobile operations over extensive areas." European states organized the education and work of military engineers to include surveying and mapmaking alongside the construction of fortifications, and in integrating these areas of expertise, engineers of French, Austrian, German, and British armies coined a "common technique of topographical representation." Engineers trained at the Royal Military Academy at Woolwich, established in 1741, received formal instruction in theoretical and practical mathematics. From 1733, midshipmen who studied at the Royal Naval Academy in Portsmouth learned astronomy, geometry, navigation, geography, the measurement of heights and distances, and the "Construction of Mapps and Charts." Illustrations and exercises in the "Manuscript of Navigation" that guided their lessons demonstrated the use of the measuring chain, plane table, circumferentor (surveyor's compass), and alidade (telescopic sight). A special unit on the "Planning of Harbours" invited them to imagine themselves locating the "most remarkable Objects all round the Bay" to fix landmarks as they charted a prospective harbor from a boat floating within it. Military surveyors brought these established methods for rendering space to North America and the West Indies during the Seven Years' War. These included a rigorous plane-table surveying method that constructed maps out of an interlocking matrix of "traverses," or measured distances. After establishing a known baseline with a chain, surveyors then sighted distant landmarks from the end points of this measured line and inferred their locations through triangulation. This procedure set aside impressionistic sketches of "an entire Country" in favor of "one quarter mile truly laid down." Such high-resolution images captured the scale at which surveyors imposed a "Series of Triangles" on the landscape through controlled observation.[19]

To achieve the purpose of guiding military movements, these maps brought together two kinds of spatial knowledge. Moving through a landscape with purpose depends on the ability to follow paths aimed at a destination. The exercise of this *route* knowledge depends on locating visible landmarks that help the navigator get from one point to another across space. Maps that picture larger spaces from above, by contrast, offer a powerful, and cognitively distinct, view of space. This *survey* knowledge shows every place within the frame in clear relation to every other. Soldiers armed with high-resolution maps could reconcile positions within the bird's-eye view of the map with the details they saw around them. By verifying landmarks against a good survey, they could eliminate the distortion of the spatial field that takes place when venturing beyond a known route. Maps that did this well promised the ability to move from one route to another—what those who study the cognition of wayfinding call path integration—across the entire network represented in the survey, freeing officers to react to the moment and change directions with confidence. Beyond its strategic utility, this emerging standard for mapping embodied an idea of precision that resembled the thick narrative descriptions of the physical and biological world created by eighteenth-century naturalists. It held out the promise of the map as a lasting record of the place, enabling imperial officials to direct action from far away, long after it was made. When Britain lavished time and money on military surveys like those that mapped Scotland from 1747 to 1755, it produced images with topographic detail dense enough to permit viewers to picture themselves within the landscape and see it in three dimensions from any vantage point.[20]

As the Seven Year's War unfolded as a contest over the control of a small number of strategically located positions, wartime maps concentrated on high-resolution representations of forts, ports, and the rivers and roads that connected them. To simulate a coherent survey of the unexamined North American interior, military engineers positioned these scattered sites across space in rough sketches. When the demands of military campaigns ended and commanders took stock of the lands their fleets and armies had conquered, they aspired to show these spaces within the proper idiom of military landscape representation. As the war drew to an end, army and Admiralty mapmakers used a common style, scale, and selection of features to map these diverse places and worked together to join information about water, coastlines, and interiors. They embedded plans of forts within a landscape divided by types of terrain, shaded with hachures (the graphic shorthand for showing elevation before the advent of contour lines), and marked their maps with straight edges that stood for cleared fields, rectangles that described the footprints of structures, and grids of town plans. **[53]**

The army's trained surveyors, their skills sharpened in wartime, formed a sought-after cadre of experts promoted to lead new expeditions to map the new empire. Assigned the task of building roads through the forest during Braddock's march in 1755, engineer Harry Gordon sent letters home to prominent generals, presenting his familiarity with America's western woods as a credential for preferment in the army's networks of patronage. He dispatched a map from another sector of the war, the hotly contested corridor between Albany and Montreal. His "Sketch of the Country" from Fort Edward on the Hudson River to Crown Point on Lake Champlain sought to convey a "true Idea of the Nature of these Places" not yet captured by previous maps. Promoted to the rank of captain by 1766, Gordon led an expedition down the Ohio River, detailing the great waterway's western navigation. An ensign under Gordon's command, Thomas Hutchins, continued mapping the new British west along the Mississippi to Illinois country. Since the capture of Quebec in 1759, Hutchins had ranged along the Ohio as a soldier in Lieutenant Colonel Henry Bouquet's punitive expedition against the Western Indians and explored the shores of the Great Lakes to provide firsthand accounts of Native trading networks. In 1760, Jeffrey Amherst ordered Lieutenant Dietrich Brehm to "Explore the country" around the Great Lakes, "taking Plans, or Sketches of the Lakes, Rivers, and Lands, and Posts which have the Face of the Country."[21]

When British forces occupied New France in 1760, the territory's military governor, General James Murray, initiated a comprehensive survey of what would become, after the formal cession in 1763, the British colony of Quebec. The impulse to map Quebec came from military rather than administrative designs. Murray expected the province to be handed back to France after the peace had been negotiated, and he wanted to gather strategic intelligence that might be useful in support of a future invasion. As Murray explained to William Pitt in 1762, with this survey in hand to reveal the intricate passages along the waterways of the St. Lawrence River valley, Britain "never again can be at a loss how to attack and conquer this country in one campaign." Murray dispatched eight army engineers to lead surveys along different sections of the river. The composite map they produced contained seventy-four separately mapped sections that, when joined together, formed an interconnected image forty-five feet long and thirty-six feet tall. Representing space at the scale of two thousand feet to one inch, these maps were among the highest-resolution topographic maps produced by eighteenth-century surveyors anywhere. The Murray maps' design as a strategic profile of the province was made clear by the addition of demographic summaries that enumerated how many men capable of bearing arms lived in each district.[22] **[54]**

Together, these surveying expeditions examined more than two hundred miles of riverside terrain. After establishing the distance of baselines with chains, they used a circumferentor to measure angles to visible objects and fix the positions on which they stood as well as those of the landmarks they observed by triangulation. These they marked on a sheet of drafting paper affixed to the writing surface of a portable plane table, and as the surveying party made its way along the river, they repositioned the plane table, took new measurements, and connected a series of these located points to draw the contour of the river's edge. Along the way, as they moved from point to point, they drew visible features of the landscape within a two-mile-long band around the rivers. They represented forests with a stylized smattering of trees and sketched farmhouses, fields, fences, roads, and churches along the strips of riverside land cleared by the French. To show how the land rose up from the banks of the St. Lawrence, they shaded their charts with hachure marks to provide the illusion of topographic relief. These expeditions provided Murray with scores of high-resolution sheet maps that recorded the geography and population of the districts under his authority. Murray's engineers reduced and recopied their survey to fit on a few sheets of drafting paper so it could be pasted together and rolled into a portable scroll. **[55]** Officers arrived in London bearing copies of this composite map, which gave senior officials a striking view of the new colony. George III received an elegantly drafted copy of the map redrawn by Samuel Lewis, hired by the Board of Trade to make several such presentation versions from field sketches and fair copies sent home from America.[23] **[56]**

These maps made clear how little of the countryside along the river was occupied and how much flat, open land close to navigation might yet be cultivated. Murray's report on Quebec, written after the preliminary treaty terms made it clear that Britain intended to keep its conquest, considered the ways in which this colony now opened up to view could be improved. Echoing a well-worn trope of French lassitude compared with British industriousness, Murray deemed the Canadians "lazy," inclined to the "Gun and Fishing Rod" rather than the plow, and "not much skilled in Husbandry." Quebec's new attorney general, Francis Maseres, was "much delighted with the appearance of the country," a verdant land penetrated by a broad and "noble" river, and thought it "more beautiful than England itself." He agreed with Murray that only the "supineness and indolence of the French inhabitants" could explain how so promising a country remained the "poorest in all America." With their firearms confiscated and their economy no longer being held back by what Murray thought was the stultifying effect of French monopolies, they would be forced to focus more intently on the "Culture of their Lands." "With a very slight Cultivation," Murray believed, "all Sorts of Grain are here easily produced and in great Abundance." He envisioned farmers, among them newly arrived British colonists, sowing fields of hemp and flax to serve Britain's insatiable consumption of naval stores. Their wives and children would while away the "long Winters, in breaking and preparing the Flax and Hemp for Exportation," the sale of which would divert them from their habit of "Manufacturing Coarse things for their own Use" and, instead, provide the means by which they could purchase clothes "of a better Sort" made in Britain. This happy cycle of production and consumption could be set in motion with new bounties to encourage the French subjects of the king to begin contributing to an imperial common good.[24]

French *habitants* had been so focused on the fur trade, Murray believed, that they had neglected improving a bountiful fishery. He projected the creation of an "immense and extensive Cod Fishery" in the "River & Gulph of St. Lawrence" capable of providing an "inexhaustible Source of Wealth and power to Great Britain." Such a program of development would necessarily entail the planting of new settlements in the "Neighbourhood of the best Fishing Places, to which the industrious & intelligent in that Branch, may be invited and encouraged to repair." To foster a fishery of such magnitude, Britain must prepare a new colonial project that the French, in their neglect, had left undone: mobilizing settlers to cultivate a "Rich Tract of Country, on the South Side of the Gulph," from which they could grow crops, raise animals, and pursue fishing from newly established shipyards, ports, and harbors. He imagined that this recolonization of Canada could begin by purchasing the estates of Jesuits "at an easy Rate, and dispose of the same to many good Purposes." Most Canadians would learn the arts of market agriculture under the new regime, he concluded, and Canada would, "in a short time, prove a rich and most useful Colony to Great Britain." When he assumed command of this conquered territory in 1760, Murray reported that "No Chart or Map whatever having fallen into our Hands" could precisely define the lands that the "French stiled Canada." His newly surveyed map of the lands along the St. Lawrence answered that omission and paved the way for the development of British Quebec. The ideal military map presented a small part of the surface of the globe—on the order of a half mile to two and a half miles per inch—in great detail. Drafted to take thoroughgoing command of territory, this comprehensive style of surveying offered the kind of information about natural resources, rivers and roads, and agricultural potential needed to direct its colonization as well. Military maps made to this standard helped secure territory in the short term, but they also proved well suited to taking possession of it over the long term.[25]

Settling Inward in West Florida

In 1763 and 1764, the British army took possession of southeastern places previously occupied by the French and Spanish: long-settled St. Augustine; the Gulf Coast ports of Mobile, Pensacola, and Apalachee; and two interior, riverside forts at Natchez and Tombigbee. In 1765, it built a new fort at the juncture of the Iberville and Mississippi Rivers. Army mapmakers sketched plans of the forts inherited from France and Spain, sending officials in London their first images of British Florida. As other soldiers mounted cannons and secured shelters, these engineers drew buildings and town plans that imagined them transformed into centers of British civility. Trained to design fortifications in neat schematic drawings, they presented these structures in sharply rendered lines, angles, and polygons. Beyond these sites, Florida's uncharted coastline faced them as an immense and opaque space whose lack of definition shielded it from the imaginations of imperial improvers. The surveyors' task was to reveal this uncertain geography, sketching forts, locating harbors, identifying inlets, and probing the rivers that flowed through them from the interior, thereby opening the lands of East and West Florida to view so that they could be parceled out, taken up, and populated with slaves by well-capitalized planters ready to sow merchantable crops.

As the Lords of Trade composed their new map of American empire in the spring of 1763, they consulted a "Plan of Forts and Garrisons propos'd for the Security of North America" and integrated its recommendations into their master plan. Attributed to imperial adviser Henry Ellis, the plan detailed a scheme for troop deployments across the continent. It designated Quebec, Montreal, Niagara, Detroit, Nova Scotia, South Carolina, Pensacola, and Iberville as "Grand Stations"—each to house a regiment of 750 British troops—and identified other frontier sites ceded by France and Spain as "Outposts," to be manned by companies of soldiers. Just as the Board visualized its general plan for British America on a copy of Emanuel Bowen's *An Accurate Map of North America,* the army used the same map as a template to illustrate the spatial arrangement of its deployments through the summer of 1767. Daniel Paterson, assistant to the army's quartermaster general, drew his "Cantonment of His Majesty's Forces in N. America" to locate these regimental stations and connect them, by red lines, to the outposts that they were to supply with troops. **[57]** Paterson's map made visible the idea that, by knowing and occupying discrete places across the continent, Britain could take possession of the whole.[26]

The "Plan of Forts and Garrisons" imagined the army as a force for counterpoise on a frontier verging on disorder. Its ring of regimental stations and outposts marked the new limits of an extensive British sovereignty in eastern North America, but at the same time, it treated the Peace of 1763 as a mere respite from future geopolitical ruptures. The plan imagined a heavily fortified band of territory between the Ohio River valley and the Nova Scotia peninsula to secure recently conquered places. Between disgruntled French and Indian subjects in the north and potent indigenous nations in the south, the army would back up claims to British authority over these ethnically mixed populations with the power to enforce "Subjection." Occupying southeastern forts not only demonstrated sovereignty but also opened a number of tantalizing possibilities for projecting British power on land and sea. From Pensacola's "Central Position" along the Gulf of Mexico, British vessels could contain and, in times of war, prey on the "Spanish Homeward bound Commerce." From Mobile, British merchants gained access to a river that "stretched far" into the country between two "large Tribes of Savages," the Choctaws and the Creeks. St. Augustine could protect East Florida "in a future War . . . against any Hostile Attempt from . . . Havana," and its mounted rangers could "scour the Peninsula, in which there are still lurking some of the Yam[a]see Indians"—British America's "old and inveterate Enemies."[27]

The plan also suggested an improved method of military colonization in which the army would position troops at the frontiers before settlers arrived to claim tracts of land. Over the course of a long and bloody history, American colonists had made a habit of "settling outwards" in a pattern of unregulated expansion that had "expose[d] the Colonies to every kind of Insult." Taking possession of these forts presented an opportunity to establish a new modus operandi for territorial empire: "erect Forts early, at the Entrance of Our Dominions, & settle inward." This strategic vision for colonization imagined spaces for settlement along river corridors confined between fortified outposts at colonial borders and well-defended port towns that provided access to Atlantic markets. Instead of allowing settlers to disperse across an undefended and undefined landscape, the plan aimed to concentrate settlement exclusively along well-charted riversides, where planters would put slaves to work, clearing and cultivating the best soils for commercial agriculture. In this way, forts could serve as seedbeds for new colonies while protecting the king's new territories "against Encroachment" and intrigue by "Neighboring Powers." Drafting maps, plans, and charts of the forts, harbors, towns, and rivers that made up these contained zones of intended settlement was the first step in initiating this improved mode of establishing colonies.[28]

British troops relinquished Havana in 1763 and sailed to Florida to raise the Union Flag over forts that Spain and France had left behind. In 1716, France had established Fort Rosalie, an outpost some two hundred miles upriver from New Orleans, in Natchez Indian territory. [**58**] A half century later (October 1766), engineer Harry Gordon visited Fort Rosalie (renamed Fort Panmure). His arrival marked the final leg of a journey that had begun at Fort Pitt and, for the first time, established the safe navigation from the Ohio River to the Gulf of Mexico, opening the prospect for new colonization throughout a vast continental watershed. As scores of soldiers around him worked to repair its wooden structures for permanent British occupation, Gordon looked out from this elevated perch above the Mississippi and beheld the "Prospect of a very large & handsome Country, in many Places cleared, diversified with gentle Risings, which are covered with Grass and other Herbs of a fine Verdure." Across this verdant landscape he imagined grape vines growing on the uplands and indigo planted in the valleys; and, struck by the wild growth of mulberry trees, whose leaves are food for the silkworm, he judged the region to be "favorable for Silk."[29]

Downriver, the army established Fort Bute in 1765 to secure an alternative point of access to the Mississippi River. The diplomats who set the terms of the Treaty of Paris in 1762 agreed to keep New Orleans in

Louisiana, defined an alternative water route between the Mississippi River and the Gulf of Mexico, and set this passage as the boundary of British West Florida. Prime Minister John Stuart, Earl of Bute, observed that since the Mississippi River had several mouths, Britain could gain independent access to it from the Gulf of Mexico by entering the passage known as the Rigolets, crossing Lakes Pontchartrain and Maurepas, and proceeding along the Iberville River until it terminated at the Mississippi. Any number of maps Bute might have consulted to verify his presumption that this was a navigable route and that British vessels could sail "between the Mississippi & the Sea [along it]" without the "Obligation of passing New-Orleans," including Jean Baptiste D'Anville's *Carte de la Louisiane* (1752), would have given him confidence in its existence. [**59, 60**] The open waterway that appeared unobstructed on paper turned out to be a stream so narrow that boatmen had to strike their masts to pass under the interlocking canopies of the trees that grew on its banks. When Lieutenant Philip Pittman navigated a flat-bottom bateau toward the river from Lake Maurepas in 1765, he switched from fathoms to feet to measure its shoaly waters as he approached the Mississippi. [**61**] At six miles out, he declared two feet of clearance, and his party left its mired vessel to make its way on foot to the "Point where the Fort is proposed." The passage was so "choaked up by wood" that soldiers detailed from Pensacola and a gang of fifty slaves, presumably hired across the river from their French masters, spent months clearing it before building the fort. Within a year, new debris clogged the passage they had so laboriously opened.[30]

"The free Navigation of the Mississip[p]i is a Joke," concluded Harry Gordon, who predicted that "no Vessel will come to Ibberville from [the] Sea." The error of the "Peace Makers" had rendered what might have been the "most valuable Colony to the Crown in N. America" into useless land, cut off from any path to the Gulf of Mexico except that which passed under the batteries around New Orleans. This quandary inspired schemes to dig a canal above the fort to provide water to the intermittently dry and perennially clogged portion of the Iberville. [**62**] British surveyors returned to this spot twice more with plans to make the Iberville's sluggish waters flow, believing an independent passage to the Mississippi was a prerequisite to any meaningful settlement along its eastern, British, side. West Florida's provincial surveyor, Elias Durnford, drafted new charts of the Iberville and two other small waterways, the Amit and Comit, in 1771. [**63, 64**] He downplayed the difficulties of making them navigable and found "Remarkably Good" and level land, stocked with hardwood trees that testified to its fertility, throughout the area.[31]

Between 1767 and 1771, Britain granted 160,929 acres of land in 101 separate tracts along the Mississippi, each of which appears on Durnford's "Map of the Mississip[p]i River." [**65**] The spatial pattern of these tracts demonstrated the attraction of military positions as sites for investment, speculation, and settlement. For eighty miles south of the Yazoo River, a few scattered plats (sketch maps of the tracts' boundaries) marked spots where bayous and streams intersected with the Mississippi. At Natchez, however, a cluster of twenty-two contiguous grants surrounded the fort site, anchored by a five-thousand-acre grant to London merchant Samuel Hannay and a twenty-thousand-acre grant to Alexander Montgomerie, Earl of Eglinton. Interspersed among the lands of these metropolitan absentees were smaller headright and patent grants of one thousand acres or less intended for resident proprietors, some of whom became actual Natchez settlers. Despite the organized migration of several groups from the east to join this paper settlement, only "ten log houses and two frame houses" surrounded the deserted fort in 1776. Predicated on the false promise of the Iberville's navigability, forty-four grantees patented more than 83,000 acres of land from Pointe Coupee to the site of Fort Bute. Their narrow, rectangular plats, stacked adjacent to one another along a sixty-mile stretch of the Mississippi, formed an almost contiguous lattice of land claims.[32]

Some two hundred miles to the east of the congested Iberville, Philip Pittman sketched a "Plan of Mobile" in 1763. [**66**] Although the harbor was "shallow & impracticable to any great extent as a port," it did come outfitted with the large brick fort that France had constructed in 1711 to enlarge Louisiana's command of the Gulf Coast in the heart of Choctaw country. French Fort Condé became British Fort Charlotte, but it overlooked the same malarial plain, home to a few streets of "Straggling houses." Although the structure was, in theory, "Strong enough to resist all Indian Attacks," the soldiers who secured it were "so ill, and Weak," that they could scarcely mount a guard. Military plans, with their perfect symmetry and clear edges, masked the reality of conditions on the ground. They pictured how such sites could be perfected as much as they recorded how they actually appeared. In his "Plan of the Fort at Mobile," Pittman not only drew the footprints of the barracks, magazines, and casements as an idealized arrangement, elevated on the fort's distinctive seven-pointed star-shaped foundation, but also included an overlay that showed these structures well roofed and outfitted with chimneys. [**67, 68**] In reality, the entire fort was "in ruins" and the barracks were "so much decayed as not to be worth repairing."[33]

What Mobile lacked in situation and structures as a fortified port town it made up for as a point of access into the interior. From the harbor, the Tombigbee River (also known as the Mobile River) traced a winding, 150-mile course to an inland fort. France had established Fort Tombecbè in 1736 to bolster the power of the Choctaws against the British-allied Chickasaws. An unnamed army surveyor pictured this compound after the British took possession of it in November 1763. [**69**] It overlooked the river from its eighty-foot-high chalk bluff as a place as well situated for agriculture as it was for the Indian trade. Archaeological remains of plentiful maize cobs and white-tailed deer bones from the site suggest that colonial-era soldiers lived off the land. Sketches of the landscape included renderings of an extensive garden, irrigated from its own well, in which French soldiers had planted "plenty of Vegitables" and laid out beds in a formal, geometric pattern. Across the river, large cultivated fields testified to the area's proven capacity for agriculture. Crops of oats and barley planted there had come to "great perfection." Connected by a "Road to the Chactaws," Tombecbè, renamed Fort York, seemed ready to serve as a trading post in Indian country as well as a center around which settlers might take up grants of proven farmland.[34]

The army pictured the space between Mobile and Fort Tombecbè on a 1765 "Draught of the River Mobile." [**70**] Composed of four sheets of drafting paper pasted together to chart the river's long and winding course, it represented every bend, turn, and juncture at two miles to the inch. The few scattered French and Indian settlements noted along its banks, some abandoned, marked spaces that awaited more intensive settlement by British colonists. The river overflowed its banks every year, preserving the "richness of the soil, which for this reason is inexhaustible," a fertility demonstrated by the thick growth of canes and hardwood trees. Anchored by two secured points at each end of a lengthy stretch of river, this place seemed ideal for "settling inward," as put forward by the plan's vision of military colonization. But for all of its agricultural promise, Fort York remained an isolated outpost among Choctaws, whose population was reckoned as "full six thousand Strong," many of whom lived in a "large Town within Musquet Shot of the Fort." They regarded Fort York as a place to deliver deerskins and receive annual presents of guns, cloth, and rum. Even if it were perfectly defended and secured, in the event of an attack by the Choctaws its soldiers would become "so many Hostages in their hands for them to exact what terms they think proper from the English." Nevertheless, by 1775 Britain had granted fifty-eight tracts of land encompassing more than fifty thousand acres along the Bay of Mobile and the rivers that flowed into it. As delineated by surveyors Elias Durnford and Joseph Purcell, these property lines clustered where the river branched around Naniabe Island, site of a deserted Alabama Indian village. **[71, 72]** As noted on the map, the "rich dark brown Soils" claimed by these prospective colonists extended in a narrow vein about two miles from the riverside and seemed fit for "Indigo, Hemp, Flax, or Corn."[35]

Diplomacy, not military power, created this new corridor for colonization. On November 21, 1763, Britain solemnized a new alliance system at the congress at Fort Augusta that attempted to bring all the southern Indians together under the common sovereignty of Great Britain. On November 22, 1763, British troops took command of Fort Tombecbè. Southern Indian superintendent John Stuart and West Florida governor George Johnstone hosted some two thousand Choctaws and Chickasaws in Mobile in March 1765 to settle a boundary line with the colony as the "first necessary step towards the Settlement" of these lands. The resulting agreement permitted the British to "settle on Tombeckby River," but only up to its juncture with the "rivulet called Centibouck." This agreement opened a plantation hinterland upriver from Mobile, granting West Florida a "tract of rich[,] convenient, and extensive territory," and accounted for the distinctive inward bulge in West Florida's boundary.

Although this cession brought less than half of the land between Mobile and Tombecbè within the colony's territory, this stretch of the river was the only part that schooners could navigate year-round.[36]

Further east along the Gulf Coast, British forces took command of the fort at Pensacola, a Spanish outpost since 1698. Its new commander, Major William Forbes, sent home a "Plan and Sections of the Fort at Pensacola" in 1764, cautioning the secretary of state that it "appears fifty times better upon paper, th[a]n in reality it is." **[73]** The image traced the rectangular outline of a vast palisade that enclosed a space roughly half a mile wide. Within it, a well-ordered village of dwellings, barracks, storehouses, and the governor's brick house seemed to inhabit an orderly grid, complete with tree-lined gardens. In fact, this "entirely rotten" stockade fort was "so defenseless that any one [could] step in at pleasure." The barracks were "nothing more than miserable bark huts, without any sort of Fire places, or windows." Pensacola looked attractive as a town plan, but visitors reported that when experienced from the muddy ground, it was a "horrid place in every respect except the Harbour." Sheltered in squat palmetto-thatched structures, the soldiers complained of bitter conditions during the winter of 1763–1764, and since there was "Scarce a Chimney to be seen," it was "no wonder they [were] cold." They were already at work digging a ditch around the fort on which Forbes planned to mount defensive pickets, illustrated in the plan's sectional drawings. Traced in yellow ink were the shapes of four defensive bastions by which this inherited stockade fort might be improved to better defend Pensacola.[37]

Pensacola Bay provided a "magnificent" natural port that possessed every maritime advantage for pursuing the "Spanish Trade." Yet the "Country for fifty Miles backward from the Sea, [was] a white Sand which will produce Nothing." Before Pensacola could "become a place of great trade" and take advantage of its "fine safe harbor," there would have to be a true town established, a "Fort built, & Inhabitants sent here for the City, for there are none at present." The Board of Trade's vision of improved colonization found another expression in the imagery of occupied forts embedded within rationally designed urban plans. West Florida provincial surveyor Elias Durnford encased old Pensacola within a new urban plan, acting on the same principles his counterparts elsewhere used to plan model cities in Portsmouth on Dominica in the West Indies and Charlottetown on St. John Island in the Gulf of St. Lawrence, among other places. [74] The site of the old Spanish fort would be Pensacola's new central square, surrounded by a grid of lots, some of which were drawn over old structures built to no clear plan. Each rose-colored lot in town corresponded to a green-stained garden lot in the as-yet undeveloped wastes to the north of town, providing each resident with a plot of land in which to grow provisions. Two eighty-foot-wide "Princip[al] Streets" emerged from this imagined Gulf Coast metropolis to "communicate" with the interior, conduits for the deerskin trade that also anticipated the growth of a plantation hinterland. Joseph Purcell's 1778 plan documented the emergence of a well-inhabited place taking form within the pattern of this new urban grid. [75] Plans for a new governor's mansion that bristled with chimneys and featured Georgian architectural touches furthered this bid to develop Pensacola into an outpost of English society. Since the seventeenth century, treatises on military architecture set a new ideal standard for the city that featured a "geometrically perfect layout surrounded and defined by impeccably designed fortified defenses." British mapmakers realized this vision of rectilinear urbanity, at least on paper, for this West Florida port town.[38]

As Pensacola took the form of an improved European city, the unwillingness of the Creek Nation to cede land around it threatened British colonization plans in West Florida. Notoriously "Jealous with regard to Land," the Creeks had their own imperial ambitions in the region, which they believed had reached an apotheosis with the departure of the Spanish from Florida in 1763. At the Pensacola Congress in 1765, the Creek leader known as the Wolf declared "all the land Round" the town off limits to settlement and threatened that as "soon as the English began to settle the lands, they would declare War, & begin to Scalp the Settlers." The Creeks agreed to a boundary that created a fifteen-mile buffer around Pensacola, one large enough to enable a modicum of settlement but that did not extend far enough into the interior to reach the "rich Soil" along the Escambia River. Before the Indian boundary was permanently fixed, cutting his government off from access to the interior, Governor Johnstone encouraged Superintendent Stuart to propose that the Creeks grant West Florida a four-mile zone around every sizable river that flowed to the sea from deep in the interior, opening the riversides of the region to British settlement. Because the British were "planters" rather than "hunters," Stuart explained to the Creeks gathered at Pensacola for the congress of 1771, they had no designs to "get Your Hunting Grounds," the vast stretches of sandy "pine Land" maintained as deer habitat between the rivers. He asked only that they give up "Lands on both Sides of the Scambia," no more than five miles back from either bank, "as far up as a Boat can go." The logic of political economy suggested that these neighboring societies, with their different uses for the land, might each find its wants met through the judicious partitioning of riversides and interiors.[39]

Johnstone and Stuart made their proposal in the knowledge that military surveyors were hard at work delineating the "hitherto unknown Coast of that extensive Country" so that boundaries could be set around these proposed corridors for colonization into Creek territory. Charged to contribute its piece to the Board of Trade's mapping initiative by surveying the Gulf Coast, the Admiralty sent George Gauld to Pensacola in August 1764. He began his assignment by charting its famed harbor. **[76]** During the summer of 1765, he sailed across the Gulf to Espíritu Santo (present-day Tampa Bay) to "examine if it [was] fit to receive Capital Ships." **[77]** In early 1766, he worked aboard the *Active* to probe the rivers that flowed into Pensacola Bay, generating a new chart that followed the Escambia, Blackwater, and East Rivers to their headwaters. Later that year, Gauld headed east along the coast, surveying every bay from Pensacola to Cape Blaise. His 1766 "Survey of the Coast of West Florida" represented his findings in a complete chart that opened this space to British navigation. **[78]** Made in the midst of a war between the Choctaws and the Upper Creeks, who sought to quash their rivals' independent access to British trade goods at Pensacola, Gauld noted the presence of small tribes that had taken refuge along these remote shores. Arrows marked the currents and actions of the tide; cross-sectional prospective views pictured "several remarkable Parts of the COAST, as they appear from the SEA," enabling safe navigation to each of the area's sheltered harbors and bays. A view of Santa Rosa Bay behind its distinguishing red cliff was crowded with live oak, cypress, cedar, and yellow pine.[40]

More than two hundred miles east of Pensacola, the Florida coast turns south with the peninsula to form Apalachee Bay, another inlet that opened onto Creek country. To claim this portion of the Gulf, Spain had planted an outpost three miles up the St. Marks River sometime before 1640 and fortified it in the 1670s. San Marcos de Apalache had been a center for the Creek Indian trade and a supplier of provisions within a larger Spanish Caribbean network that included Veracruz, St. Augustine, and Havana. Spain abandoned San Marcos in 1704 in the wake of Anglo-Indian attacks throughout the region, but it reestablished the settlement and rebuilt the fort in 1718. British sea captain Thomas Robinson visited in 1754 and praised the "excellent and centrical position of this fine port," which opened up onto "its own river" and whose "interior parts, as far as the *Apalachean* mountains," might in British hands "carry on more commerce than all the other settlements in *Florida* put together." Venturing "into the country," Robinson found abundant mulberry trees and "exquisite grapes" growing wild. The first British map of the site revealed that Spain had been in the process of reconstructing San Marcos on a more impressive scale, replacing the small wooden fort with a larger triangular structure that pointed south toward the Gulf.[41] [**79**]

So difficult was navigation to San Marcos de Apalache (renamed St. Marks) that its isolated contingent of British soldiers came close to starvation as they waited for the provincial schooner *Dependence,* dispatched from St. Augustine with supplies and provisions, to round the Cape of Florida. Envisioned as an outpost of sixty soldiers by the "Plan of Forts and Garrisons," a skeleton force of some twenty men garrisoned St. Marks Apalache in the late 1760s. The Admiralty's George Gauld and the army's Philip Pittman teamed up to chart the land and water around the fort in 1767. [**80**] One could read the history of this imperial marchland in their annotations, from the ruins of the "Old Spanish Fort" and the site of the Creek town of Talahassa to the swath of devastated "hurricane Ground" that recalled the storm in the late 1750s that drowned twenty men and weakened Spain's hold on this outpost. The surveyors categorized the visible landscape into pine barrens, savannahs, ponds, and swamps, and marked the point distant from the coast at which the "land begins to be pretty good." The determination of the Creeks to maintain their command of the Apalachee interior made this geographic information moot and foreclosed any plan to colonize these lands.[42]

George Gauld was a "Mathematical Genius," enthused George Johnstone to the Board of Trade. Not only did his charts reflect remarkable "Exactitude," but he also stitched detailed sketches of small places together, creating elaborate annotated charts at a more "general Scale" so that this coastline could be comprehended as a whole. Viewing them "opened a new Light to us, both respecting the Nature of the country, and the Importance of it, which exceeds the most sanguine Expectations." By examining the manuscript charts Gauld brought back to Pensacola, Johnstone could see the natural harbors in which new port towns might be sheltered and from which settlers could advance upriver to follow his vision for development. In 1763, British advisers also anticipated that from a "good Port within the Bay of Mexico," situated for "commerce as well as command," Britain might also "bridl[e] the power of Spain" and flood its markets with manufactured goods. Lord Adam Gordon had already eyed Espíritu Santo, surveyed by the Spanish in the 1750s but never occupied, for its strategic position. In "any future War with Spain," a naval station there could "[cut] off all communication between La Vera Cru[z], and Cuba," because the trade winds pushed ships heading out of the Gulf so that they "c[a]me almost within sight" of its harbor. Gauld's surveys located these sites of command along the Gulf of Mexico, anticipating their settlement. At the Pensacola Congress of 1771, however, the Creeks rejected the idea of extending the territory of West Florida in noncontiguous spokes up the rivers and leaving an Indian hunting ground between them. Upper Creek headman Emistisiquo made it clear that other Indians might negotiate to cede interior land to the British, as did the Choctaws when they decided to "give Lands on the other side [of the] Tombeckby River," opening a riparian strip for colonial settlement above Mobile. But "on this side of it," he insisted, the "Land belongs to us." The Creeks refused to accommodate invasive British settlement above the tidewater and, by so doing, rendered the military's elegant and precise surveys of West Florida's bays and rivers irrelevant as a prospectus for colonization.[43]

Deputy Indian superintendent David Taitt traced the boundary these negotiations established on a 1772 manuscript map. [**81**] In the lower right-hand corner, one can find Pensacola's distinctive urban grid, built around the refurbished, star-shaped stockade fort. Just a few miles along the Escambia's course, at Boundary Creek, he drew the red line that hemmed Pensacola in and delimited its future. The boundary line followed the old road to Mobile before widening along the course of Bayou Roch to open part of Mobile's Choctaw hinterland up for settlement along the Tombigbee.

Samuel Lewis, hired to make presentation maps of the new empire for the king, gathered together drafts by provincial surveyor Elias Durnford into one image in 1772. **[82]** Drawing from a stockpile of charts of newly surveyed rivers and index maps that located rectangular plat boundaries along their banks, his image of West Florida was a plan of its major rivers—the Mississippi, Iberville, and Mobile—with the "Situation and Extent of the Lands granted by the English thereon." Lewis's plan imagined the site of a "New Town" upriver from abandoned Fort Bute. Those who came to live within the grid of eighteen blocks—flanked by armed redoubts and centered on a church—that defined it would find clear sailing to the Gulf of Mexico along the Iberville River, its recurrently clogged channel to be filled with flowing water from a "Proposed Cut." From the site of this planned Mississippi River port town upriver to the Yazoo River juncture, 107 British proprietors had claimed tracts of land by headright, purchase, and township grant, ranging from fifty to nearly fifty thousand acres. Seventy-eight more claimed tracts from Mobile to the Choctaw boundary and around the Bay of Pensacola, largely in increments of five hundred and one thousand acres. Beyond the yellow and green ink washes and the smattering of trees and bluffs with which he accentuated these riversides, Lewis left the colony's interior blank.

Two years later, at the request of Superintendent John Stuart, he drafted a new map that pictured these prospective rivers of settlement with the broader geography of the Southern Indian District. **[83]** Although this image lacked the record of colonial possession of its predecessors, it opened the southeast along its major waterways, offering the king a way to see his new Gulf Coast colony as a continental space populated largely by Native subjects. Just as Creek hunters helped make the eastern portion of this space productive by delivering deerskins to Augusta along the Coweta Path to be sent downriver to Savannah for export, this map of rivers depicted a vast interior, intersected with red-inked paths and marked with designated tribal hunting grounds, which promised the same for its western reaches. It left the lands above tidewater east of Pensacola—part of the Creek Nation—unmarked. An almost imperceptible dotted line traced the boundaries Stuart had negotiated with the Creeks and Choctaws, setting off a relatively small portion of the coastal zone for colonization, a limitation clarified by Joseph Purcell's 1779 map of the formal boundary survey of the line separating West Florida and the Choctaw Nation.[44] [**84**]

Pensacola's harbor was "so very fine," observed James Grant, "that it took every Body in." As a failed aspirant to the West Florida governorship, Grant had once been deluded by the idea that Pensacola would figure as the capital and center point of a wealthy and expanding British colony. He counted himself "very lucky" for getting posted to St. Augustine instead. When it became clear that the Creeks were adamant about refusing land cessions, its fortunes constricted to a patch of "barren sand, incapable of producing either Corn or provisions." Although West Florida's negotiated Indian boundary opened to take in significant tracts of land as it approached the Mississippi River, the limitations of the colony's territory for settlement, planting, and navigation dashed its colonizers' expansive dreams.[45]

At the very large scale of feet to the inch, British engineers sought to master Florida's spaces with precisely rendered plans of military fortifications and strategic routes along its rivers. Britain's scheme to define spaces for settlement between ports and forts in West Florida came to an end almost before it could begin when, as part of a general retrenchment of American forces, the army abandoned forts at Natchez, Iberville, Tombigbee, and Apalachee in 1768 and 1769. During their brief histories as British outposts, each of these places was in need of repairs, improvements, and, above all, healthy armed men. As troops withdrew, the idea that these fort and port sites could anchor vectors of plantation settlement within the territories they defined based on their environmental potential alone went with them. A few companies remained to defend Mobile and Pensacola, indispensable harbors whose occupation demonstrated British sovereignty in the region. Britain's attempt to recast colonization along Florida's Gulf Coast revealed how thoroughly all of those who participated in this enterprise were steeped in the idea, central to the Board of Trade's vision, that each place in the new empire could be prompted to produce the most lucrative commodities that could thrive in its distinctive environment. No matter how remote, ill defended, or vulnerable these isolated outposts were, there was a British visitor who inspected the land, imagined its riverbanks thick with plantations and fields flourishing with valuable cash crops, and penned a report or drew a map that testified to its promise as a rising colonial site. A lack of geographic information about Florida before 1763 meant that there were no contravening ideas about place and space to obstruct the logic of Britain's plan for West Florida. Only when rendered at a general, continental scale that presumed the navigability of mapped rivers and downplayed the spatial claims of different Indian powers in the region, did Britain's military plan and its visualization on Paterson's "Cantonment of His Majesty's Forces in N. America" map seem plausible in "this *Western remote out of the way place.*"[46]

A new spatial understanding of empire formed around the military experiences of the Seven Years' War. As Britain battled France for control of North America and the West Indies, its military surveyors generated maps, plans, and charts of tropical harbors, frontier forts, and frigid coastlines. These images captured the strategic essence of remote American places, but this new abundance of cartography underscored Britain's fractured knowledge of the hemisphere. Wartime mapmaking revealed British America with startling new precision, generating a scattered collection of high-resolution images of discrete places. This rush of cartographic production during the war changed the way British people, at home and abroad, understood American geography. New maps conveyed a sense of America's continental extent, put names to distant settlements, traced the courses of waterways beyond the tidal zones, and described the peoples and edges of Indian country. As readers consumed accounts of defeats and victories, perusing views of the places over which armies fought, they saw remote frontiers as worlds unto themselves in which the visual confusion of mountains, watersheds, and coastlines began to take on a recognizable order. These maps, no matter how accurately they might depict battle sites, fort plans, harbors, routes, and portions of coastline, were by-products of the exigencies of the military moment, inked on sheets of paper scattered in private and official collections across Britain and the colonies after the war. Even if they could have been brought together in the same place, they offered an uneven image of America from which to govern it. Unverified maps became tokens that stood for the largely unknown places around which senior officials imagined the new empire taking form.

A soon as Parliament approved the terms of the Treaty of Paris, Secretary of State Charles Wyndam, Earl of Egremont, sent a map (now lost) to Prime Minister George Grenville, marking new forts and colonial boundaries "according to the best ideas I have been able to collect." The army and Admiralty created new general maps of North America, envisioning it as an integrated place defended by forts, defined by an Indian boundary line, open to settlement in the northeast and southeast, and linked across the Atlantic Ocean through key ports. Yet officials in charge of this continental vision for defense, commerce, and development could not see the land at the continental scale with confidence. During the war, army plans and Admiralty charts pinpointed strategic places—forts and ports—from which force could be projected along extended frontiers. After the peace, the "Project for a Military Establishment in North America"—described in the "Plan of Forts and Garrisons" and represented on Paterson's map—sought to overawe new French subjects, back Indian diplomacy with "Authority," ward off French "Encroachments," and better "retain the Inhabitants of Our An[c]ient Provinces in a State of Constitutional Dependence upon Great Britain."[47]

The Board of Trade initiated its General Survey of North America in 1764 to create an imperial archive of images that were internally consistent with one another and capable of generating maps that could be reduced and enlarged to represent American space at different scales. Integrating the broad scale of the best general maps (which pictured the whole but failed to capture the accurate relations of the parts) with the focused scale of high-resolution military maps (which rendered small spaces precisely but did not represent these in relation to one another) became a key objective of imperial planning. British geographic ignorance of the interior, for which there were no good "charts or accounts," meant that officials could scarcely see this space as a coherent territory, much less control it. Creating maps that met such a standard for representation demanded countless miles of travel by "able and skilful surveyors," whose journeys and measurements within this territory would provide the raw data by which a new map of the continent could be constructed—river by river, valley by valley, and town by town—until the "whole of this country," and especially "that which lies between the great mountains and the Mississippi," was laid out on paper.[48]

rly as No. 51.
g contiguous to George Town makes it more valuable than 51 or 52.
d for Fishing or Agriculture; Large Vessels may go up Cardigan River.
situated for Farming and Fishing; Boughton River is shallow, and
igable only for small Vessels, and that with difficulty.
y good for Fishing and Cultivation; the Carrying Place from the Bay of
tune to St. Peters proving thro' it a great Advantage.
ry good Place for building small Vessels; at present an established
hery here.
Coast is very shallow, and but an indifferent Township.
ll situated for Agriculture and Fishing; plenty of Fish being off Bear
ve at the beginning of the Season.
rly as No. 63.
inland Lot in St. George's Parish.
HIS County Town Lot is very convenient for Trade and Shipping; and has
Mutually secured by Batteries on each Side of the Town.
Shipwreck Point
Shipwreck Cove
Short Point
Beaver Point
Britain Pond
Savage Harbour
St. Peter's Bay
River Morrel
Bedford Bay
Tracadi
Oyster Cove
Hills R.
Fortune
Boughton River
Cardigan River
George Town
Lot
Brudenel R.
Brudenel Pt.
Montague R.
Livingston Bay
Sturgeon
Murray
St. PATRICK'S PARISH
St. GEORGE'S PARISH
St. ANDREW'S PARISH
St. JOHN'S PARISH
BEDFORD PARISH
KINGS COUNTY
35
36
37
38
39
40
41
42
43
48
49
50
51
52
53
54
55
56
57
58
59
60
61
62
63
66
Hillsborough River
Dern Cove
Mermaid Farm
Lot
PORT JOY
Pownal Peninsula
Pownall Point
POWNALL BAY
Observation Cove
Goose Id.
Bacon Point
Bacon Cove
Point
St. Peter's Id.
H
BOROUGH
BAY
Vernon River
Great Afsention Village
Orwell River
Orwell Point
ORWELL BAY
Prim Village
Prim Id.
Pinet Village
Jenyns River
Prim Point
Gascoyne Cove
Jenyns Point
Flat R.
Belle River
Burnt Woods
Wood Islands
Indian Rocks
dry at low Water.

THREE

Securing the Maritime Northeast

The Treaty of Paris promised to bring clarity to the maritime northeast, a region that had been the focus of so much oppositional cartography, diplomatic contention, and bloodshed before 1763. It declared all of the "countries, lands, islands, places, and coasts" that comprise modern-day Labrador, Newfoundland, and the Maritime Provinces (now known collectively as Atlantic Canada) to be unambiguously British. The Gulf of St. Lawrence, formerly the entryway to New France, had interrupted Britain's claims to the North Atlantic coastal plain. Since 1713, the borderlands separating Nova Scotia and Acadia, inhabited largely by Native Americans, had devolved repeatedly into violent confrontations. With the exception of two small, hardscrabble islands, St. Pierre and Miquelon, the lands and waters from the Great Lakes to Newfoundland became unequivocally British—in the eyes of European statesmen, if not to their thousands of indigenous and French inhabitants.

In its plan for massing settlers on the land in optimal configurations, the Board of Trade called for the state to reshape America in two ways: by undertaking colonization directly in territories earmarked for settlement, and by prohibiting settlement in places where it was deemed detrimental to British interests. The Board's commissioners pictured a delimited settler society in the maritime northeast designed to exploit the region's "obvious Advantages" for commerce. For a century, New France had been configured expansively, as French colonists pursued a lucrative fishery in the Gulf of St. Lawrence and a thriving fur trade with Native hunters and trappers across the Great Lakes Basin. The Board of Trade delimited British Quebec to a rose-colored band that faced the Atlantic and reached only as far into the interior as the waterways that flowed into the St. Lawrence River. In pale yellow, the Board's map designated

Newfoundland and the Labrador coast as northern fisheries that Britain was now positioned to exploit more effectively. As Quebec shrank in this new order, Nova Scotia grew. The Board's commissioners recommended establishing the "true & just" boundaries of Nova Scotia to fill the space that British cartographers had long presented as a natural part of the Crown's domain. They envisioned the colony encompassing an "extensive . . . Line of Sea Coast to be Settled by British Subjects" from Maine's Penobscot River to the Bay of Fundy, across the length and breadth of the peninsula from Cape Sable to Canso, and along the western shores of the Gulf of St. Lawrence from Chaleur Bay to St. John and Cape Breton Islands. To shore up its formal sovereignty over the St. Lawrence River, "that Great Inlet to the Heart of North America," and capture the valuable fishery in the Gulf, the Board urged the "Speedy Settlement" of the islands that guarded this maritime entrance to the continent. After decades of frustrated efforts, Britain paused before what seemed to be an empty geographic canvas to consider methods to populate and develop the maritime northeast.[1]

Into this moment, ripe with possibilities for resurgent colonization, stepped Samuel Holland, with a plan for the comprehensive survey of North America as the means of realizing this vision of a well-settled northeastern frontier. As a young Dutch officer who had served alongside the British during the War of the Austrian Succession (1740–1748), Holland had presented two of his town plans to the Duke of Richmond. His expert draftsmanship secured him a lieutenant's commission in the Royal American Regiment, a unit stocked with experienced European military engineers, many of whom assumed leading positions as surveyors and mapmakers in British America. Promoted to captain during the Seven Years' War, Holland analyzed seized French charts of the St. Lawrence River alongside James Cook and led the effort to map its riversides under James Murray. In 1762, he sailed for London bearing copies of his composite map of the completed survey of Canada, which he presented to King George III and the Board of Trade. [**85**] This two-sheet reduction of the enormous "Murray Map" modeled a new standard for colonial cartography, informed by the best military practices for representing coasts, terrain, rivers, fields, towns, and landmarks in high resolution. This "accurate Map . . . of the settled parts of [the] . . . Colony of

Quebec" testified to Holland's "great knowledge of the northern parts of America" to the Board of Trade, which invited him to devise a plan for an expansive survey that could map all of British America with comparable accuracy.[2]

With this exemplar image in hand as a demonstration of how such maps could open distant lands to view for the purpose of "dividing, laying out, and settling" them, Holland proposed a General Survey of the American Colonies at a scale of one mile to the inch in 1763. He suggested creating new maps and charts of the entire Atlantic coast, from the "Channels & Harbours" to the "Rivers and Lakes." The Board adopted Holland's framework and urged its approval on the grounds that "no time should be lost in obtaining accurate Surveys of all Your Majesty's North American Dominions, but more especially" of those territories just acquired from France and Spain, whose lack of British inhabitants and considerable "natural Advantages" called out for rapid occupation. In 1764, the commissioners launched the General Survey to map "all of his Majesty's Territories on the Continent." They appointed Samuel Holland surveyor general for the Northern District and William De Brahm, another continental European military engineer, surveyor general for the Southern District. An imagined line extending west from the Potomac River divided the continent "as nearly equal as possible" into two great surveying districts.[3]

Each surveyor general commanded a deputy and several assistant surveyors and had at his disposal the services of Admiralty vessels, teams of ensigns, and an array of surveying and mapmaking instruments. Funded by a recurring annual parliamentary appropriation, the General Survey was an ongoing initiative with no set end point save the distant moment when every place in eastern North America might be fully mapped. The Board instructed its surveyors general to observe the "greatest precision and Exactness" in the creation of their maps. The location of every place required a measurement of latitude and longitude "settled by just astronomical Observations" that took the magnetic variation of the compass from true north into account. The information contained by the maps was meant to be a comprehensive record of place that assessed the land's capacity for colonial occupation. Each was to "convey a clear & precise Knowledge of the actual State of the Country," not only its boundaries and location (its "Limits, Extent, Quantity of Acres" and "Principal Rivers and Harbours") but also its capacity for development (the "Nature of it[s] Soil and Produce, and in what Points capable of Improvement"). To begin, each surveyor was to focus on preparing new colonies acquired in 1763 for the arrival of settlers. In the north, Holland would map the islands and chart the western coasts of the Gulf of St. Lawrence. His counterpart in the south would undertake the mapping of the new province of East Florida to "accelerate" the establishment of a plantation society in a virtually unknown country. Their work combined terrestrial surveying of coastal lands with hydrographic surveying of the waters next to them, establishing the knowledge needed to divide the land up for occupation and secure a safe navigation to and from new colonies. As the commissioners fielded schemes for "making Settlements upon these Coasts," the maps and reports produced by the General Survey were "to be in great Measure the Guide" by which their proposals would be judged.[4]

As Holland and De Brahm began their work at opposite ends of the mainland, the Admiralty initiated complementary expeditions to complete British coverage of North America's coastline. Along the Gulf Coast, George Gauld examined the "coasts and harbour" between New Orleans and the western Keys. James Cook and J. F. W. Des Barres charted Nova Scotia, the eastern shores of the Gulf, and Newfoundland. Army expeditions along the Ohio River, around the southern shores of the Great Lakes, and into Illinois country sketched out a general geography and ethnography of the interior on a handful of new maps. The superintendents of Indian Affairs, north and south, mapped the location of an Indian boundary line as it emerged between 1764 and 1774 as the result of negotiations with Native nations. The Board of Trade organized this collaborative surveying project, communicating with other agencies and disseminating manuscript maps and reports among them.[5]

The Board of Trade directed the surveyors general to create and submit maps at the uniform scale of two miles to the inch. This standard set a resolution that offered close views of terrain at a scale large enough to show the location of individual structures, the layouts of towns, and the distinct edges of fields. Although the surveyors reduced such detailed images to create new general maps of provinces and regions, these two-mile-to-the-inch maps captured space at a foundational level. Such maps commanded the authority of systematic, firsthand observation and could be redrafted to represent strategic spaces in ways that matched unanticipated needs of military commanders, imperial administrators, colonial governors, and new settlers for reliable maps of specific places. Holland took up his charge as the northern surveyor and spent more than a decade charting the Gulf of St. Lawrence and the adjacent coasts of New England.

Lessons from Canso

A cluster of islands just off the coast of the Nova Scotia peninsula took on outsized significance in Britain's quest to define its North American empire before the Peace of Paris. Canso marked the northernmost frontier of a band of coastal land stretching north from South Carolina, which Britain regarded as a territorial empire by 1713. As geographic promises laid out in treaties fell apart, the Board of Trade learned important lessons from the struggle to keep Canso. To truly possess the sparsely settled borderlands of the maritime northeast, Britain would need to understand its complex coasts and rocky terrain; parcel out the land based on this knowledge; and plant colonists who could farm, fish, and defend the edges of empire by their physical occupation of it. Colonial fishermen prized the sheltered channels between the Canso islands as a place to sojourn, not to settle. The Board of Trade initiated some of its earliest settlement schemes to pull this and other remote outposts out of the economic orbit of New England and populate them from London.

The maritime northeast featured scattered settlements and overlapping spheres of practical power. It was not so much shared by the French, British, and Wabanaki (the loose confederation of indigenous peoples in the region) as it was intersected by these groups as they engaged in farming, fishing, hunting, logging, and fur trading, sometimes in collaboration, sometimes independently, and sometimes in competition with one another. By the close of the seventeenth century, European mapmakers had rendered portions of the coastlines that matched these fragmented interests with precision, but the "problem lay in connecting all the pieces." Piecemeal cartography reflected the discontinuous exploitation and occupation of the region, and by seeking to map it all, Britain expressed its bid for exclusive command.[6]

This contested region came into focus as a strategic problem over the course of Queen Anne's War (1702–1713). Far from Boston and Montreal, French and English colonists ventured beyond fortified areas of settlement and into spaces inhabited by Abenaki and Mi'kmaq Indians that fell between clearly delineated colonial jurisdictions. New England townspeople along the upper reaches of the Connecticut River, Acadian farmers entrenched along the Rivière au Dauphin in northwestern Nova Scotia, timber harvesters in the woods of Maine, and itinerant fishermen encamped on the shores of Newfoundland's Avalon peninsula were all drawn away from well-settled colonial centers to exploit alluvial soils, productive marshlands, pine forests, and abundant shoals of fish in this vast, resource-rich zone on the edges of New France and New England. Those who lived in these detached settlements bore the brunt of the violence when war broke out in 1702. At its conclusion in 1713, each side sought to shore up its claims to territory in this contested region through treaty diplomacy. Britain gained title to Acadia, which it renamed Nova Scotia, extending its claims to the eastern coast of North America to the Gulf of St. Lawrence. France was awarded the large islands in the Gulf and began constructing a fortress at Louisbourg on Île Royale (Cape Breton Island). The Treaty of Utrecht declared the sea at the edge of these opposing territorial empires as a space open to the fishermen of both nations.[7]

When mapmaker Herman Moll shaped the mainland empire into a new form, seemingly defined by nature for British occupation, he marked its edge at "Canceaux," where the interposition of the Gulf of St. Lawrence broke the extent of British sovereignty before it could reach Newfoundland. Despite his map's certainty that Canso belonged to Britain, this place was in fact a contested outpost under perpetual threat from the French and their indigenous allies. This fishing port drew hundreds of vessels from the distant port towns of New England (in 1720, ninety-six ships arrived to take cod), but only a handful of people remained as permanent winter residents. How could strategic lands be populated in places where settlers were reluctant to take them up of their own accord? How should the state oversee colonization to mitigate the dangers that came from claiming underpopulated lands? By thinking through the implications of Canso, British officials forged a new commitment to state-sponsored colonization and conceived of new ways of implementing it.[8]

Armed with the treaty's guarantees of British fishing rights, New England merchants sent vessels to Canso. The New Englanders' arrival at the "best and most convenient fishery" on the coast met with determined resistance. The French "instigate[d]" the Indians to attack the king's "subjects fishing at Canço" and, by impeding them, cemented control over the valuable *pêche sedentaire,* the regional fishery anchored by permanent settlements at Île Royale, Île St. Jean (St. John Island), and other places around the Gulf. The British retaliated by seizing French vessels and raiding Acadian villages. The French complained of the "depradations at Canceau" and submitted for British review a "Coloured Map" that pictured the port "separated from the mainland of Nova Scotia by a small arm of the sea." As "an island," they claimed, it belonged to France, along with all the other Gulf islands, by the terms of the treaty. The British categorized Canso as an integral part of Nova Scotia as they attempted to secure sovereignty over it in diplomatic negotiations from 1720.[9]

As the Board of Trade worked on a new general plan for colonial reform in 1720, news of the attacks on Canso revealed Britain's struggles to defend its interests in a region in which its hold on territory was so tenuous. The "daily disputes" with the French "concerning [th]e fishery at Canço" made it imperative to fortify, settle, and defend it, but when pressed to find a chart of the place, neither the commissioners nor the Admiralty could find "any maps of this country" that could locate it. Such a glaring deficiency in geographic knowledge was the shock that initiated a new policy regarding surveying and mapmaking. It convinced them that Britain should send an "able person from hence to take a survey, and make exact maps of all the several Colonies from North to South, which the French have done for themselves, from whence they reap great advantages whilst we continue in the dark."[10]

Thomas Smart, captain of the Admiralty ship *Squirrel,* brought home a rough sketch of the islands in 1718 drafted by Cyprian Southack, an experienced New England pilot and privateer. Southack published a version of this manuscript in Boston in 1720 to inform wary ship captains about this perilous fishing station marred by "French Intruders." **[86]** Only by 1732 did Captain Thomas Durell complete a more rigorous and detailed survey of the islands for the Admiralty, more than a dozen years after the first conflicts over them. Southack compiled and published the results of his coastal explorations as *The New England Coasting Pilot* in the mid-1730s. **[87]** On one chart, he marked the "East Boundaries of Nova Scotia" as mainland British America's ultimate frontier, facing Île Royale across the narrow channel known as the Gut of Canso. Inscribed in text across this territorial edge of empire, he narrated his own decades-long exploits in a region riven by violence. Just below this description, he located "Canso Road and Harbor" in the first legible, widely circulated image of this strategic outpost, denoting its navigational hazards and demonstrating British occupation of the islands with icons of hearth-bearing houses. South of the peninsula, he marked out the three large fishing banks that drew New England captains to Canso despite the dangers.[11]

As they reacted to the crisis at Canso, the Lords of Trade turned to another underpopulated district, the Maine coast. As surveyor of the woods, Colonel David Dunbar was empowered to take possession of large stands of white pine trees, the only source of masts for oceangoing sailing ships, and secure them for the Crown. With the Board's support, he bound this mercantilist mission into a settlement scheme that aimed to plant a new colony on the Pemaquid River. When his party of soldiers arrived in 1729, they established Fort Frederick, traced a grid of urban lots beyond its walls, and cleared a tract of nearby land for planting—accomplishments Dunbar documented on a map sent to the Board. **[88]** The Pemaquid colony challenged the interests of merchants who profited from illicit logging in the district as well as those of the "Great Proprietors," who held title to millions of acres in Maine by virtue of "several Antient Claims" granted in the 1620s. Advocating for these New England interests, Massachusetts governor Jonathan Belcher used his political influence to scuttle the Pemaquid scheme. Dunbar abandoned his quest to populate this coastal frontier with Scots-Irish immigrants in 1731. Under the sway of New England elites who regarded Maine as their eastern frontier, such lands, observed the Board, remained "waste and uninhabited," claimed by colonists who had never legitimized their paper titles by carrying them into "actual Possession & Culture." The commissioners could not displace the New Englanders' geographic understanding of this space as a "maritime border zone," centered in Boston and organized around the commercial imperatives of "New England seafaring," but they attempted to redefine it. "All the Lands from Canço to the River Kennebeck," they argued, should be regarded as space that needed to be settled and defended in organized townships from London.[12]

After the Board tried and failed to colonize Maine, the commissioners reflected on the ideal process by which new colonies should be formed. Although they castigated New Englanders as selfish interlopers in the maritime northeast, they regarded the New England town system as the "best Method to be followed" to populate the northern and southern frontiers of North America. Rather than permitting settlers to disperse across territory as independent planters, Massachusetts approved petitions from organized groups of families to which it gave a "Tract of Land from Seven to Twelve Miles Square" in a formal corporate grant "called a Township." The Board's commissioners admired the organization of such settlements, particularly the provision that they must be built around a town that supported a church, a minister, and a school, and praised their rules for granting lands to inhabitants based on their numbers and needs. Such townships not only provided "some sort of Security against the Savages" but also formed civil, self-governing polities within the larger political order of the province.[13]

The Board of Trade began planting its own versions of these townships, which blended the practical communalism of the "New England way" with a lost medieval ethos of obligated ownership. The traditional English township, constituted by a community of free burgesses, bound its inhabitants together in ways that constrained "private right" in the use of land for the public good. Historians have understood the "Scheem . . . for Settling Townships" in the South Carolina Backcountry as a regionally specific solution to the problem of defending the Lowcountry's black-majority plantation society, but it was in fact part of a continental conception imposed from Whitehall. From 1730 to 1775, the Board of Trade organized the granting of townships in West Florida, East Florida, Georgia, South Carolina, New York, Quebec, Nova Scotia, and St. John Island. Although the specific terms of acquisition and ownership differed in each place, most shared common features: the Board approved metropolitan petitioners as New World landowners only after the commissioners had vetted their means, characters, and intentions; the Privy Council, bypassing provincial governments, granted the land in large tracts, typically twenty thousand acres, directly to the proprietors, who were required to settle and develop these tracts or risk forfeiting them; and township recipients paid for the privilege of gaining title to so much land with regular quitrent payments that not only acknowledged the Crown's enduring sovereignty over and interest in this land but also provided a direct fund for governing new colonies that did not depend on the legislative appropriations from colonial assemblies. Out of the failure to settle Maine in 1729–1730, the Board of Trade codified a new mode of colonizing that would, it imagined, populate uninhabited territories across the continent.[14]

At the beginning of King George's War in 1744, French and Mi'kmaq forces descended on Canso Island's town, burnt it to the ground, and imprisoned its inhabitants. This notorious attack undermined Britain's possession of Nova Scotia and troubled the imperial idea of an integrated mainland empire. In 1746, Massachusetts agent William Bollan saw this part of the world as Herman Moll had drawn it in 1715, as a "British American Empire . . . extended upon the Sea Coast from Georgia to Newfoundland saving the Break made in it by the French having possession of Cape Breton." To illustrate his published plea that the Crown seize Île Royale, Bollan included a map of North America that reasserted the idea of imperial dominion stretching—settler colony by settler colony—in a continuous "Chain" of occupation, capped by a contested Acadia. [**89**] His copy of Jacques-Nicolas Bellin's *Carte de L'isle Royale* presented the location of this rupture in high resolution. [**90**] Only the narrowest of sea-lanes—the Gut of Canso—separated British Nova Scotia from Cape Breton and its menacing fortress at Louisbourg. The great island occupied a maritime "Center-point" for the western Atlantic. It was only a week's sail from Louisbourg to New England, Quebec, or Bermuda, and the commercial fleets that served the West Indies followed the Gulf Stream north, close to its shores, before heading east across the Atlantic.[15]

Not only did Île Royale disrupt the sweeping vision of an exclusive British dominion over Atlantic North America, but its hungry population of soldiers warped the political economy by which these two rival empires were to inhabit this shared space. Every early modern military outpost needed a nearby community of farmers to provide for its material wants, and Louisbourg's demands brought it into an illicit alliance with the so-called neutral French settled in Nova Scotia's Annapolis River valley, which tarnished their standing as subjects of the British king. In addition, this demand supported a "destructive Clandestine Trade" by which "near an Hundred Sail of decked Vessels" from New England arrived at the French fort laden with cargoes of trade goods ranging from planks and onions to butter and cattle. Stung by the destruction of its Nova Scotia fishery, Massachusetts gathered an invading force of some four thousand men at Canso that "wrested" Louisbourg "out of the Hands of that haughty, perfidious, and insulting Nation" in 1745. The busy hands of skilled surveyors among the invading force populated Britain's imperial libraries with numerous images of the town, fort, and harbor as it appeared in the mid-1740s. **[91, 92]** With the Union Flag flying above its ramparts, eager merchants imagined an exclusive continental "Fishing Coast" stretching from the "Bank of Newfoundland, to the Southernmost Part of Georgia" once France was expelled from the Gulf of St. Lawrence. By taking possession of Cape Breton Island, one New Englander imagined that just as the small number of colonists who first landed in Massachusetts were "now multiplied into an incredible Number of Inhabitants," Britain could likewise settle a "Chain of Towns from Lewisburg to Canso; from thence to Annapolis Royal," the Acadian heartland in Nova Scotia, and "so on, to Casco" Bay on the thinly settled coasts of Maine. The previous century's "amazing Enlargement of the British Dominions," with its "Encrease of People, Trade, and Towns" across the mainland, suggested that with the French gone, Britain could finally take possession of the maritime northeast.[16]

By returning Île Royale to France at the peace talks of 1748, Britain nullified Massachusetts's great victory and inhibited the expansion of a greater New England into the Gulf of St. Lawrence. Although Britain had claimed Nova Scotia for more than three decades, there was "not so much as one English Family settled there, beyond the Walls of the only Garrison in this extensive Country." To secure Nova Scotia, Halifax's Board of Trade undertook an extraordinary step in the history of British colonization: shaping a colony's development as a wholly controlled extension of an "imperial state under Crown-in-Parliament rather than as an autonomous dependency of the crown."[17] [**93**]

As the Board of Trade began these efforts to implant a true settler colony on the Atlantic shores of Nova Scotia, it also defended Britain's "Right to the extended Boundaries of Nova Scotia." The British-French commission created in 1748 to determine the boundary between French Acadia and British Nova Scotia bogged down in an interminable exchange of claims and counterclaims, backed up by reams of old charters and maps. As French forces began taking up positions across the Bay of Fundy at the heads of the Saint John and Kennebec Rivers in present-day New Brunswick and Maine, the Board gave commercial cartographer Thomas Jefferys access to manuscript maps and charts to launch another salvo in this "map war." The product of this private-public partnership was *A new map of Nova Scotia and Cape Britain* (1755), an image that ran roughshod over French claims and marked off all the territory from Canso to the St. Lawrence River as British space. [**94**] Published with an accompanying pamphlet that cataloged the image's sources as well as its systematic regard for geographic precision from the best materials available, Jefferys's map drew together Britain's episodic efforts to map Canso, the Nova Scotia Coast, and the Gulf of Maine into a unified statement of authority. As this map asserted Britain's definitive sovereignty over Nova Scotia, the Board's efforts to populate its new province faltered. As one settler reported, of the "five thousand" brought to colonize, "scarce five hundred" remained by the mid-1750s.[18]

As war with France destabilized all paper claims to territory in the northeast in 1755, Thomas Pownall visualized, on a hand-drawn map, a process of colonial expansion that could back up bald assertions of sovereignty with a plan for more extensive settler occupation of North American frontiers. [**95**] A Halifax protégé, Pownall was lieutenant governor of New Jersey, soon-to-be governor of Massachusetts, and brother of Board secretary John Pownall. Like Herman Moll and Thomas Jefferys, he drew "Black Lines" of jurisdiction for the "several English Colonies & the British Territories up to the River St. Lawrence & the Great Lakes," setting the Atlantic coast of North America aside as a natural space of British sovereignty. Pownall's sketch map pictured lands inhabited by colonists and slaves in a wash of ink, red for the "New England Settlements . . . settled in Townships" and green for the rest of the visible mainland that was "settled in scattered Farms" from New York to northern Virginia. "These are the English Frontiers to [the] Sea," Pownall wrote by way of annotation. Picturing occupied territory in this way, as a contiguous form that sprawled across colonial boundaries, produced an unprecedented image of empire, prescient in its representation of colonized space as a "continuously expanding domain." Pownall's population map matches, in content and conception, the visualization produced by the first rigorous spatial reconstruction of colonial settlement, geographer Herman Friis's 1940 dot maps, still the standard source for images of early American territorial expansion.[19]

New Englanders would continue to expand toward the St. Lawrence River on their own, "taking Possession of & settling [the] Country" by granting new townships, which Pownall judged "as Good a Form as can be devised." Behind the occupied regions of Virginia, Pennsylvania, and New York, however, Pownall traced a grid of vast squares and urged a plan of "Military Townships in the Manner of New England." Command of the Great Lakes by boat would make the granting of a bounded Iroquois homeland, marked off by a dotted green line south of Lake Ontario, palatable to those loathe to acknowledge Indian dominion. Pownall imagined an expansive American empire securing this "Frontier to the back Country." His plan differed from the New England town system in that it was to be controlled and monitored from London and would feature a fort constructed by the army at the center of every township instead of a church. This vision faltered when it came to envisioning a means by which Britain would take command of territory to the north and east of New England. The sum total of British occupation beyond the coasts of Maine could be shown by a few scant brushstrokes in Nova Scotia. It remained unclear how so much blank territory might be credibly colored in red and green in the future.[20]

During the Seven Years' War, Canso remained a flashpoint in Britain's quest to occupy its northern frontier. In 1754, Mi'kmaq raiders "murdered twenty one English Fishermen at Canso" and "carried their Scalps" to the French fortress at "Cape Breton[,] where they were well received, and 'tis said rewarded." In the western part of the province, British troops began deporting Nova Scotia's French inhabitants, deemed an internal enemy, in a coordinated program of early modern ethnic cleansing. As army detachments descended on Acadian settlements, military surveyors scouted the landscape. As they debated whether the "Luxuriance of Mar[s]h and pasture Ground" along the Shubenacadie merited a comparison with the Severn or Connecticut Rivers, they gathered notes for a ten-page report on the landscape's potential for planting, defense, navigation, and mining for the Board of Trade. Ensign Winckworth Tonge sketched the watercourses between the Bay of Mines and Halifax, describing a route through the Mi'kmaq-controlled interior that might allow British Protestants to leave their enclaves along the Atlantic coast and fan out across the peninsula. **[96]** Charles Morris, Nova Scotia's provincial surveyor, codified wartime coastal and river surveys into general maps that showed how Nova Scotia's Atlantic outposts connected to these new interior spaces for farming and fishing, left vacant after the forcible removal of the Acadians.[21] **[97, 98]**

During the war, the Board of Trade established fourteen new townships in Nova Scotia. By 1763, the province was settled by 7,794 colonists, one-third of whom lived in Halifax to serve its military bases on a site surrounded by miles of "High Mountainous, Rocky Land, Incapable of being Improved." The rest were scattered across the townships in clusters of a few hundred people who had cleared too few acres to feed themselves. Most of them had departed overcrowded New England towns to build farms, ply their trades, and fish from the southern coast. Beyond the "rich & fertile" reclaimed marshlands cultivated by the Acadians, good land was hard to find in Nova Scotia. When settlers, eager to claim a proffered bounty for developed acres, burned the trees and brush they had cleared, these fires raged out of control, destroying their new fence lines and revealing a soil "covered with a bed of Stones." The lands of short-lived Lawrenceville Township were nothing but "Rocks covered with Moss." Instead of residing in nucleated towns, farmers dispersed to cultivate scarce patches of fertile soil, and still found themselves unable to secure a level of subsistence they might have enjoyed in New England. Unable to make enough fodder to feed their cattle, settlers bought "their Hay from the Massachusetts at Excessive Prices." New England fishermen entered the Bay of Fundy by the hundreds to take cod every summer, outcompeting Halifax merchants for the profits of the fishery, but making no attempt to establish a port within Yarmouth Township, a tract that faced Boston across the Gulf of Maine. Instead of a prosperous settler colony capable of sustaining its own growth, Nova Scotia remained in the 1760s and 1770s a "Province too much dependent on New England."[22]

The bloody history of Canso taught the Board of Trade three powerful object lessons about taking possession of contested American territory. First, the early eighteenth-century diplomatic compromises by which France and Britain were to divide territory and share the seas proved unworkable. As long as France remained ensconced at Louisbourg, from which it could support Wabanaki raids on scattered British settlements throughout the region, Britain's formal title to Nova Scotia would remain an empty assertion. Second, to truly reap the economic benefits of the fishery, Britain would have to mobilize settlers to clear, cultivate, and improve lands around these waters. Although the British had no rival when it came to "planting themselves in America and breeding there," a generation of colonists had not been able to establish a hold over Nova Scotia. The maritime northeast had to be shielded from the influence of self-interested New Englanders who undermined its long-term settlement by treating it as a place in which to plunder, smuggle, and profit. And third, the Board needed unmediated access to information about the land in order to direct colonization from the outset. The long, slow acquisition of geographic knowledge of Nova Scotia and the Canso Islands brought this institutional incapacity to control territory from a distance into sharp relief.[23]

Discovering the Gulf of St. Lawrence

The Peace of Paris settled the conflicting claims that made this region such a volatile eighteenth-century space. By its terms, France renounced "Nova Scotia or Acadia," "Canada," and the "island of Cape Breton, and all the other islands and coasts in the gulph and river of St. Lawrence." However, the treaty's establishment of exclusive British sovereignty could not entirely erase the region's earlier history as a place of overlapping interests and activities. Although France relinquished the right to settle colonies on northeastern lands, it retained the "liberty of fishing" in its seas, secured by a long-standing practice of drying cod on the coasts of Newfoundland. When the Board of Trade ranked the economic benefits of the new American territories acquired in 1763, it put the northern fishery first as the "most obvious Advantage from the Cessions." It celebrated the unification of an "Exclusive Fishery," by which Britain claimed all the riversides and islands close to some of the world's richest fishing banks, as if the nation had awakened from a bad mercantilist dream that had lasted since the Treaty of Utrecht, when the great islands of the Gulf had been "Dismembered from Nova Scotia and ceded to France." Britain now claimed a vast maritime zone that had once loomed as a vulnerable, unsecured frontier. For a century, France had developed the Gulf as the center of a lucrative fishery and the point of access for its fur trade with Great Lakes hunters and trappers. Surging British trades in fish and furs were now poised to make use of this space. Since the conquest of New France in 1759, enterprising fishermen had begun taking "Whales, Seals, [and] Sea Cows" in the St. Lawrence River.[24]

Along with other masters of British naval vessels in 1759, James Cook responded to the order that all ships at sea survey the islands and coasts they visited in the final years of the war. Cook's portfolio of "draughts and observations" of the region demonstrated his "genius and capacity" as a surveyor. After the peace, the Admiralty assigned a squadron to the government of Newfoundland to police Britain's newly acquired Gulf fishery and, prompted by the Board of Trade, appointed Cook to undertake a systematic survey of the eastern coasts of the Gulf of St. Lawrence. In April 1763, he began by answering an urgent call from London to chart the coasts and map the interiors of St. Pierre and Miquelon off the coast of Newfoundland's Burin Peninsula. **[99]** To make the French right to fish in the Gulf of St. Lawrence meaningful, Britain ceded the tiny islands of St. Pierre and Miquelon to France "to serve as a shelter to the French fishermen." They remain France's only North American territories to this day.[25]

The Board of Trade presumed that the islands were "Utterly Incapable of Producing Provisions" to sustain a resident population. Cook arrived to assess the capacity of the islands for settlement and navigation on June 15, 1763, the same day as incoming governor François-Gabriel d'Angeac arrived to take possession of them. As Captain Charles Douglas delayed the impatient Frenchman for two weeks, entertaining him aboard the *Tweed* at the additional expense of £50, Cook completed his survey, and the British "Deliver[e]d up [the islands] to the French" and sailed back to Newfoundland. Beneath his image of these "Barren and Mountainous" islands, dotted in places with small firs, Cook wrote an extensive remark that judged St. Peters to have "one of the worst Harbours about Newfoundland" and dismissed Langly as a swampy waste. Miquelon might have enough pasture to keep "several hundred head of Cattle," but nothing larger than a "fishing Shallop" could safely cross the bar to enter Dunn Harbor. Although some officials feared that St. Pierre and Miquelon might offer a refuge for Acadian exiles or become an outpost from which unscrupulous traders might smuggle French sugars into New England, Cook's survey and chart documented their desolateness.[26]

From the minute observations of these small places, Cook imagined the challenge of taking possession of Newfoundland as a whole, documenting the history of British fishing along the complicated topography of this vast island to undermine claims of exclusive French fishing zones. His "Sketch of the Island of Newfoundland" pictured it in historic terms as a place of contending French and British use rights and in cartographic terms as a place of scattered geographic knowledge, overwhelmed by lacunae and imprecision along an extended coastline. [**100**] He color-coded its shores to show the long-standing British fishery on the southeastern coast from Placentia Bay to Cape Bonavista (in red), the more recent occupation of dispersed harbors (in dark blue), and the northern and northeast shores (known as the French Coast or the Treaty Coast, "Where the French are allowed to fish") in light blue. Much of the rest, and virtually all of the eastern shores of the Gulf of St. Lawrence, remained shaded in yellow as lands and waters for which there was no reliable chart.[27]

For two centuries, reported Leonard Smelt—an army engineer sent to survey the defenses of the British settlement at Placentia in 1751—the "Banks and even [the] Coast" of Newfoundland had been "open to all comers," serving as the territorial base of operations for a multinational European fishery. Because the island "abound[ed] with many excellent Harbours, the most of which [were] equally convenient," it was "impossible to defend any one of them by fortifying another." Rival fishermen, threatened by an outpost of soldiers, might simply uproot and replant their "houses, wharfes, stages, flakes & everything necessary for curing their Fish" on another site. Far from establishing an exclusive domain, a fort would neither "secure the possession of the Island or Fishery, nor protect any district of the country, excepting its own small circuit," he concluded. Britain could not take effective possession of Newfoundland by planting settlers or fortifying harbors, but it could amass a new body of geographic knowledge of its coasts and, by doing so, open them to enterprising ship captains. After declaring St. Pierre and Miquelon uninhabitable, Cook turned next to the long-contested passage between Labrador and Newfoundland, charting coves and harbors that had long been haunts for French fishermen. Cook promoted York Harbor not only by sounding the depths of its bays and providing sailing directions illustrated by a prospective "View of the Coast of Labradore" but also through extensive handwritten remarks that praised its anchorages, its convenience for gathering wood and water, and its nearby seas, teeming with "Cod and Seals." [**101**] Before Cook arrived to map it, Newfoundland was a vast, contested, and scarcely inhabited province, known only through a number of "extreamly erroneous" charts in circulation. When he completed his surveys in 1767, he produced what he thought to be a "perfect good chart of Newfoundland and an exact survey of most of [the] good harbours."[28]

For four years, Cook surveyed Newfoundland's western coasts aboard the *Grenville* during the summers and wintered in London to convert his sketches and survey notes into finished charts. He brought this painstakingly acquired information together on three charts that illuminated the northern, western, and southern coasts of Newfoundland, which previous mapmakers had so "Doubtfully described" before 1763. He secured the Admiralty's permission to publish these images, which proclaimed Britain's intentions to establish its own residential fishery in the Gulf. Along with the charts, "fishing adventurers" could purchase printed sailing directions that would bring them safely to sheltered anchorages. Beyond the bleak shores of St. Pierre and Miquelon, Cook's southern chart revealed the contours of a rocky coastline; marked the flow of offshore currents; and inscribed the waters around bays, coves, points, heads, and capes with hundreds of numbers, each one representing a point sounded in fathoms by the lead line. Leaving his cabin to examine Newfoundland beyond the shore firsthand, Cook reported expansive forests and an abundance of river salmon, beaver, and bears behind some of its bays. He marveled at the "uneven craggy barren surface" of the land in the southwestern part of the island and, from the head of White Bear Bay, beheld a landscape of "barren rock . . . covered with white moss and watered by a great number of ponds." His chart opened up the "great Bays of Placentia, Fortune, and Despair" to view at a time when British officials feared smuggling from the French islands, the migration of French-allied Indians across the Gulf, and illegal attempts by French fishermen to settle on Gulf shores. **[102]** Along the Straits of Bellisle in the north, "all which coast we had not the least knowledge before," Cook's high-resolution insets showed anchorages and entry points for Chateau Bay, Old Ferolle Harbor, and Quirpon Harbor, all beneath the gaze of two sober, fur-clad Inuit seal hunters.[29] **[103]**

His 1768 *Chart of the West Coast of Newfoundland* stretched more than five-and-a-half feet long and presented the coastline from the Bay of St. John to the Bay of St. George at three miles to the inch, a scale that rendered it legible to mariners for the first time. **[104]** Prospective views of the rocky highlands beyond the shore pictured it as it would have appeared from aboard a fishing boat, inviting captains to come to anchor at scouted locations. At the scale of one mile to the inch, the locations of "beaches and places fit for stages and other conveniences for landing and drying fish" became visible on insets of selected harbors. By 1768, more than twelve thousand men in 389 ships made up Britain's Newfoundland fishery, 151 more vessels than had ventured to these contentious coasts in 1764. Although most of them continued to ply the waters close to England's historic fishery along the Avalon Peninsula, a few dozen "fishermen adventurers" followed these charts into the Gulf of St. Lawrence. After honing his skills in the maritime northeast in the art and science of joining coastal hydrography and terrestrial surveying, Cook applied this "running survey" method aboard the *Endeavour* to bring the South Pacific into view from 1768.[30]

The Board of Trade supported Samuel Holland's branch of the General Survey with two assistant surveyors, a draftsman, a small company of soldiers, and advanced surveying equipment, to be transported and assisted by a ship manned with a crew of forty. The Admiralty purchased an English merchantman called the *William*, outfitted it with guns, and rechristened it as the armed ship *Canceaux* under the command of Lieutenant Henry Mowat. On April 21, 1764, Holland, Mowat, and their men boarded the *Canceaux* at the Deptford Naval Yard in London and set sail for America. After a rough crossing and a perilous voyage into the Gulf and down the St. Lawrence River, Holland established the General Survey's office in Quebec City. The surveyors embarked in September for St. John, one of the "valuable Islands lying within that Gulf" that the commissioners identified as the General Survey's first object, to be completed with the "most pressing Expediency." When the *Canceaux* came to anchor at Port-la-Joy harbor, below the small garrison stationed at Fort Amherst, the island's landscape—illustrated on a 1764 army map—bore the marks of past French settlement. **[105]** Before British forces invaded St. John in 1758 and began deporting its inhabitants, French colonists had established fifteen small villages scattered around the island's bays and inlets, their numbers swelled by hundreds of refugees fleeing the British army in Nova Scotia. By the time Holland arrived, fewer than three hundred French settlers remained.[31]

Holland began his survey of St. John determined to represent it comprehensively. Although the first French colonists had arrived there four decades before, the island remained largely uncultivated. A few hundred farmers had managed to clear fewer than 12,000 of its 1.4 million acres, leaving a string of hardscrabble farms clustered along the banks of the central Rivière du Nord-est (renamed Hillsborough River). From his boat, the French colonel who assessed the island in 1751 could see "log houses of the settler[s] rising among the stumps of the recently felled trees and the strong, if patchy, harvest waving over the yet unlevelled and unfenced fields." Rarely extending from the banks of the rivers and bays "more than one-farm deep" into the interior, the remains of the French settlements of Île St. Jean displayed the inadequacies of unregulated colonization. The French had left most of its land in woods and waste; they failed to generate a marketable agricultural surplus that could draw new settlers to the island; and they had not invested in the ships and harbor facilities that would make it possible to exploit the fishery, by far the region's most valuable potential enterprise. Acadia's skilled marshland farmers, who supplied most of St. John's pre-conquest population, were reluctant to clear hardwood forest when they found few familiar wetlands ready to be diked and desalinated. They kept growing their favored mixed crops of wheat and peas despite the fact that oats, barley, and rye would have been better suited to the island's soils.[32]

Once ashore, Holland's men salvaged materials from an abandoned barn to build a crude shelter for themselves and their telescopes. He named this spot, just south of Fort Amherst, Observation Cove, headquarters for the General Survey on St. John. In February 1765, Holland dispatched four teams, each consisting of a surveyor, three soldiers, a sailor, and an Acadian guide, to different portions of the coast. By dogsled and snowshoe, they began the coastal survey of St. John, situating their plane tables along the shore and sending soldiers to stand with raised flags to provide the surveyors with visible landmarks so that they could measure their angles. Taking advantage of the ease of travel down frozen rivers, these teams ventured into the interior of the island and extended their emerging coastal picture of its shape and topography to include riverside lands. When the ice melted that spring, Mowat dispatched small boats from the *Canceaux* to transport the surveyors and soldiers around the island by water to complete their survey. As the midshipmen waited for orders to pack up the gear and move down the coast, they sounded the island's waters along each stretch of shore, recording depths and hazards as part of the Admiralty's complementary surveying effort, producing charts of the island's major bays.[33] **[106–108]**

The laborious process of plane-table surveying brought Holland's men into sustained engagement with the landscape. As they moved their plane tables along the shorelines of their survey zone day after day, they scanned the viewsheds before them in search of landmarks—including abandoned windmills, ragged lines of sand hills, and their own pitched tents—that allowed them to fix the distances and proportions of the landforms they observed. This mode of surveying was a disciplined way of seeing, and it forced the eye to take in every portion of the coast systematically in order to draw its shape accurately. The field book of assistant surveyor Thomas Wright records how the General Survey's definitive image of the island's northwestern coast, assigned to him as a survey area, materialized out of this process. In dense script across twenty-eight pages,

Wright recorded a long column of observed "Angles" in degrees, minutes, and seconds, and associated these figures with the lengths of measuring chains to define a sequence of connected lines, oriented in geographic space. Wright plotted his measured lines, lettered like geometric diagrams to correspond to the data in his tables, across twenty-one sketch maps to form a coastal outline for part of St. John Island. [**109**] Over the twenty-six days of work recorded in his field book from June 7 to July 10, 1765, Wright's party produced, on average, seven miles of surveyed coastline per day, and measured a total of 846 segments that, when joined together, stretched just over 203 miles. In the dead of winter, such work was tedious and uncomfortable—and occasionally fatal—but especially as winter turned to spring and the island's vegetation came to life, it offered an ideal opportunity to reflect on the land's agricultural potential. Although Samuel Holland himself surveyed only a portion of the northern coast, from Orby Head to East Point, his report to the Board of Trade included natural and agricultural comments on each of the townships he laid out on his maps, suggesting that his assistant surveyors had provided him with notes and observations for the entire island.[34]

By the end of the summer of 1765, these coastal and river surveys were complete, and in a tent set up as a drafting room, the surveyors fought off swarms of mosquitoes to compile their large-scale field surveys into a general manuscript, "Plan of the Island of St. John," divided into sixty-five townships, three towns, fourteen parishes, and three counties. [**110**] Ringing the coast and extending along the rivers was a band of stylized green woodland, punctuated by hachured cliff sides and the forms of mudflats at low tide. Although most of the interior remained an unobserved blank on this map, the information his surveyors brought back to Observation Cove permitted Holland to divide it. Conforming to the Board of Trade's instructions to apportion the colony into twenty-thousand-acre townships, he drew long rectangular polygons across the island, angled and arranged to fit the topography of "Mountains, Rivers, and other natural Limits" and divide the entire prospective cadastral landscape of the island into roughly equal-sized tracts. This rectilinear mosaic prohibited future landowners from engrossing the best lands along the "Sea-Coast, Navigable Rivers," and other desirable spots and instead divided the land into "Oblong Squares, extending from the Sea-Coast up into the Country," so that every township might have a "due Share and proportion of what ever Advantages the Island affords." Each township took in some part of the observed coast before it extended into the unknown interior. Following the Board of Trade's ideal civil framework for a new colony, every township fit into a designated parish, and every parish sat within one of three preordained counties. Throughout the continental mainland south of New England, new frontier jurisdictions metastasized haphazardly to take in the latest migrations of settlers and the claims of speculators. By contrast, before the first British colonists stepped ashore on St. John, Holland's survey imposed a clear jurisdictional order on the land.[35]

On this and other manuscript maps that summarized his survey's findings, Holland appraised each township as to its convenience for prosecuting the fishery, the qualities of its soils for farming (ranking the land "bad," "indifferent," "pretty good," "good," and "very good"), and its proximity to three planned towns. **[111]** Tables of this assessment appeared next to the image of the divided island on the maps he sent to London. Townships that terminated at the island's north shore were deemed "well situated for fishing" in the teeming banks of the Gulf, and those blessed by geography with natural harbors—especially townships 5, 33, and 39—seemed ideal sites on which to build wharves and fish flakes. Many of the townships he ranked "very good" for agriculture featured lands already cleared by French settlers, a developmental advantage that, along with the presence of preexisting houses and barns, promised early plantings of large crops in the first seasons of settlement. Others had marshes that would "make fine Meadows" for harvesting hay, a trait shared by four adjacent townships of Halifax Parish on the low-lying southwestern coast.

Some areas seemed scarcely worth the benefit of having, such as township 7, with "indifferent Lands and Woods and no Fishery," perched upon a section of the sea coast that was "Steep and Rocky." With neither good lands nor a harbor, township 20 would "be of little Value until the Neighbouring Townships are Settled." Even those without the best natural endowments for crops, cattle, or fish would "in Time become of Value" because of their situations in relation to navigable rivers, prospective roadways, and one of three planned market towns—Georgetown, Charlottetown, and Princetown. Holland's survey of St. John produced maps of the new colony that not only made use of rigorous empirical methods to represent its geography as it was but also scrutinized its natural endowments to imagine a colony that could be rapidly settled and fully peopled by commercial farmers and fishermen and their families.

The Board of Trade issued a call for proposals to develop St. John on May 9, 1764, while the *Canceaux* was still making its first westward crossing to Canada. With the results of the survey at hand by 1766, the Board took up these "Petitions & Memorials" for township grants and advertised in the *London Gazette* to announce the day on which the petitioners or their agents should report to the Plantation Office in the Treasury Building to be interviewed. There, they made their cases that they stood ready to populate the island in return for the privilege of taking possession of such large portions of it. After questioning them to discover the "Intention & Ability of each of the Petitioners to make Settlements upon the said lands," each petitioning individual or partnership wrote his name (or names) on a paper ballot. On Friday, July 23, 1767, the petitioners gathered again before the Board to witness the drawing that assigned the townships located on these 1765 manuscript maps. Holland recorded the results of this process—adding the names of St. John's new proprietors to the table of township data—on a handsome presentation map in 1767, which stamped selected townships with a seal of productive approval: "F" for fishing and "A" for agriculture.[36] **[112]**

The first chosen received one of the least auspicious tracts. Although Philip Stevens served as secretary of the Admiralty, his office failed to shield him from the impartiality of this process. His first-drawn ballot entitled him to township 1, easily located on the map as the tract that encompassed the point at North Cape, a rugged, windswept portion of the northwestern coast in which "Lands as well as the Woods [were] bad[ly]" covered by "small Spruce," a tree that grew on sandy, subprime soils. Yet the township might have been redeemed in his eyes as a long-term investment: it lay adjacent to an "exceeding[ly] good" cod fishing ground in the Gulf, which promised a source of revenue on this picturesque tract of sandy woods and high cliff sides. Stevens had only to walk a block or two from the Admiralty to the Treasury Building to compare his township with that of his counterpart, Board of Trade secretary John Pownall, who drew number 13 in the ballot. [**113**] Holland proclaimed this "one of the best Townships on the Island," ideally situated for the fishery from its location on sheltered Richmond Bay. French colonists had cleared 750 acres of prime land within the township and left behind twenty-four houses and barns (as well as a church). If Pownall gloated over his good fortune at acquiring such a valuable holding in 1767, he might have been less sanguine about the fact that, five years later, he would owe six times as much in quitrents as did Stevens for the privilege of becoming a St. John proprietor. Under the terms of the free and common socage status under which most British American land was granted, each landowner was liable for annual payments to the Crown, usually set at 2 shillings for every 100 acres of land every year, but few paid it. New provinces such as St. John, lacking a legislature and an entrenched creole elite, provided a clean-slate advantage that allowed the state to experiment with new land and taxation policies. Settling new postwar colonies after 1763 offered an opportunity not only to direct colonization from London but also to set this new colony on a course toward financial self-sufficiency and political dependency that it had long wished, and long failed, to impose on the mainland.[37]

The plan to collect quitrents on St. John also involved a scheme of graded taxation keyed to the economic value of the land. Elsewhere in British America besides New England, colonists retained the right to choose the best lands they could find within a properly constituted jurisdiction; often, they transgressed this legal restriction by venturing beyond the bounds of parishes and counties and simply squatted on land that they expected would be later ratified by a formal grant. On St. John, the state selected who received what township by random ballot and distributed all of St. John's land at once, a system designed to preempt the machinations of land speculators and encourage the comprehensive development and peopling of the island. To compensate those who received "bad" or "indifferent" townships for the limited commercial potential of their lands, the Board tailored the land tax system to impose the lightest burdens on those with the worst lands and make those who gained prime places pay the most. Those with poor lands paid the standard rate of 2 shillings per hundred acres; those with middling properties paid 4 shillings; and those with "very good" land, as well as those whose lands seemed well situated for the fishery and agriculture, paid 6 shillings. Within four years of taking up a township, each proprietor or partnership had to entice at least thirty-three people to settle on its lands; within ten, a full one hundred people had to be in residence. Failure to pay these quitrents or attract settlers meant risking the forfeiture of these lands in their entirety to the Crown.[38]

Who were these men who aspired to become the great landlords, planters, and fishing magnates of St. John? As the officials who organized the work of the Admiralty and the Board of Trade—the very agencies behind the surveying, development, and defense of British America—Secretaries Stevens and Pownall were well positioned to take personal advantage of their offices by claiming land in the Gulf of St. Lawrence. But they were by no means unique among proprietors as well-placed metropolitan colonizers. Military officers composed the largest group of township holders, figuring as proprietors or co-proprietors of thirty of St. John's sixty-six granted townships. Granting land to military men who had served in the Seven Years' War was already part of the Board of Trade's plan to settle new lands acquired in 1763. The Royal Proclamation of 1763 commanded governors of old and new colonies alike to grant, without "fee or reward," tracts of land ranging from fifty acres for every "private man" to five thousand acres for those with the rank of "field officer," a privilege extended to naval officers as well. Such a provision did not entitle them to these twenty-thousand-acre townships, however, and these military officers—including an admiral, a vice admiral, a commodore, and numerous current or former lieutenants, captains, and colonels—had to make credible claims that they were prepared to settle and develop these lands like everyone else.

Military officers were not only the largest category of proprietors on St. John; they were also the most likely to be in a partnership, suggesting that wartime experiences had nurtured well-connected, entrepreneurial cadres of men eager to seek their fortunes in the postwar empire. The entire officer corps of the 78th (Highland) Regiment of Foot banded together to claim township 42 (a "pretty good" tract suitable for agriculture, distinguished by a mostly burnt woodland, a ruined village, and a "grove of Cypress shrubs"), having proved its worth at the capture of Louisbourg in 1758, the Battle of the Plains of Abraham in 1759, and the surrender of Montreal in 1762. Other officers partnered with merchants (who served as proprietors or co-proprietors for nine townships), a banker, and others whose occupations went unnamed; such partners could promise capital as well as connections among London ship owners and suppliers to help the new townships gain settlers and market commodities. Senior officials made up the next largest group of proprietors, involved in nineteen townships. These included not only Stevens and Pownall but also Shelburne's secretary, the Lord Advocate of Scotland, the governor of Newfoundland, Quebec's surveyor of the woods, Quebec governor James Murray, and surveyor general Samuel Holland. Eight members of Parliament became St. John township owners in 1767, as did one peer of the House of Lords.[39]

The Board of Trade's scheme for settling St. John displaced independent colonists as the primary agents for New World colonization. In their place, trained surveyors documented the land's value, and a coterie of distant landlords, entrusted by the state with a regulated monopoly on real property, took on the risks and anticipated the rewards of settling a new colony. Convinced that the Gulf's "Islands are able to contain Millions of People" if colonized systematically, the Board commanded Samuel Holland to amass a comprehensive body of geographic knowledge about them to supplant the practice of opening the province to colonists and hoping they would discover, by trial and error, the best way to develop these new lands. [**114**] Predictably perhaps, this bid to orchestrate the colonization of St. John from London failed to achieve its objectives. Beneath the rectilinear shapes of Holland's map, the first "Settlers were distressed in not being able to ascertain the just bounds of their Lands." The colony's governor complained that Holland's precise township boundaries were "merely imaginary, except upon the Map."[40]

Envisioning Order on Remote Coasts

After surveying St. John, Samuel Holland trained the General Survey's imperial gaze on nearby Cape Breton Island, which faced British Nova Scotia across the Gut of Canso. [**115, 116**] After deporting the residents of Louisbourg, laying waste to its once formidable fortifications, and massacring some of the island's Mi'kmaq inhabitants, General Jeffrey Amherst and his forces departed, leaving this former French stronghold to be mapped and repopulated to the design of the Board of Trade's vision. The scouring of ancient glaciation had formed more than two million acres on Cape Breton into bands of hills and mountains interspersed with barren plains and narrow river valleys. This volatile geology left behind a vast saltwater lake that divided the island and gave it an inner shoreline every bit as complex as its rugged Atlantic- and Gulf-facing coasts. If left as a wasteland so close to the continent, an uninhabited Cape Breton would, the British feared, become a haven for smugglers and a target for resurgent French ambitions in the region.[41]

In the fall of 1765, Holland and his men began their two-year ordeal surveying Cape Breton. Despite the fact that it was "Almost Intirely Surrounded by Inaccessable Rocks and Clifts," "exposed to the Surf (or breaking of the Oceans)," and inundated by "Frequent Foggs" that obscured the visual observations so crucial to plane-table surveying, Holland produced a plan of the island by the summer of 1767. [**117**] He had recruited Mi'kmaq guides to lead his surveyors by snowshoe and dog sled into the interior of the island during the winter and endorsed their request that Britain provide their community with a Catholic priest in exchange for their services. Such willingness to acknowledge the wants of the island's indigenous inhabitants also gave Holland a way to forsake Cape Breton's northern highlands for settlement within the idiom of the Board's plan. "[S]tiled the savage or hunting country" on his map, and deemed "fit for nothing else, being without harbour, or rivers, & very mountainous," these lands could be "put to use" to afford a "trade in furs, with the Indians in return for English manufactures."[42]

Holland's map of the island's windswept coasts "evinced the truth" of his claim that no "Part of North America can boast of a more advantageous Situation for Commerce and Fishing." Taking advantage of the abundance of "Good & Safe Harbours" close to north Atlantic banks, fishermen would settle and defend the island in times of war, and produce a valuable peacetime bounty of fish, fur, whale bone, and train oil. Although no more than one hundred British vessels plied these waters in the years after the Peace of Paris, Holland imagined a resurgent fishery, harvesting teeming shoals of cod on an even larger scale than the French had achieved. His surveys not only established the depths of bodies of water, the contours of coastlines, and the topography of terrain but also attempted to describe the capacity of places for development with predictive value for imperial planning. With his maps and accounts of the French *pêche sédentaire* in hand, Holland calculated that pursuing the fishery from the island was worth "upwards of £130,000" to Great Britain and could employ 7,536 men and 653 fishing vessels. This was aside from the opportunities for mounting whale hunts and fishing for the salmon, mackerel, and herring that abounded along the island's coasts and bays, trapping beavers "found in every Lake and Pond" for their pelts, harvesting timber, mining coal, and quarrying for marble. Such prospects suggested an "inexhaustible Fund of Wealth" that could repay investments in attracting settlers, laying out tracts of land, and building towns. Industrious British fishermen, unlike their indolent French predecessors, might also be expected to "bestow some Labor on their Farms, clearing them for culture."[43]

As he had done on St. John, Holland partitioned the arable part of Cape Breton into a grid of townships, projecting lines of division that followed bearings on his compass. The Board of Trade sent Samuel Holland into the Gulf of St. Lawrence to understand which places along its expansive coastlines and among its innumerable islands possessed the natural capacities that might make them good candidates for colonization. It charged him to examine, analyze, and represent them, in text and image, and, once he judged them "fitted for Inhabitancy and Cultivation," to conceive of a "Plan of division as may be best calculated for the Settlement of them, having always for [his] principal Object, the Carrying on of the Fisheries." But despite this work preparing Cape Breton for settlement, officials delayed its colonization for the better part of a decade. Just as they had fended off the Earl of Egmont's bid to claim all of St. John, the Board of Trade and the Privy Council rejected a proposal from Charles Lennox, Duke of Richmond, to take possession of the island in 1764 as a privately held province. They fought off another metropolitan proposal in 1766 to establish a monopoly to extract its substantial coal reserves on the grounds that encouraging such an industry went against the grain of mercantilist principles and amounted to a dubious "Innovation, in the system hitherto pursued in the Regulation of the Colonies." Finally, in 1774, Colonial Secretary William Legge, Earl of Dartmouth, heeded the advice of Nova Scotia's provincial surveyor, Charles Morris, who advised that the "whole island of Cape Breton should be reserved" as forested repository for the Royal Navy, to supply it with black birch for shipbuilding and black spruce for ship masts. Dartmouth ordered this remote province of Nova Scotia, home to a few hundred Acadians and Atlantic fishermen, formally closed to new grants.[44]

At the other end of the Gulf, 250 miles to the northeast of Cape Breton Island, John Collins, Holland's assistant surveyor for the province of Quebec, laid out an orderly framework of lots along the sheltered coves of Chaleur Bay in 1765. **[118]** At the scale of four hundred feet to the inch, he modeled a settlement system for Grand River, Bonaventure, Port-Daniel, and Paspébiac, the last of which centered on the fixed grid of a model town. **[119–122]** All modestly occupied sites of the French fishery before the war, Collins found a handful of tracts with enough good land to suggest that newcomers might raise cattle and plant crops of vegetables and flax when they were not harvesting fish from the Gulf. Although "well Situated for Carrying on a Cod Fishery," these lands were, in general, mere spaces to "cure the Fish taken." Few settled these coasts until the British transported a group of loyalist refugees to Chaleur Bay in 1785.[45]

Other places mapped by the General Survey seemed too forbidding to colonize from the outset. Thomas Wright charted Anticosti, a large island that occupied the strategic mouth of the St. Lawrence River, delineating every cove and cliff along more than three hundred miles of its rocky coast. **[123]** One man in his party died on this "barre[n]" island during the "excessive severe" winter of 1766–1767, and Wright declined to mark off the lines of a prospective settlement scheme to apportion its "1,699,840 acres of very indifferent land." Plans to develop the Gulf's northern shore, including the mainland district of Mingan and its adjacent Seven Isles, foundered when this "extensive coast" was revealed to be but "poor barren rock that can never be turned to any account in the way of husbandry." The Magdalen Islands, like the desolate isles of St. Pierre and Miquelon charted by James Cook, appeared small, rocky, and barren. Despite their reputation as an ideal site from which to conduct the walrus fishery, any potential inhabitants would be hard pressed to find enough good land with which to feed themselves. After these islands were meticulously charted by Peter Frederick Haldimand, who died after falling through an ice-covered lake on Cape Breton in 1765, their promise as a colonial outpost was archived, along with his image.[46] **[124]**

From 1770 to 1775, Holland conducted the General Survey for the Northern District from Portsmouth, New Hampshire, and focused its efforts on the vast and complicated stretch of coastline from Gardiners Bay, Long Island, to the Saint John River in modern-day New Brunswick. Before he sent his assistants to their separate survey areas, he described his vision for the region's development to Colonial Secretary Wills Hill, Earl of Hillsborough. Emboldened with rumors that Britain was poised to detach the province of Maine from Massachusetts as part of the general reorganization of the empire, Holland proposed boundaries for a new colony to be called "New Ireland," which he illustrated with his "Sketch of the Country between New Hampshire and Nova Scotia." **[125]** Armed with reports of colonists laying waste to fine stands of white pines, claimed by the Crown as sources of ship masts, Holland drew new boundaries to make a "separate government of the whole Territory" that could better organize an expected surge of postwar settlement. Although Hillsborough gave no sanction for such a move, sure to have provoked a confrontation with colonial proprietors and legislators in Massachusetts, Holland's proposed colony followed from the Board of Trade's premises about American empire. His red boundary line encompassed a sparsely inhabited space, misgoverned from distant Boston, which possessed vital natural resources. Perhaps Holland imagined that with his direct knowledge of its landscapes, he might gain a royal appointment to govern the colony that his map defined and named. His idea for the colony not only completed the transplantation of geographic nomenclature from the British Isles to America, adding a "New Ireland" next to New England and New Scotland, but also imposed a rational jurisdictional plan for an ambiguous region, filling in the space beyond the limits of two other "regular" colonies closely administered from London—Quebec and Nova Scotia—whose boundaries Britain had changed in 1763.[47]

Holland and his surveyors opened Maine to imperial scrutiny. Families migrating north from long-settled parts of New England after 1760 arrived in the territory seeking riverside sites with good water transportation to the Atlantic for farming, fishing, and lumbering. Especially north of the Kennebec River, this coast was a hydrographic maze composed of an "infinite number of Islands and numberless Harbours and Bays" that demanded more time and attention than any other sector in the General Survey's Northern District. He and his fellow military surveyors brought their skills to bear on the task of representing this space between land and sea in ways that revealed it as a place for settlement and navigation. North of Cape Ann, their charts pictured the town of Ipswich surrounded by hills shaded to give the illusion of three-dimensional form; they traced courses of shallow creeks that meandered through the great marshes around Plumb Island Sound. [**126**] Assistant surveyors James Grant and Thomas Wheeler pictured Piscataqua Harbor at the especially high resolution of two thousand feet to the inch, zooming in on the growing town of Portsmouth, New Hampshire, to show the layout of its streets; the network of roads that drew in its hinterland; and even the outlines of fences and the locations of farmhouses, mills, and churches scattered along its bay side. [**127**] Where James Grant's and George Sproule's survey areas met between Casco and Penobscot Bays, their overlapping manuscript charts laid bare the complexity of this "drowned coast," formed more than ten thousand years earlier, when rising sea waters submerged an area scoured by retreating glaciers, turning mountains into islands and valleys into bays. [**128, 129**] Holland verified their surveys with "Astronomical Observations" to fix every "Geometrical Position" in proportion to points of latitude and longitude. Sproule's 1772 "Plan of the Sea Coast from Cape Elizabeth, to the entrance of Sagadahock, or Kennebeck River" recorded numerous marks of settlement by drawing tiny red rectangles that located structures within pale gaps of color that suggested patches of cleared land. [**130**] To the north, where Grant, Thomas Wright, and Charles Blaskowitz charted the mountains of Mount Desert Island and the sheltered coves of Passamaquoddy Bay, signs of human habitation and development appear fewer and further between right up to Fort Frederick and the mouth of the Saint John River in neighboring Nova Scotia.[48] [**131, 132**]

With its "Multitude of Harbours, and Retreats from Shipping," Maine was, Holland proclaimed, a "Country richly calculated for the purposes of Navigation, Agriculture and commerce, if good Soil, deep and secure Havens, with plenty of Ship Timber are any Recommendation." The men who measured the new frontiers of British America saw themselves as agents of empire. Holland supported a request by his assistant surveyors, military men without "landed Property," for large grants of Maine land. Not only did they know these coasts better than anyone else, but the hardships they had suffered to map them for the modest pay of five shillings per day had demonstrated their loyalty to king and country. The Board of Trade's surveyors would make ideal colonial proprietors, certain to give "an Air of Policy & Civility to at least one Spot in an immense Tract of Country where every Man seems to act, as he holds his Lands, by his own Pleasure and Coveniency." To give his surveyors a "mark of distinction" and to promote their "attachment to the Right Honorable Board" of Trade, Holland designed a green uniform for the General Survey. He requested that his men be outfitted with "leather caps" (better suited than conventional hats for their "excursions thro' woods, & by water") and that these caps be embossed with the words "Trade & Plantations" over an image of the Board's emblem: two ships under full sail next to an island on which Britannia sits, with her olive branch, spear, shield, and liberty cap, surrounded by bales of goods.[49]

Their work advanced the imperial interests embodied in this image by opening the places they surveyed to regulation. So frequently was Henry Mowat and the armed ship *Canceaux* diverted from the work of sounding depths off the coast by the task of "Watching for Smugglers" that the "Naval Part" of the General Survey began to lag behind the "Land Part." As Holland began his New England survey in 1770, John Wentworth, governor of New Hampshire and surveyor of the woods, reported to Hillsborough that Maine settlers were already "cutting & ha[u]ling mast or white pine trees out of the King's Woods." He had personally marked and seized some of these illicitly felled masts as contraband along the remote Androscoggin River, "where the saw mills are." So far, the "many scattering Settlers on this extended Coast" had not yet plundered the large stands of "Mast Pine Trees growing some little distance into the Country," but the British could not expect that colonial lumbermen would refrain from cutting them down out of "virtue or forbearance." They must be "compelled by fear or stimulated by Interest," Wentworth concluded, or the "amazing Quantity of Timber fit for every naval purpose"—a supply that Holland called a "Grand Magazine of Masts"—would be lost. The charts of the General Survey exposed these sites of trespass and smuggling to view for the better protection of the empire's resources.[50]

Holland's surveyors captured "every Variation of the Ground" along hundreds of miles of coastline at the high-resolution scale of four thousand feet (three-fourths of a mile) to the inch. After these surveyors copied figures from their field notes onto clean manuscript charts, Holland dispatched these images to London, giving the Board of Trade's commissioners piecemeal glimpses of the region. These "detached Pieces" that he sent home from "Time to Time" were proof that the General Survey's laborious work had created a durable empirical image of the maritime northeast with a degree of "Fidelity & Accuracy" that was unmatched, with every place "properly delineated agreeably to Nature" and "Nothing being left to the Imagination." To be useful as tools to police and regulate these coasts, however, this information had to be systematically organized on new general maps. In 1768, Hillsborough stressed to Holland the ultimate purpose of this work: filling a metropolitan archive with a comprehensive body of geographic knowledge about America. Although he was happy to receive so many detailed charts, Hillsborough reminded Holland that, in the near future, the "Surveys transmitted by you, and by the Surveyor for the Southern District[,] should be reduced to one uniform Scale, and Copies made of them," one for the collection of the king and the rest to be distributed to the "different Offices" that made up the imperial state—the Colonial Office, the Treasury, the army, and the Admiralty—to serve as a permanent record of place. When the Earl of Dartmouth assumed the office of colonial secretary in 1772, he marveled at the charts that Holland had "already transmitted," praising them as "far more correct & accurate than any that have been before made" of this "very large and important district of North America." He, too, pressed the surveyor general to have them "collected together & united into one general Plan." Holland reported delays in completing the survey of a coast that was "nothing but a Chain of Bays, Harbours, & Inlets, interspersed with a Multitude of Islands," whose complexity prevented him from drafting such a "general Map on projection of these Northern parts."[51]

Holland never created a "General Map under one General Scale," uniting the entire output of the northern branch of the General Survey within a single frame. But the surveyor general worked toward this "comprehensive Idea of our Surveys" by compiling a map of Maine that drew many of these high-resolution images into a composite image in 1772. With "A Plan of the Sea Coast from Cape Elizabeth on the West Side of Casco Bay to St. Johns River, in the Bay of Fundy," Holland presented the secretary of state for the colonies with a map that combined and reduced the "large survey to the scale of two miles to an inch." [133] He anticipated that many more such images would someday "compose the Atlas of all We have surveyed, so soon as this service is finished," he promised, but in the meantime, he hoped that Dartmouth would "be pleased at seeing in one View, so great & valuable an Extent of Country."[52]

In the mid-1770s, Grenada's colonial governor George Macartney reflected on the new geographies of Britain's relationship with the world after 1763. The places that made up this "vast empire on which the sun never sets and whose bounds nature has not yet ascertained" had to be clearly understood before they could be governed for the benefit of the nation. He reflected on the importance of St. John Island in particular, and urged that its nascent settlements receive "extraordinary encouragement." The island's innovative quitrent system had failed to support the costs of government into the 1770s, and its financial difficulties initiated a protracted campaign for reform that pitted British officials, the island's absentee proprietors, and its struggling class of settlers—approximately one thousand had arrived on the island by 1775—against one another. The Board's attempt to make St. John self-sufficient by encouraging agriculture to complement the fishery was likewise proving hard to mandate. "Fishing [was] so enticing an Employment" to the "common people" that they could "scarcely refrain from it . . . to raise themselves a little bread." They would need imports of provisions as well as constant supplies of manufactured goods to focus on this profitable pursuit, but this economic dependence was a good thing, Macartney concluded, for it guaranteed that the island "must ever continue in due subjugation to Great Britain while her Navy commands the American Seas."[53]

New Englanders were now attempting to commandeer the trade of St. John "in so irregular and unfair a manner"—by gouging the residents for desperately needed provisions and goods—"as rendered it exceedingly prejudicial to the settlements on the Gulf." Always in search of "present profit" rather than the long-term goal of sustainable planting, they should not have been allowed to extend their commercial sway over the colony or to serve as the mercantile link between the northeast and the West Indies, which was so vital to both regions. Now that the "infatuated colonies" of the mainland were close to open rebellion, Britain should try to establish St. John as a new commercial center for the northeast in order to wrench it permanently out of the hands of New England merchants. Macartney was prescient in his view that the coming war with America might favor St. John as an "Asylum" where displaced loyalists might resettle. By the end of the century, some 4,300 people lived on the island, concentrated in the same pockets of good riverside farmland and north-side fishing harbors that the French had colonized so haltingly in the first half of the century.[54]

The Board's scheme for St. John's development continued after the disruptions of the war, but settlers pressed for outright ownership of tracts belonging to absentees who had failed to settle their townships or remit their quitrents. The Crown reclaimed more than two million acres of undeveloped township land in Nova Scotia, but in St. John, proprietors leveraged their political power to retain ownership. By 1798, Samuel Holland had drawn only 136 residents to his own township (lot number 28 on his map), a meager yield from the pamphlet he published to "induce Inhabitants to settle on his Estate." In addition to these twenty thousand acres, the surveyor general had acquired three thousand in New York and twenty-four thousand in Vermont. After commanding British troops in New York, he continued his work as Quebec's provincial surveyor, preparing lands in what would become Upper Canada for loyalist resettlement in the 1780s, and accepting a long-standing appointment to Quebec's Legislative Council. His United States lands lost to independence, he acquired vast new landholdings along the St. Lawrence River.[55]

By the close of the eighteenth century, disgruntled St. John tenants, among them resettled North American loyalists used to liberties that came with landownership, expressed their discontent with the Board of Trade's lingering experiment in reformed colonization. By working the land, they had acquired a stock of practical knowledge that called for erasing Holland's linear township boundaries and empowering them to shape viable tracts more attuned to the landscape's resources. Instead, St. John's land system encumbered their efforts by demanding that they improve this severe land for the benefit of others, pay cash rents, or be dispossessed. They found their woodlands cut off from the "marsh meadows" on which they could graze cattle and produce the manure they needed to amend poor soils; what crops they managed to grow they defended against plagues of mice, capable of consuming all their grain before it could be reaped; and they raced against the constraint of a short growing season to plant and plow the land before punishing winters set in.

Farmer Joseph Robinson, born in South Carolina and resettled as a loyalist to St. John, recalled the history of Carolina's negligent Lords Proprietors, forced to surrender to the Crown their vast grant earlier in the century. Once Britain changed the "mode of colonization" to favor liberal headright grants, the province "parceled out" lowcountry lands to the "settlers according to the number of their families," granting tracts at minimal cost to colonists whose investments in slaves and improvements to the land brought new "happiness, population, and wealth" to a rising planter class. In St. John, by contrast, oppressed tenants spent their labor to no purpose in a "wilderness island" and had to "drag out their existence . . . in a state of low spirits, in much want, misery and distress." Despite their demonstrated loyalty to the king, they were trapped at the bottom of landed hierarchy in a new-forged peasantry as "subjects to other subjects." Even this cold and infertile isle could be redeemed, Robinson proposed, by granting colonists tracts of "100, 200, and 300 acres," encouraging them to put their hard-won environmental knowledge to use and animating a "lively spirit of industry" among them to improve a languishing colony.[56]

Long after the last legal remnants of the Board's land system disappeared on St. John (renamed Prince Edward Island in 1798), the province's roadways and borders, skewed at an angle of 15½ degrees from true north, still reflect the geometry by which Samuel Holland solved the puzzle of how to apportion its lands into townships. With its crenulated shores and numerous islands, the Gulf of St. Lawrence contains some 12,500 miles of coastline when measured at high resolution. The Board of Trade's failed attempts to survey and settle the maritime northeast offer a case study in the mistakes made by an overreaching state that became enamored with its access to empirical knowledge but could not marshal it effectively. Given the vastness of American spaces, Britain simply did not possess a means of seeing them clearly, falling short of the mark of supplanting the knowledge of agricultural landscapes that colonists had built up over generations through direct experience.[57]

South Carolina

No. Wt. fork of Long Cane Cr.

Post Bo. Le.

Post

Pine

Black Oak

Course So. Wt. 50

Indian Land

e William Bull Esquire —
day of June 1765. I have
een the province of South
the Upper Middle and
hich Line is Represented

We The
were pres
at Devise
and agree

Signed Seale

FOUR

Marking the Indian Boundary

On October 7, 1763, King George III issued a royal proclamation that set Britain's American empire on a new legal foundation. In a surprising revision of previous understandings of Indian land rights, the proclamation declared the unpatented interior of the continent to be the property of indigenous states and ordered colonial governors to grant no new tracts of land to Europeans where such claims might prevail. This effort to contain the expansive energies of settler societies seemed doomed from the start. No law could contain a people bent on taking land from those they viewed, with unconcealed malevolence, as having no right to exist on it. Given how little Britain did to defend Native lands against encroachment, the proclamation appears to have been a promise that it never intended to keep. Despite the king's high-minded rhetoric, its provisions seemed only to quiet Indians' fears before their lands could be gradually acquired in a process of dispossession by treaty. Even before the American Revolution brought the forces of settler society into wholly unfettered power on the mainland, this bad faith can be seen in the renegotiated boundary lines that pushed the colonies' western borders deeper into Indian country. Even as the proclamation was being decreed, speculators vied for "lands west of the mountains, forming companies and dispatching surveyors, as if the Proclamation Line did not exist." In an important sense, however, such a line did *not* exist, except as a prospective division drawn across the surface of paper maps. No one could locate this geographic abstraction in the real spaces of eastern North America. It took Indians, British officials, and colonists more than a decade to set a tangible course for this boundary across the continental interior.[1]

To begin to understand the process of geographic negotiation and assess what it meant for the future of empire, settler society, and indigenous autonomy, we should set aside the notion that Britain's attempt to reconcile the zero-sum game between Indians and colonists over American land began and ended with the proclamation. Reconstructing this moment requires us to see three lines, not one. Each of these lines has a history that reveals a different moment in the British quest to use new knowledge to reshape American spaces. The proclamation put forward the idea of a line. The notion of separating colonies from Indian country had deep roots in the Board of Trade's agenda of reform. As part of its broader campaign to reorganize the empire after 1763, Britain sought clear boundaries between each of the westward-facing colonies and the indigenous nations in eastern North America. The idea of using the Appalachian Mountains as a natural boundary had surfaced before, most notably during the Albany Congress of 1754. But it was at the Treaty of Easton of 1758 that the line took hold as the centerpiece of a new Indian policy. The proclamation put this idea of a line across the mountains into law, ordering colonial governors to halt any new grants of land beyond the heads of rivers that flowed into the Atlantic Ocean and on any lands anywhere that might be claimed by the Indians. Across a map purchased in a London print shop to celebrate British America's territorial extent after the Treaty of Paris, the Board traced the first image of this line by hand and submitted it to the king. The Board's line diligently separated two great river watersheds: one that flowed into the Atlantic, and one that flowed toward the Mississippi River.

A second line sought to turn this abstraction into a surveyed boundary. Britain's superintendents of Indian affairs convened a series of diplomatic meetings to determine its location, beginning with an inaugural Indian congress at Fort Augusta, Georgia, in 1763. Augusta began a process of negotiating and surveying a general boundary that crossed the continent from the Mississippi to the Mohawk River. In nine congresses that followed, negotiators sought to hold white settlements where they stood and affirmed hunting grounds as legitimate components of Native territories. As this negotiated line followed the demographic pattern of how the land was occupied, it departed from any clear, natural definition and instead traced a course through the human geography of eastern North America. As Native headmen stood and spoke on behalf of their nations before the superintendents, the colonial governors of bordering provinces, and one another, they put forward spatial definitions of homelands that were distinctive to their historical experiences as Choctaws, Creeks, Cherokees, Delawares, and Iroquois.

At the Fort Stanwix Congress of 1768, this negotiated line's position began to erode as Indians and colonists agreed to open more land for settlement in anticipation of the boundary's final ratification. The boundary that imperial reformers hoped would keep rival societies in a state of equipoise with one another changed as fast as it could be surveyed and mapped. A third line reveals how the boundary was renegotiated from 1768 to 1774. The last unbounded segment of the line between the Cherokee Nation and Virginia witnessed the most dramatic change from the vision of the proclamation. Instead of honoring Indian land rights universally, Virginia's expansion ran roughshod over the claims of Ohio River Indians and opened the interior to unprecedented new colonization.

In 1763, the Board of Trade drew a red line across the surface of a map. By 1774, a surveyed boundary circumscribed the colonies and set them off from Indian country. Over the course of a contentious decade, the challenge of locating, surveying, and mapping this boundary absorbed the efforts of scores of Indian leaders, colonial governors from New York to West Florida, and the offices of the two superintendents of Indian affairs, north and south. Teams of British surveyors, accompanied by Indian counterparts, entered the woods with lists of agreed-to landmarks. With their instruments and instructions, they hacked a path across the depopulated marchlands of empire, places marked in recent memory by clashing aspirations and acts of violence, and turned the idea of a line into a visible division established by firsthand observation and documented by maps sent to London. The collapse of British government in America during the Revolution meant that almost as soon as it was created, this line lost the power to express the idea of Indian national integrity, much less enforce it on the ground. The threat it posed to the interests of colonists encouraged the resistance that led to independence, as Americans mobilized against the image of bounded, diminished colonies surrounded by places of imperial control. The promise of a fixed and regulated frontier, though put forward in apparent good faith by the Board of Trade's idealistic inner circle, served as a useful fiction for indigenous actors as they used opportunities for negotiation to gain goods, erase debts, and attempt to shield their towns and hunting lands against the imminent prospect of settler invasion.

The Imagined Line, 1721–1763

The idea of a line first germinated in the deliberations of the Board of Trade as a way to bring the mainland territories into a regulated imperial system. It began as a "statement of policy principle" and evolved into an "expression of a newly self-conscious British imperial view" that restructured Indian affairs in the spirit of comprehensive reform. Above all else, it was to be a jurisdictional division that delineated the polities on either side of it. It held that Native people were inhabitants of nations that possessed fixed territories, a precondition to negotiate with agents of the Crown as legitimate states. At the same time, imperial reformers sought to close the open geography of mainland colonies in order to transform them from empires in miniature—each with its own foreign policy and schemes for expansion—into bounded provinces with fixed places on a map of North America. The success of wartime negotiations with northern Indians at Easton in 1758 convinced the Board's commissioners that the key to pacifying North America was through a formal process of treaty diplomacy focused on creating a general boundary.[2]

A half century before the Proclamation of 1763, the Board of Trade identified the open mainland frontier as a critical strategic concern. Its commissioners saw colonial charters that extended without limit to the "South Sea on the West" as a framework for disorder, because they invited colonists to encroach on Indian land and thus provoke their retaliation. Merchants, officials, and speculators "Scattered along the Sea coast of North America" competed to claim the interior beyond the settlements as a range for their private interests, pushing their frontiers west in waves of violence that either "Ext[i]rpated the Natives" or "dr[ove] them farther into the Mainland." Expansive plantation settlement along with predatory trading practices triggered South Carolina's devastating Yamasee War in 1715. In the aftermath of this strategic disaster on the southern frontier, the Crown assumed control of the distressed colony from its proprietors in 1719. In 1721, the Board of Trade laid out a "System of Reformation" for Indian relations that aimed to contain selfish monopolies and fraudulent speculations that had thrown the interior into a state of perpetual war and squandered the promise of commerce to secure a self-sustaining peace.[3]

By 1721, the Board had embraced the concept of Indian possession as well as the need to establish diplomatic relations with Indian states, and although it did not recommend a "boundary of your Majesty[']s Empire in America," it named this possibility for the first time. Taking in a "view of the Map of North America" as a whole, the Board saw the Appalachians—this "chain of Mountains, that run from the back of South Carolina, as far as New York"—as a bulwark against the threat of French attack. Instead of unleashing settlers across the mountains, it advocated a system of western colonization. Once passes were discovered, the army could build forts to anchor a revitalized Indian trade "far westward upon the lakes and rivers." By channeling the unrivaled ability of British manufacturers to supply Indian wants "at honest & reasonable prices," a newly regulated trade could offer them fair dealing rather than the "unreasonable avarice" that had provoked frontier violence in the past. To make this vision of security and prosperity a reality, the Board pressed colonial governors to "make treaties & alliances of friendship with as many Indian Nations as they can." Transparent diplomacy, contained settlement, and open trade would convince them that they had "but one King and one interest."[4]

Indians understood how to picture space in the terms dictated by European cartography. Their descriptions provided mapmakers with credible sources for remote places, which filled in what would have otherwise been blanks on New World maps. When asked to show the way, indigenous guides created impermanent sketches on dirt and bark that impressed colonists with their verisimilitude to the topography of the real world as experienced by travelers. When Indians represented their own countries on their own terms, however, they often imagined regions as blank planes populated by abstract shapes, which made relationships among corporate groups the most salient features of North American geography. The Catawbas presented such a map to South Carolina's governor in the early 1720s. **[134]** Painted on a deerskin "by an Indian Casique," it described the southeast in the idiom of "the path," a geographic idea at the center of Native conceptions of space throughout eastern North America. On it, the Catawba Nation appears as a composite of eleven communities, each encompassed by a circle, separated by space, and linked together by strong double lines. At the center of this complex stood the town of Nasaw, the largest circle commanding the most connections, the most prominent of which was the "English Path." This path widened as it approached a maze of right angles meant to stand for Charlestown's urban grid. It became the road that colonists called the "Broad Path" as it reached the peninsula and King Street within the bounds of the city. An icon of an oceangoing ship, with its curved hull and a rigged mainmast topped by a flag, made it clear how South Carolina's capital was also the place at which Indians gained access to goods imported from the Atlantic economy. At the periphery of the Catawbas' world sat the square outline of Virginia, as well as circles that registered the presence of the Cherokees and the Chickasaws.[5]

The Catawbas, some of whom had joined in the attacks on British traders during the Yamasee War, presented this image of themselves at a critical juncture in Britain's relationship with Native America, asserting their desire to become indispensable intermediaries between Carolina and the indigenous interior. Its configuration of paths described a wished-for world in which even the smallest Indian group remained distinct, its circle surrounded by empty space between occupied sites. At the same time, every Indian circle and British square remained linked by routes of commerce, kinship, and affiliation. It was this idea of the path—and the tension between separation and connection that this abstract geometry of circle and line worked to reconcile—that Native speakers returned to again and again at the Indian congresses of the 1760s and 1770s. The purpose of speaking at these congresses was to clear, straighten, and "whiten" the paths of interdependence that had become obstructed by misunderstanding, convoluted by suspicion, and bloodied by violence. These paths were the literal roads along which Indians and Europeans traveled and traded as well as symbols of the presumption of mutual goodwill and a promise to resolve disputes through negotiation.

Unlike European maps of nations that placed all land within one jurisdiction or another, these Native American network maps positioned individual polities as separated entities rather than adjoining territories. The space between the circles were borderlands, often depopulated by warfare, claimed as deer hunting grounds, and incorporated into the territorial aspirations of surviving nations. Overlapping claims to these unsettled forests, marshy wastes, and rugged mountains made these in-between woods dangerous places, better to pass through quickly and arrive at the threshold of a circle that promised welcome. This was an archipelago of Indian corporate islands separated by contested space, not a vast "indigenous commons" to which these polities had stable attachments that gave each a clear claim to sovereignty. As much as Indians in 1763 wanted the British to set a boundary that would stop encroaching settlers, their relational idea of territory was not congruent with the absolute divisions encouraged by the law of nations and represented on European maps.[6]

A proposal to separate Virginia from Iroquoia along the "high Ridge of Mountains which extend all along the Frontiers" circulated without clear resolution during negotiations at Albany, New York, in 1722, and Lancaster, Pennsylvania, in 1744. The "Plan for a General Union" presented at the Albany Congress in 1754 advocated for terminal western borders along the ridgeline of the "Alleghenny or Apalachian mountains" to mold Atlantic-facing provinces into "more Convenient Dimensions." Benjamin Franklin, Thomas Pownall, and other reformers anticipated imperial schemes that would take command of regions within the interior, planting "Protestant subjects . . . westward of said Mountains in convenient Cantons," like the Board of Trade's townships in South Carolina, Georgia, and Nova Scotia. The Albany scheme "came to nothing," however. Far from ending provincial expansion, the meeting witnessed large cessions of land to Pennsylvania. Yet the notion of bounding colonies to soothe Indian fears that they would be "driven from their Lands" persisted. Reformers who embraced the idea of a new Indian policy in the 1750s shook their heads at a century and a half of frontier clashes. They criticized the ad hoc diplomacy that made "Indian Affairs" a byword for greed, fraud, and the pursuit of shortsighted gain over long-term security. In the face of French expansion and the "wretched footing" of Britain's standing with the Indians, the Board's first superintendent of Indian affairs, Edmund Atkin, advocated a "General Plan" for the management of Indian diplomacy and trade that would take charge of all "our Colonies on the Continent."[7]

The idea of a line persisted as British officials considered sweeping new policies to shape the mainland colonies. This was evident in 1755 when Sir Thomas Robinson, secretary of state for the Southern Department, proposed to set a new boundary with France at the mountains as part of a compromise to end the Seven Years' War. Although British hawks scuttled any capitulation to French claims, the notion of using the mountains to establish a general colonial boundary—an idea that had gestated for years within the deliberations of the Board of Trade—now entered into broad circulation at the highest levels of government. As British forces prevailed in the North American interior after 1757, commanders in the field worked to shore up these gains by convincing northern Indians to put an end to their attacks against frontier settlers. Wartime negotiations in Easton, Pennsylvania, in 1758 made the idea of the mountains as a boundary the keystone of Britain's new negotiated relationship with Native peoples. Pennsylvania governor William Denny promised the northern Indians that colonists would not "settle westward of the Allegheny Hills" and that they would "not make Plantations in that Country[,] as it would take away their hunting Country and drive them too far back." These specific agreements with Iroquois, Delawares, and Shawnees established a model that Britain later used to imagine a line extending across the continent.[8]

Experiences of encroachment did not entirely explain why eastern Delaware fighters had wreaked such violent havoc on the frontier during the war, but their anger at fraudulent treaties that brought British settlers beyond the mountains was "one Reason why the Blow came harder." Headman Teedyuscung described his goal for the Easton agreements as a search for clarification after the spatial confusion brought about by rapid colonial expansion. The Delaware leader felt like a "Bird on a Bo[ugh]; I look about and do not know where to go; let me therefore come down upon the Ground, and make that my own by a good Deed, and I shall then have a Home for Ever." The idea of mountains as a visible feature of a contested landscape appealed to Indian as well as British desires for clarity. Mid-eighteenth-century maps accentuated the "regular curve" of the "Allegéni Ridge" as a decisive western edge to one sector of the settled seaboard. **[135]** Cartographer Lewis Evans, creator of the century's most influential map of the middle colonies, believed it to define the "antient maritime Boundary of America." Eons ago, he speculated, when the seas had subsided to reveal the coastal plain, this original shoreline left behind a rocky mark that cut across Pennsylvania and divided North America's watersheds. Charles Thompson copied the shape of Evans's Allegheny Ridge to illustrate how wantonly this natural boundary had been violated by dishonest land deals, causing the "alienation of the Delaware and Shawanese Indians from the British Interest."[9] **[136]**

The northern Indians' understanding of these mountains as a border with British America was no hoary conception handed down through the ages but rather a new idea that reflected the recent surge in settlement across the northern frontier and the common desire to put a stop to it. Specific understandings of territorial limits held by individual Indian nations with particular colonies were no longer up to the task of protecting any of them. When General Edward Braddock told six Ohio River chiefs in 1755 that the "English Should Inhabit" their lands once the French were driven out and that "No Savage Should Inherit the Land," he suggested to them that British sovereignty over the trans-Appalachian west meant their certain dispossession. The idea of a line through the mountains helped Britain recant this vision about the future of the Ohio country as a place without Indians and replace it with a credible alternative that corresponded with indigenous ideas. After British forces took Fort Duquesne and renamed it Fort Pitt in 1758, a Delaware leader advised General John Forbes "to go back over the mountains and stay there." At Easton, a representative of the Ohio Iroquois people known as the Mingos told the Delawares to "look towards the Mountains" and pledged an alliance of resistance to such encroachment "if you see English Brethren coming over the Mountains."[10]

With the English "settling so fast" around them, the Indians were "pushed back, and could not tell what Lands belonged to them." Surreptitious incursions into hunting lands, once a buffer between Native Americans and Europeans, undermined Indian expectations that they could claim any space as their own for long. The new proximity of settlers, revelations that many independent transactions had eroded the scope of collective hunting lands, and pressure by provincial governments for large cessions at formal treaties combined to make Indians fear for the integrity of their homelands and react to the unexpected appearance of white surveyors, settlers, and hunters as a violation. Britain's objective for the Easton agreement was to "ease the Minds" of northern Indians who had been "very much disturbed" by the boundary confusion of the recent past. Forming a barrier that could be mapped across the mountains promised to establish a territorial marker that would insulate indigenous societies from perceived threats to lands they regarded as necessary for their survival.[11]

Before they signed off on the Allegheny ridge division proposed at Easton, the Delawares made a final bid to reclaim lands in the Delaware and Susquehanna River valleys claimed by Pennsylvania on the colonial side of the mountains. They sought clear documentation of "how far [Pennsylvania] purchases extend[ed]" and for this information to be laid before the king and circulated throughout the colonies so that private frauds could be revealed. Perhaps remembering how misleading maps had suggested that the Indians were giving up less than they actually were in the Walking Purchase of 1737, they produced their own sketch of a "large tract" around Wyoming on the eastern branch of the Susquehanna River. This lost Indian map suggested permanent "Boundaries fixed between you and us." Teedyuscung chalked an image of this land on the conference table to reiterate the Delaware vision of a "certain Country fixed for our use" at Wyoming, big enough not to be "pressed on any Side." When English negotiators asked the Iroquois to affirm their role in authorizing the sale of these lands at previous conferences, representatives from the Six Nations demurred: "they would have nothing to do with the land on this," the English "side of the hills, for they had no right to it nor ever claimed any." The Iroquois had in fact claimed the right to barter away these and other lands inhabited by subordinate Indians in 1736, 1749, and 1754. But in an act of intentional forgetting, the statement laid down a principle that sought to forge consensus among groups with diverse histories of dispossession in negotiations with colonial officials. The idea of a single line between Indians and Europeans following the ridgeline of the mountains imposed geographic clarity on multilateral diplomacy.[12]

Maps helped broker this British-Indian peace by representing land visually and thus putting doubts about the locations of places in English legal documents to rest. Delaware headmen "arose and said they did not Rightly understand that Paragraph relating to Lands" in the deed proffered by Pennsylvania governor William Denny at a subsequent meeting at Easton, held to confirm the agreement. The text that described where this land was "left Matters in the Dark," and they did "not know what lands he meant." Did he mean to keep only the eastern lands on the "other side of the Mountains"? If so, they would assent to it once they had seen the original deed and authenticated it by recognizing their own marks on the paper. Maps that showed the renegotiated dividing line across the mountains sealed the deal. Pennsylvania gave up its claim "Westward of the Allegheny or Appalaccin Hills," displaying an indenture of release that renounced its 1754 purchase from the Six Nations. Affixed to this written instrument of exchange was a "Draught of the said Land" that the province still claimed east of the mountains. This map made clear that, regardless of how the text of the document might be mistranslated, mistransmitted, or otherwise misunderstood, Pennsylvania had given up its western claim, and the Delawares had lost their lands in the east, including their hoped-for homeland at Wyoming. The Indians "express[ed] their Satisfaction" with the agreement and "particularly with the Limits as described in the Draught annexed to their Confirmation Deed." The British prepared a new "draught of the Boundary line" to represent the final Easton agreement and sent copies of this map to the Delawares, Shawnees, Wyandots, Six Nations, and the governor of Pennsylvania, so that all could see how the mountains now established the province's new limits.[13]

As soon as the British army had pushed the French out of the "fine fertile" region between the "Mountains and the Ohio" in 1759, Virginia's lieutenant governor, Francis Fauquier, urged the Board of Trade to put in place a system for granting those lands. His colony's prospective settlers as well as its "Speculative Gentlemen" had already imagined this place as a new colony, which they named, evocatively, the "Lands on the Waters of the Mississip[p]i." The Board of Trade rejected Virginia's settlement plan because, the commissioners explained, it violated the agreements at Easton by which Britain had "solemnly relinquished to the Indians all the Land to the Westward of the great mountains, and Engaged not to make any Settlement upon it." Although the

Easton treaty reworked the terms of a land deal that had involved only Pennsylvania, the Six Nations, and the Delawares along a comparatively small stretch of the northern frontier, Britain elevated this agreement to the level of an international treaty that bound all colonies to respect Indian land rights, a broad interpretation later enshrined in law by the Proclamation of 1763. As Virginians "fixed their Eyes" on lands along and beyond the Ohio River, the Board appeased Indian fears of territorial erosion at the expense of their desire for provincial growth. The Privy Council approved the Board's recommendation of December 2, 1761, to forbid all governors to "pass Grants of or Encourage Settlement upon any Lands within the said Colonies which may interfere with the Indians bordering thereon." It rejected New York's attempt to grant land within the notoriously "exorbitant" Kayaderosseras Patent along the Mohawk River. The colony had treated the Iroquois with "Cruelty and Injustice" by defrauding them of their "Hunting Grounds in open Violation of those Solemn Compacts by which they had yielded to Us the Dominion, but not the property[,] of those Lands." Only after Britain "awakened to a proper sense of the Injustice and bad Policy of such a Conduct towards the Indians" and reaffirmed promises to keep their land secure at Easton did the "horrid Scenes of Devastation . . . ceas[e,] and the Six Nations and their depend[e]nts became at once from the most inveterate Enemies our fast and faithful Friends."[14]

Sir William Johnson, the northern superintendent of Indian affairs, endorsed the idea that a "certain line should be run at the back of the Northern Colonies, beyond which no settlement should be made, until the whole Six Nations should think proper of selling." Johnson sought a "remarkable boundary"—something more than a mere property line between lands already purchased and those that remained to be so—across which speculators would be barred from buying the land piecemeal, putting it "beyond the power of disposal" by individuals. If such a line preserved a large, continuous tract of hunting land, it could keep the Iroquois "contented and satisfied." By generalizing the Easton boundary to apply across the mainland, the Privy Council sought to "ascertain the British Empire in America." Britain's desire to know American territory in this definitive way in the aftermath of the Seven Years' War meant, first and foremost, making the ownership of contested lands certain and determining under what jurisdiction they fell. The king issued these new restrictions to "give equal Security and Stability to the Rights and Interests of all [of his] American Subjects," Native as well as European. Asserting these powers over frontier land was a right Britain claimed as the only authority that stood dispassionately above the self-interested scramble for new lands to preserve the "peace and security of Our Colonies and Plantations upon the Continent of North America" as a whole.[15]

The Proclamation of 1763 gave continental scope to Easton's regional agreement by proposing a common boundary that spanned the British mainland. Those who planned the new American empire from London sought above all to "conciliate the Affection of the Indian Nations" as the most important end of this new system of Indian affairs. Preventing any "Incroachments on the Lands they had reserved to themselves for their hunting Grounds" anticipated Indians' volatile reactions, rooted in the psychology of territoriality, which would bring them to war. The very idea of "encroachment" binds together notions of disordered space compounded by unwelcome change over time. It is a specific kind of fear that expresses alarm about intrusions into spaces vital to survival: those surrounding bodies, houses, and homelands. The implicit boundaries that encroachments violate are often inchoate and unspecific, revealing themselves in terms of the context of the violation. Encroachment's impact compounds the losses of ceding ground with the insult it offers to a person's or a people's "social persona." It is a refusal to acknowledge the fundamental character of who they are by not respecting the places in which they live and on which they depend. Ohio Indians attacked Pennsylvania and Virginia settlers during the war not only because they had "settled on their lands" but also because their "Complaints on this Head [were] not regarded" by British officials. The sound of firearms, the sight of surveyors with their instruments, the appearance of smoke from a new cabin at the edge of the woods, or an overheard rumor of a secret land sale to Europeans might trigger a sense of encroachment without specifying a consistent idea of the boundary that had been crossed. Fear of encroachment is a progressive anxiety because it anticipates future violations that accumulate over time until the original boundary is erased altogether and the territory it once defined is lost. The ultimate purpose of establishing a frontier line was to make this implicit sense of a boundary into a real border that all could see.[16]

This idea of conciliation served as a guiding principle for diplomacy, not only for the Algonquians and Iroquoians who lived near the Ohio River but also as a general mode of approaching all Indians living in parts of North America and the West Indies—including the Creek country of the southeast and the Carib-occupied areas of the Ceded Island of St. Vincent—to which Britain claimed title by the Peace of 1763 but over which it did not yet exercise direct rule. As colonists molded their rage over wartime attacks into a portrait of irredeemable Indian barbarity, imperial reformers offered a sympathetic analysis that understood those "merciless Devastations which have greatly distressed the Western Frontiers" as rooted in reasonable fear rather than irrational savagery. Indian speakers at the congresses expressed fears of their lands being dismembered through cadastral surveys, and British authorities listened as they recounted histories of fraud, squatting, and speculation that had displaced people who once lived in the river valleys of the coastal plain, who had been forced to take up new lands in the interior and now found that with every generation came a new threat of displacement. By setting indigenous land beyond the limits of colonial jurisdictions, it could no longer be scouted, surveyed, and converted into tracts, at least not those deemed legitimate in the eyes of the law.[17]

The Board of Trade urged the king to "fix upon some Line for a Western Boundary to our ancient provinces." Its "Report on Acquisitions in America" proposed an interior of protected Indian "hunting grounds, where no Settlement or planting is intended, immediately at least, to be attempted." Later instructions to the governors would list specific geographic landmarks for new interior boundaries for the mainland colonies. In the meantime, the commissioners provided the king with a map "in which these limits are particularly delineated." The Board drew a red line across a copy of Emmanuel Bowen's *An Accurate Map of North America*, tracing a course around the rivers that flowed into the Atlantic Ocean, and coloring in the bounded mainland in a wash of aquamarine. As the Board's secretary, John Pownall, explained, limiting settlers to land along Atlantic rivers kept them within the "reach of navigation." It was a "happy coincidence" that the Indians wanted placating, because this strategic necessity provided an opportunity to enforce an optimal geography for imperial political economy. Americans that moved across the Appalachian divide to "remote settlements" would lose the "interests and connections" that bound them to Britain. Fixing this limit to "colonizing" would keep them within Atlantic circuits of exchange, whereas beyond the "Heads of those Rivers, which run into the Atlantic Ocean," interior "Settlements can have little or no Communication with the Mother Country, or be of much Utility to it." The Board's line marked the threshold beyond which colonists could not be expected to serve their own economic interests as well as Britain's. Its band of red ink disentangled two continental watersheds to create distinct spaces for two different kinds of societies, one Native American and the other European. Once marked by the Board and sent to the king, this line represented the "Certain fixed Limits" alluded to in its June 8, 1763, report, which envisioned the reconfigured empire. It became the imagined border just beyond the "Heads or Sources of any of the Rivers which fall into the Atlantic Ocean" in the king's proclamation of October 7, 1763. It suggested a place to begin to determine the "precise and exact boundary and limits of the lands" reserved to the Indians, where "no settlement whatever shall be allowed," as set forth in the 1764 plan that laid out the terms by which British superintendents were to treat with the Indians. With the mainland colonies confined to the watersheds of rivers that flowed to the Atlantic, the "imaginary Boundary Line" that French cartographers had drawn on "Maps to all our Colonies on their backs Westward" could be made real, even though the French were gone. After the war, Britain embraced what it had once feared. Beyond it, the Board of Trade imagined an "Indian Country, open to Trade, but not to Grants and Settlements."[18]

The idea that North America could be divided by a line on a map made sense only within the logic of the map itself. To make the continent legible on his *Accurate Map,* Bowen represented its topography with a series of symbols. The Board's line followed his image of a mountain range made from hundreds of repetitions of the same icon—each a tented peak fronted by its own tiny foothill—a graphic shorthand that created the illusion of linear ridges. In fact, the Appalachians are a two-thousand-mile-long complex of ranges, valleys, and dissected plateaus spread across a territory larger than the inhabited coastal plain. This vast cordillera, so named because it resembles a great braided rope, stretches across the eastern continent at a forty-degree angle from true north, a skewed course that bedeviled the idea of a north–south ridge line cleanly bisecting the eastern continent into a colonial coast and an indigenous interior. These mountains widened out to the southwest so much that Creek and Cherokee towns actually fell to their east, on the colonial side of the map's proposed line; they converged on the coast in the mid-Atlantic in formations that bend along an east–west axis, their "lofty peaks over topping one another" to form a range called the "Endless Mountains." This meant that Iroquoia was as much to the north as it was to the west of a rapidly peopling British region. The Board had already conceded the presence of Virginia settlements "beyond the great mountains" by extending its red line to the eastern banks of the Ohio at the juncture where the Kanawha River joins it. So ancient is this river, which begins as the New River in the North Carolina piedmont, that it had carved a course through the heart of the Appalachians, confusing the very idea that the British could distinguish North America into distinct hydrological zones. No map could disentangle such geographic realities with a line drawn in London, particularly because so little was known by survey about the interior.[19]

The Board of Trade's line could only be as precise as the map on which it was drawn. Emanuel Bowen was a practiced compiler of information first published on other maps, and his main source was the era's most sophisticated representation of the continent, John Mitchell's *Map of the British and French Dominions in North America* (1755). This image was itself compiled from previous maps, written reports, and surveys taken at different scales, and although Mitchell attempted to catalog every known place at mid-century, his map did not establish the location of any particular place in relation to any other with certainty. Although "many Copies of Mitchel's Map" circulated in the colonies, according to rival cartographer Lewis Evans, "nobody pretends to look into them for any Place on our Borders." Inland from the settled coast, its credibility fell apart the closer its details were examined. Other, less careful mapmakers filled the frames of their maps with images of an interior country they did not really know. Although there were "no Mountains whatever" below the 34th parallel, the Appalachians were shown to extend south in "almost all Maps," which notoriously showed "these mountains other than they are." Beyond the last township, parish, or county in which provincial surveyors had laid out land in measured tracts with compass and chain, beyond the furthest reaches of colonial boundary lines that they had extended into the west to sort out conflicting provincial claims, no place was known well enough to draw a line with enough precision to locate territory on either side of it. Britons who passed through this "mountainous Country" estimated the distance between "one known Mountain to another with a Compass" to locate landmarks they observed along their route. Such rough methods made the images on even the best maps of the west little more than approximations. Although they believed they were separating rivers and tracing the ridges of mountains, the commissioners of the Board of Trade drew their red line across a terra incognita.[20]

The Royal Proclamation of 1763 took the framework for empire devised by the Board of Trade, with all its geographic uncertainties, and made it law. In the months leading up to the proclamation, the Board urged the king to express his "determination to permit no grants of Lands" in the west and to use its map to set "fixed limits" for the mainland colonies. Although they had received no formal report of the uprising later known as Pontiac's Rebellion by the time it was drafted, these "late complaints of the Indians, and the actual Disturbances" they had caused, urged its immediate public proclamation. The king declared the unpatented west off-limits to new colonial settlement, arresting the legal process by which the North American provinces had, since their establishment, enlarged their dominions and extended the reach of their jurisdictions. Even east of the mountains—in fact, wherever "any Lands whatever" might lay unclaimed—colonists were to assume them to be "reserved" to the Indians as their property. The proclamation defined the Indian country in the negative, as the sum total of territory "not having been ceded to or purchased by" the Crown. In its early drafts of the proclamation, the Board considered and rejected language that would have limited this Indian country to territory actually "occupied by the said Indians" or claimed only by "Indians in Alliance with Us." In the end, however, it opted for a much broader definition of indigenous possession that demanded the most restrictive injunction against new grants of land. As a matter of law, the proclamation stilled the relentless machinery of the colonial land system and validated the Indians' ownership regardless of who they were or how they used the land.[21]

Except as a band of ink on the paper surface of a few maps, however, the "Proclamation Line" did not exist. The October 1763 issue of the *Gentleman's Magazine* published the text of the king's proclamation along with a map that simplified North American geography for a mass audience. **[137]** It compressed a complex topography of ridges and valleys into a single mountain chain, along which a single line could be drawn. In bold letters across the Mississippi River watershed, mapmaker John Gibson labeled the new British west as "Lands Reserved for the Indians." Gibson's line put North Carolina's western boundary past the Cherokee towns and projected Georgia into Creek country. These were not signs of his ignorance about the finer points of North American geography that had appeared on Bowen's more detailed map, because Gibson himself had engraved its copperplate, carving every bend of every river by hand with the edge of his steel burin. For this image, he pictured the proclamation's prospective division in broad strokes, scaling a continental geography to fit within the modest, six-by-nine-inch dimensions of a widely distributed octavo magazine. Other maps published in London in 1763 took liberties in drawing lines between the colonies and Indian country. Readers of the *Royal Magazine* saw a dotted line demarcating safety from danger, beyond which "Scalping Tribes of Indians Inhabit"; a dashed line separated territories claimed by Britain before the war from those ceded by France and Spain at the peace. **[138]** Nothing on this map indicated a place for Native polities within the boundaries of an enlarged British North America. From 1758, Britons read Edmund Burke's "History of the Present War" in *The Annual Register* to keep up with the exploits of the king's armies and fleets in America. After turning the page on Burke's final installment in 1763, they beheld the "extent of our empire" by viewing it on a specially engraved image, *A New map of the British Dominions in North America*, which promised to incorporate new boundaries dictated by the Peace of Paris as well as by the proclamation. **[139]** Within them, Indian place names fell within the jurisdictions of colonies that extended to the banks of the Mississippi. Such an image asserted long-standing aspirations for continental grandeur more than it imagined how new subjects in new territories might inhabit a stable space within the empire.[22]

These and a few other uncertain lines drawn across North America make up the whole visual record of what historians have called the "Proclamation Line." The maps on which they appear document a moment in which the notion of a divided North America had been conceived but before the actual location of any boundary had been determined. There was no self-evident border with Native America. If ordered to locate Virginia's frontier, reflected Francis Fauquier in 1767, he would have been "at a loss where to have begun." As the Board of Trade reconceived Britain's Atlantic empire in 1763, this line remained unfixed in cartography as well as in public consciousness. To give meaning to the proclamation's endorsement of Indian land rights, the idea of this line through the mountains, drawn across a map of North America at the scale of the continent, would have to be brought down to earth.[23]

The Negotiated Line, 1763–1768

Britain attempted to impose an explicit boundary on an uncertain, bloody, and contested frontier. This line was to be no mere notion but rather a real location, affirmed by a process of negotiation that described, marked, and mapped it through a collaborative process of surveying. At ten formal diplomatic congresses (and many smaller meetings) from 1763 to 1774, Cherokee, Creek, Choctaw, Chickasaw, Catawba, Delaware, and Iroquois negotiators—along with thousands of Native spectators—joined the Indian superintendents and the colonial governors to come to an agreement on each section of the boundary. The first such gathering was held in early November 1763, at Fort Augusta, Georgia, just as news of the proclamation reached America. At this frontier fort, which served as the primary place of exchange for the Creek deerskin trade, southern superintendent John Stuart set the precedents by which the "consent and concurrence" of Indian leaders would be required to "ascertain and define the precise and exact boundary," beyond which "no settlement whatever shall be allowed." Following rules that were codified in 1764's "Plan for the Future Management of Indian Affairs," Indians were only permitted to sell land to the colonies collectively, when it was offered by "principal Chiefs of each Tribe" at a "general meeting" over which the superintendent would preside. Such cessions became binding only after the lands were "regularly surveyed by a sworn surveyor," accompanied by Native representatives who joined him to walk the boundary as it was marked. The document that ultimately legitimized this transaction was an "accurate map" produced from this survey, to be filed along with a deed in the relevant colonial land office. Only after each Indian nation had set its "boundary line between their Country and the settlements of [British] subjects" in this transparent way could the violence of the past be forgotten and a peaceful future made possible by reciprocity begin. The negotiated line was, first and foremost, a jurisdictional boundary meant to end encroachment by taking western lands out of the purview of provincial governments and including them within the domain of Indian states. As such, it created two distinct spaces for civil order, in which each side would exercise judicial authority to resolve disputes on its own terms. Although it served to separate two kinds of territory under the common sovereignty of the Crown, it was also meant to operate like an international boundary across which a free and open commerce could take place, a hallmark of independent nations in amity with one another. As the key to securing land, organizing trade, and administering justice, this "boundary line being established by solemn compact with the Indians" became the focal point of British-Indian diplomacy in the generation before the American Revolution. More than any other initiative, British reformers believed that surveying this line could help Indians imagine viable futures for their societies next to the colonies and under the distant authority of the king.[24]

John Stuart opened the Augusta Congress with a speech intended to begin this new era. One of several agents charged with the task of putting the Board of Trade's vision into practice in the 1760s, this former Scottish Indian trader was appointed to oversee a fractious southern frontier. His "Map of Cherokee Country," like Samuel Holland's much-praised map of Quebec, was a patronage performance that demonstrated an original knowledge of a contested American place. **[140]** His appointment to high office, from which he dictated Indian policy to six colonial governors, reflected the Board's intention to empower those men who knew America best to administer the new empire. Stuart's manuscript map conveyed new information about the locations of Cherokee towns, nestled in four regional clusters in the river valleys of southern Appalachia. He represented "Cherokee Country" as a place of Indian habitation without corporate boundaries. Distant hunting grounds to the northwest and southeast show the range of seasonal migrations in pursuit of deer but confer no clear claims to national dominion. Stuart noted sites of "Abandoned plantations" and those vacated settlements "destroyed by the Cherokees" on the borderlands with Virginia and South Carolina. Stuart had captained a South Carolina militia company during the Cherokee War (1758–1761), and this map, as a documentary artifact of that conflict, saw Indian towns in strategic terms—as targets as well as threats. The map followed a convention of European cartography by representing Indians ethnographically rather than nationally—in this instance as a population whose 2,920 enumerated arms-bearing men appeared in a table at the bottom of the map. Stuart annotated the road by which he had "escaped to Virginia" in 1760, after his capture at Fort Loudon. As he fled along the Holston, he learned that the river was "Navigable into the Settlements of Virginia" and put this observation on his map, suggesting a corridor through which Indians and colonists could anticipate mutually beneficial exchange across a recognized boundary once the war was over. As the southern superintendent, Stuart became the Cherokees' most powerful advocate within the empire, tasked with putting this wartime view of space aside to help forge a lasting peace.[25]

Speaking before some eight hundred Native listeners, many of whom had traveled hundreds of miles to hear what he had to say about the future of Indian societies under British rule, Stuart chose his words carefully, working from a script of conciliation prepared, sometimes verbatim, from reports that Henry Ellis and other advisers had submitted to the Board of Trade and the secretary of state. He described a political geography in which Britain's exclusive sovereignty in eastern North America would not mean Indian dispossession. To counter their suspicions that the "English entertain a settled design of extirpating the whole Indian race, with a view to possess and enjoy their Lands," he asked them to imagine an empire of commerce in which they could flourish as participants. Stuart met with the southern Indians at Augusta to "quiet their

Apprehensions, and gain their good Opinion," as they learned that Britain would stand unrivaled as the sole European imperial power east of the Mississippi, as stipulated in the Treaty of Paris, signed just weeks earlier. Before the meeting began, Stuart dispatched an emissary to assure the Creeks that the British "don't want any of your lands" and demanded (and received) a signed affirmation from the southern governors that was read aloud to the waiting Indians at Augusta. It declared: "your lands will not be taken from you."[26]

Stuart's address expressed the Board of Trade's imperial vision. It opened with a visible demonstration of a common British Indian policy. Four governors stood behind Stuart and deferred to him as the voice of the king and the "mouth" of the southern provinces through which they all "utter[ed] the same words, at the same time." Unlike the negotiations of the past between "one nation of Indians, with one governor," this "general" congress was itself a demonstration of the new order, in which the interests of individual colonies and tribes would be subsumed into a coordinated program of trade, justice, and settlement. For years afterward, Indians remembered Stuart's speech at Augusta as a touchstone moment. Like many others who gathered on the banks of the Savannah River to hear it, the Cherokee headman Saluy (also known as the Young Warrior of Estatoe) remembered its promise of an American continent ruled by Britain, in which Indians would persist as recognized nations—a relationship he described as a "straight path." Instead of demanding new territory as a conquering empire, Britain surprised North American Indians by inviting them to enter into negotiations to lay out a boundary that promised to end encroachments everywhere.[27]

In the early 1720s, when the Catawbas presented their deerskin map to South Carolina governor Francis Nicholson, the Board of Trade was just beginning to make the reform of Indian relations a priority of empire. Forty years later, as the Board worked to contain settlement and treat with Indian nations at diplomatic forums, the Catawbas no longer figured as Britain's gatekeepers to the Native southeast. By the time they gathered to hear John Stuart in Augusta, their lands were surrounded by settlers. In 1762, North Carolina officials had "caused Surveys to be made, and Lines to be run over the Hunting Grounds, and over burial Places of the Catawbas, to the great Alarm and Disquiet of that friendly

Nation of Indians." The Privy Council ordered the "immediate Establishment of a certain Line of Partition" between the Carolina colonies and Catawba country that would forbid white settlement there, prevent the sale of any of it as property to Europeans, and put it outside of any colonial jurisdiction. When the Indians confronted settlers who advanced up the Catawba River to claim this land, "they say they will continue to do so unless we show them a paper to restrain them." With their numbers reduced to about five hundred people after the 1759 smallpox epidemic, the Catawbas came to Augusta having already agreed to draw a formal, legal cordon around a fifteen-mile square tract that settlers would surround on all sides.[28]

Creek speakers dominated the Augusta proceedings with a proposal that located every twist and turn in their lengthy boundary with Georgia, while the Catawbas were "last to be introduced, last to be addressed, and last to speak." Their confined reservation was ratified without much ceremony or discussion, and South Carolina sent provincial surveyor Samuel Wyly into the backcountry to map its boundaries later that winter. [**141**] He measured out a box around 144,000 acres and, in doing so, added a square-shaped notch that revised the border between the Carolinas and fully encompassed the Catawba Nation within South Carolina. The proximity of hundreds of white settlers to this plot had "spoiled" a meager hunting ground, driving away "buffaloes and deer" for "one hundred miles every way." Under such constraints, the Catawbas retained their integrity as a people but lacked the means of independent survival. After Augusta, John Stuart considered them inhabitants of the government of South Carolina who had an "absolute depend[e]nce upon it." Wyly called his image of the Catawba Nation a map, but it resembled a plat—a boundary sketch that every prospective landowner had to submit to the provincial land office in order to obtain a formal grant of land. These boundaries offered a measure of legal protection from encroachment but at the price of giving up the social and economic autonomy that was a prerequisite for true nationhood. After Augusta, the Catawbas possessed a tract of land in South Carolina, surrounded by other tracts.[29]

By witnessing the fate of the Catawbas, hundreds of Native observers at Augusta reflected on the specter of dependency that came with the loss of hunting grounds. Other encompassed Indians brought similar plights to the attention of participants at other congresses. As the Iroquois line with New York took form under the leadership of northern superintendent William Johnson, the Cayugas agreed to "collect our People together and not to live so dispersed." They asked to be supplied with resident blacksmiths as they contemplated a transition to a life of European-style farming, confined to towns and fields. The agreements at Fort Stanwix in 1768 left the Mohawks "within the Line," a position that prompted them to begin the process of selling their tracts so that "they who ha[d] so little left" and lacked the ability to make a living from the land "may not lose the benefit of the sale of it." The Mahicans of Stockbridge, Massachusetts, watched as settlers cleared the land around them, an experience they said was like standing in the shadow of a "very large Tree, which has taken deep Root in the ground, whose branches are spread very wide." Native people who lived within the pale of an expanding settler society had little choice but to live in the shade of its protection as dependents. At every congress, the settlements by which encompassed groups like the Mahicans, Catawbas, Cayugas, and Mohawks turned their territory into tracts within provincial land systems presented object lessons to the Creeks, Choctaws, Chickasaws, Cherokees, and non-encompassed Iroquois, who commanded the diplomatic power to carve out sustainable national spaces through the boundary-setting process initiated at Augusta.[30]

John Stuart drew "A Map of the Southern Indian District" in 1764 to illustrate the accomplishments of the Augusta Congress and represent Native societies within bounded territories. **[142]** When the Board of Trade's clerks first unrolled this manuscript map, they saw novel shapes of Indian nations filling the spaces of the southeastern interior. Establishing that Indians possessed the civic capacity to negotiate their borders was a critical point to imperial administrators, because these societies often seemed to lack the characteristics of functioning states. The problem, as Thomas Pownall put it, was the "original natural form under which the Indian Country lay, being that of a Forest" through which Indians hunted, traversing the land as "wanderers" instead of planting it as "Settlers." Following Emer de Vattel's influential vision of a balanced world order among states organized around the need to protect property, Pownall observed that because indigenous societies "never had any idea of property in Land," they had neither developed a constitutional order to protect land rights nor elevated an executive "Stateholder" capable of enforcing law. Two centuries of "Trade, Treaties and War," however, encouraged what once seemed like crude hunting societies to become the kind of states that could engage in international diplomacy with legitimacy. British officials used the congresses and the process of boundary making to cultivate this nascent authority.[31]

Those willing to look could see that Native Americans pursued objectives that were a "Simple and Plain" reflection of a "true National Interest": securing safety, supplying wants, and insisting on "fair Usage" with others. A long history of white aggression had driven the Indians to take up arms against settlers to "vindicate their Natural Rights." A boundary with Indian country, by ending such provocations, would reveal their commonplace humanity. Recognizing the stature of Indian nations in diplomatic negotiations redefined them as civilized within the law of nations, which otherwise denied savage societies the ability to exercise sovereignty. Rituals of respect on display at the congresses demonstrated a British willingness to follow Indian protocols, coming as they did at the close of a long history of "quasi-diplomatic engagement" by which British officials had dealt with Indian states in the past. That each of the ten major congresses took place at British outposts close to indigenous lands reinforced the idea that the Indian nations entered into negotiations as partners whose preferences mattered. Still, British officials could not entirely shake the sense that negotiations with Indians were something "contemptible in Comparison to the Affairs of Civilized Nations." In place of centralized authority capable of imposing agreements, Indian government was forged by consensus among towns and clans. Every overture generated "Jealousies, and Tumults, and Counterworkings" because, instead of unified nations under stable leadership, the decentralized "Nature of Indian Government" presented a chaos of interests. Perhaps British commentators, whose government's ministries changed hands so rapidly in this era, should not have been the first to throw stones at the disorderly character of Native governance, but they nevertheless observed that Indian states were like "so many united Republics": these polities seethed with factions—fronted by "Leading Men," each of whom required "particular Attention"—beneath the surface of unity presented at treaty talks. The procedures for negotiating the Indian boundary helped bolster the credibility of enterprising headmen, who had to return to their communities with evidence that their concerns had been heard and their interests protected.[32]

At Augusta, negotiators followed the logic that the territorial needs of populations on either side of the line should determine where the boundary between their societies should be set. This principle dictated that the line would be drawn to respect colonial settlements where they existed, even if some questioned the legality of the means by which settlers had acquired their tracts. For Indians, this process offered an opportunity to define a national domain that included towns and fields as well as vast deer hunting lands. Before the first formal speech at Augusta, one of the Creek participants complained that "it was their hunting season, when they should have been in the woods providing for their families." By the congress's end, hunting grounds were recognized as an integral part of the Creek Nation. By recognizing "Lands claim'd by Indians as their hunting Grounds," Britain also set aside the standard for legitimate claims to sovereignty that required the land's improvement—demonstrated by intensive agriculture and town building—as signs of its permanent, physical occupation. The "Country on the Westward of Our Frontier quite to the Mississippi," explained the secretary of war, Viscount William Barrington, was "intended to be a Desert for the Indians to hunt in & inhabit." Restricting settlement to one side of the line preserved deer habitat on the other, and Britain's first act of "strict Justice" designed to demonstrate its benign intentions was to make it possible for Indians to live off their lands. No longer defined as unimproved woods and wastes, British negotiators recognized the role that forests played in sustaining Indian societies.[33]

Building fences, clearing trees, and planting fields had, for more than a century, imposed a "geographic separation of English cultivated settlement from indigenous 'waste' lands" that, in most colonists' eyes, made the interior appear as a vast wilderness open to European colonization. The boundary line begun at Augusta set apart two societies, one seen as devoted to hunting and the other to planting, but did so in a way that rejected the legal doctrine of *terra nullius* and its assertion that uncultivated lands could be claimed by those who first arrived to improve them through agriculture. At every congress, British negotiators upheld deer habitat as part of legitimate national domains, affirmed by past conquests and long use. Although the "American Savages" possessed "more Land, than they know, what Use to convert it to else, but to hunt upon it," this less intensive use for the land provided an indispensable foundation for national independence, providing not only the "necessaries of life" but also a valuable commodity—deerskins—that they needed to engage in trade. Indians who "should be considered as His Majesty's Subjects" could also be counted, alongside their counterparts in the colonies, as consumers and producers who "contributed to the Trade of Great Britain."[34]

Stuart's map brought the Southern Indian District into sharper focus as an inhabited place. Scores of named towns occupy the bends of delineated river systems. Its orderly, populous, and bounded nations challenged conventional views that pictured Native societies in vague relation to the territories that surrounded their towns. Two red lines of latitude drawn across the map—at 36°30′ and 35° north—represented the charter boundaries of Virginia and the Carolinas. Where these hit the new borders of the "Cherokee Nation," Stuart's lines suggested, the claims of those colonies to western lands ended. On the map, this nation took on a natural coherence as a network of mountain valleys along the southern Appalachian divide. Its towns occupied the headwaters of two great river systems, one that flowed toward the Atlantic Ocean and converged on the Savannah River, and the other that flowed toward the Mississippi watershed and converged on the Cherokee River. What made this map different from Stuart's earlier "Map of the Cherokee Country," from which he copied several details, was that the Cherokees now possessed a contiguous extent of territory and a place within a system of legally recognized states. The map reflected the Cherokees' national aspirations in the new imperial system as well as Britain's interest in affirming them.

Stuart's map not only made Indian nationhood visible but also recorded the first segment of a specific negotiated border between the Creek Nation and the colony of Georgia. The "old boundary Line by the Creeks," traced in red on the map, confined the colony to the coast and to a narrow band that stretched upriver along the Savannah. The "New

Cession by the Creeks" made at Augusta opened a substantial body of "lands to the westward, which may be settled by the white people," between the Savannah and Ogeechee Rivers. This new Creek-Georgia boundary was important not only because it was the first negotiated section of the general North American boundary but also because its position reflected the history and aspirations of the Creek Indians. The Creeks saw the southeast as a range for their own colonial ambitions. Creek warriors had depopulated Spanish Florida in the early eighteenth century, burning villages and enslaving thousands of Timucua and Apalachee "mission Indians." After Spain ceded Florida to Britain in 1763, the Creeks claimed virtually all the land between the Savannah and Alabama Rivers as part of their historic conquered domain. At the congresses of Augusta (1763), Picolata (1765), Mobile (1765), and Pensacola (1771), their headmen remained "tenacious of their hunting Ground" and held steadfast to the principle that their nation possessed the entire southeast interior, confining British colonies to the "sea coast as far as the tide flows." The line they negotiated with Britain during the 1760s followed the contours of the coastline from the edge of Choctaw country east of Mobile and across the Gulf Coast. It hemmed in the British outpost at Pensacola and reserved most of the Florida peninsula as a hunting ground for those Creek-affiliated Muscogulges, later known as the Seminoles, who lived there. Only as it crossed Georgia did the line turn inland in recognition and acceptance of the tracts that colonists had already occupied. Although the Board of Trade enlarged the colony's boundaries in 1763, the leverage that Creek negotiators brought to the congresses sharply limited the extent of future settlement in this well-regulated model colony.[35]

Three rivers—the Savannah, Ogeechee, and Altamaha—defined the shape of the Georgia colony as its first settlers arrived to claim it in 1733. Georgia's founder, James Oglethorpe, promoted this idea by sending accounts of exploratory voyages up these rivers to hired mapmakers in London. Their maps shaped an idea of Georgia as a place with abundant riverside land to entice new colonists. After 1751, with the repeal of the prohibition on African slavery in the colony, new planters contended with one another to acquire large tracts of tidal river swampland ideally situated to grow rice, which became the colony's staple commodity. As the Seven Years' War began, the colony's provincial surveyor, William de

Brahm, put forward a plan to fortify each of these rivers against the threat of French, Spanish, and Indian attack so that this valuable new plantation frontier could be defended. [**143**] He labeled the space between these fortified positions the "inhabited Part of Georgia." Nowhere on this map did he indicate that the Lower Creeks claimed much of this land as their own hunting ground or that they had previously agreed to give up only a small part of it to the British.[36]

Charged by Governor James Wright to provide a common geographic reference at the Augusta Congress, De Brahm and fellow provincial surveyor Henry Yonge drew a second map that pictures a much different, and much smaller, Georgia in relation to the Creek Nation. [**144**] This "Map of the sea coast of Georgia" circulated among Indians and British negotiators as they debated where the line should be located; when they reached an agreement, they traced it across the map. This negotiated boundary divided the three rivers that had defined Georgia's domain in British understandings for years, leaving most of the land between the Ogeechee and Altamaha Rivers in Creek country. Where the "Map of the inhabited Part of Georgia" assumed British title to all the territory it represented, "A Map of the Sea Coast of Georgia" adhered to the principle that only land formally ceded by the Indians was open to settlement. It displayed the restrictive line of the Creek grant to Oglethorpe in 1739, one based on the principle that the British should claim only those lands whose adjacent rivers were affected by the ocean's tides. The new boundary set at Augusta cut across two of the colony's three great rivers to make room for an expansive Creek Nation.[37]

The sharp bend in this Creek-Georgia line marked the inflection point between that part of mainland North America that the British had already colonized and that part which they intended to colonize in the future with new rules designed to keep overseas territories connected to Atlantic currents of trade, culture, and authority. Georgia's southern coast and the new colonies of East and West Florida were to be developed within these limited bounds as Atlantic colonies, shorn of continental interiors. The Creeks came to Augusta trading rumors that the British would demand lands that ran up the length of all three of Georgia's great rivers. If some had heard the geographic language of the Proclamation of 1763, which prohibited new grants "beyond the Heads or Sources of any of the Rivers which fall into the Atlantic Ocean," they would have had reason to be apprehensive. If Britain had in fact opened the full length of these rivers to new grants, the boundary line would have reached more than three hundred miles into Creek hunting territory and close to the Chattahoochee River, along which the Lower Creek towns were settled. Creek negotiators offered a compromise that split the difference between these two conceptions of where the boundary should be. Between the edge of the tidewater and the headwaters of Georgia's rivers, "land above the rocks"—that is, above the fall line—should "remain unsettled." This redefinition of southeastern rivers put the visible edge of the coastal plain—where water tumbled across a geological fracture in its path to the sea—into service as a visible marker for the proposed boundary. Georgia was not to extend beyond the point at which a boat could no longer be rowed from the coast. Such a line left the most forward plantations within the settled zone, along with a modicum of yet-to-be settled land. Between the Creek road to Charlestown and the Ogeechee River was to be a "line for the white people to grow between."[38]

Once they had agreed to the path of the boundary in the abstract and affirmed it with all the trappings of diplomatic ceremony, British surveying parties marked the line on foot and horseback, accompanied by contingents of Indian observers. These linear manuscript images, some composed of sheets of drafting paper glued together to form thick scrolls, depicted swaths of previously uncharted space to fix the boundary within the landscape. **[145]** The provincial surveyors who led these expeditions adapted the skills they had learned measuring out tracts of land for prospective settlers, who paid them to produce the plats they were required to submit in order to obtain a formal land grant from the Crown. An everyday colonial plat showed how its property lines corresponded to real spaces by noting where the surveyor had planted posts and notched trees as he measured its bounds with compass and chain. By looking at these signs of survey work on the ground, a prospector seeking to register new tracts could envision where these lines ran and see what lands had already been claimed. The same techniques and images made it possible for colonists and Indians to know when they had reached their boundary. At the very least, experienced surveyors could be expected to recognize and understand where the boundary was, and colonial governors prohibited them from passing beyond it. In places where the thresholds suggested by natural landmarks became ambiguous, those passing through spaces so marked could see where the line ran. To make it clear that the North Carolina-Cherokee boundary passed through Mount Tryon, colonial commissioners and Indian headmen carved their names and marks on trees at the summit.[39]

These maps documented how both sides fulfilled their obligations under the agreements signed at the congresses. In a handwritten paragraph at the bottom of his map, surveyor John Pickens affirmed that he had indeed "marked out" part of the "Boundary Line Between the Province of South Carolina and the Cherokee Indian Country," and that he had done so in the "presence of the Headmen of the Upper Middle and Lower Cherokee Towns," who had also signed the map to affirm that they had taken part in the survey that created this line, understood the image that represented it, and agreed to its course. [**146**] Negotiators at the Fort Prince George Congress in 1765 had brought Pickens and the Cherokees to this place by agreeing that the line should pass through Devises Corner, a small trading post on a bend of Long Cane Creek. This outpost sat astride the well-traveled path between Keowee, one of the Cherokee Lower Towns, and Fort Prince George, and was at the center of two notorious episodes of frontier violence. Pickens's image faces east from Indian country to show the line cutting through the network of creeks and rivers that surrounded this central landmark. The Cherokees considered Long Cane Creek to be their boundary with British Carolina, but with so many tributaries extending from its main branch, where the creek began and ended was open to interpretation. As new immigrants surged into the area by the hundreds in the 1760s, settling "upon and between Navigable Streams" to claim land as "private Property," their presence convinced the Indians that this unwritten understanding of the creek as a boundary had been violated. Cherokee negotiators recalled that in their own childhoods, they had passed through this place before "white People began to settle thick in the country." By mid-century, they were now "debarred" from traveling beyond it, with no end in sight to the erosion of their territorial claims as new settlers followed the scent of good river bottomland deeper into the interior. In the midst of the Cherokee War, this encroachment emboldened a party of young men to attack the Long

Cane settlements in 1760, killing more than a hundred settlers. Another attack in 1763, this time by Creek Indians living in the Lower Cherokee towns, killed fourteen more.[40]

The chief object of this part of the boundary was to clarify how it crossed the creeks along a contentious fifty-mile stretch. Before Pickens drafted his map in 1766, Cherokee travelers passed through Devises Corner and saw signs of new settlements that seemed to spread in every direction at once. After 1766, they found a blazed path marked by sunk posts and notched poplars, pines, chestnuts, and black oaks; this path created an objective basis for orientation that allowed Indians as well as Europeans to see each waterway, even those too small to merit names, and every log house clearly located in relation to the line. The experience of surveying this portion of the boundary shaped an idea of space that Cherokee participants took back and described to their townspeople. This jointly surveyed map helped turn perceptions of a chaotic, multidirectional encroachment into a pattern of settlement that seemed to stretch along the colonial side of the line. With this new image in place to attest to the boundary, the Cherokees believed they could tolerate the intensive occupation at Long Cane, and even a few settlements that strayed over the line. They were now part of a coherently mapped river system that could be seen as a delimited space.

In 1768, the Board of Trade presented a map of an almost-completed boundary to the king. **[147]** Its particular course was much different from the line through the mountains drawn across Bowen's *Accurate Map* in 1763. This negotiated line, attuned to the demographic contours of human geography, intersected frontier trading posts, crossed rivers, and followed the perfect lines of compass bearings. It confined East Florida and West Florida, the "new established Colonies to the South," to "very narrow limits," but allowed the "middle Colonies (whose state of population requires a greater extent) . . . room to spread." Before this universal boundary was established, each colony had made its agreements with the tribes, creating a patchwork of discontinuous and scattered borders. Now the commissioners put forward "one uniform and complete line" between the "Indians and those antient Colonies," disrupting the provocative pattern of "extensive settlement" that had previously taken place "without the consent of the Indians."[41]

This map summarized five years of negotiation aimed at the "fixing of a Boundary between the Settlements . . . and the Indian Country." Across the Southern Indian District, a "Boundary line ha[d] not only been established by actual Treaties with the Creeks, Cherokees and Chactaws, but also, as far as relates to the Provinces of North and South Carolina, been marked out by actual Surveys, and has had the happy effect to restore Peace and Quiet to those Colonies." In the north, negotiations with the Iroquois at Johnson Hall in 1765 had worked out the basic terms of a "line of separation," which awaited ratification. The map "endeavoured to trace those [lines] with as much accuracy as the general Map of America will admit of." All that remained to do to complete this great work of diplomacy was to set the boundary between the Cherokee Nation and Virginia and convene a new round of Indian congresses to solemnize the agreement.[42]

The Eroding Line, 1768–1774

Beginning in 1768, as segments of this established boundary were renegotiated and resurveyed, the course of the line, so close to being fixed in place, changed. As it did, Britain's plan for North America began to unravel. The superintendents found their authority weakened after the Board of Trade returned control of Indian trade to the colonial governors. Americans lobbied in London to secure huge tracts of land, and official interest in schemes for new colonies elevated the hopes of projectors who sought to build private empires along the Ohio and Mississippi Rivers. This revision of Indian policy left its mark on new maps, which recorded the transfer of millions of acres to Britain. New boundary lines angled into the interior, enlarging the empire's stock of unsettled lands. Some have argued that once speculators gained a hand in revising the negotiated boundary, Britain broke the proclamation's promise of indigenous autonomy in eastern North America. This process of boundary erosion began when the Iroquois ceded a vast tract of land south of the Ohio River at the Fort Stanwix Congress in 1768.[43]

Northern superintendent William Johnson came to Fort Stanwix in November 1768 charged with the task of finalizing the northern Indian boundary. The map he brought with him is now lost, but it showed the part of the negotiated boundary that passed through territories claimed by the Six Nations. "Here is the Map of which I spoke," announced Johnson to an audience of more than three thousand Indians, "where all that Country which is the subject of our meeting is faithfully laid down . . . clear and plain" for all to see. Johnson drew their attention to the "Line here described" on the map, a provisional boundary that showed what had already been agreed to at previous negotiations. This line ended at "Owegy," a gateway town that led into Onondago territory from Pennsylvania, leaving Iroquoia's border with New York open and undefined. Iroquois negotiators saw what Johnson intended for them to see: that the "way to our Towns lay open," leaving their nation exposed and vulnerable. They joined in the superintendent's project of securing a "continuation of that Line" to establish a basis for Iroquois security. Viewing their world through the medium of European maps had already changed the scale at which the Iroquois conceived of their nation. "We have large Wide Ears," a speaker at the Johnson Hall Congress the previous spring announced, and "we can hear that you are going to Settle great numbers in the heart of our Country, and our Necks are stretched out, and our faces set to the Sea Shore. . . . Our Legs are long, and our sight so good that we can see a great way through the Woods" to behold the "Blood you have spilled and the fences you have made." Indians articulated their fears of encroachment in metaphors that put the British on notice that they perceived landscape change in the region that both societies shared. Their access to paper maps helped create these capacities and envision their own national spaces as part of a continental order.[44]

New York surveyor Simon Metcalf marked his colony's new border with the Six Nations after the conclusion of the Fort Stanwix agreement, closing the open space that made Iroquois headmen so uneasy. **[148]** Where the preliminary boundary left off, this map's new segment of the line began. From Owegy north, the Iroquois assented to the extension of the line to keep the towns of the Oneidas and Onondagas, two founding members of the Six Nations, safely within it. Unlike the Creeks, the Iroquois lived close to large settler populations and thus did not have the luxury to imagine their own national space as a territorial empire stocked with abundant hunting lands. At Fort Stanwix, they listened to the Algonquian Nanticokes plead, without success, for the freedom to cross into the province of New Jersey to visit kinsmen on the other side of the line after the boundary was set. Mobility was central to Iroquois identity, especially the capacity to circulate among the tribes that composed the larger confederacy. Just as the Creeks used their own stories as conquerors of the southeast to gird their national aspirations, the Iroquois recalled the travels their ancestors had undertaken to found the league a century before as they formed a new national space in talks with Britain. Completing the northern boundary at Fort Stanwix preserved a defining geographic value by holding off encroachment from New York to the east and from Pennsylvania to the south.[45]

The Iroquois boundary with New York followed the Board of Trade's principle that the colonies should be confined to territories that settlers already occupied along with a small supply of new land into which they might expand for perhaps another generation. These new limits on the colonies aimed to keep settlements within an Atlantic sphere of influence, promoting their deeper integration into an economic and political world organized from London. What they did not anticipate was that Indian leaders, in collusion with officials, would use the opportunity of boundary setting at the congresses to give up more land than this required. Instructed to terminate the western boundary of Virginia at the juncture of the Ohio and Kanawha Rivers, Johnson instead asked for and accepted a grant of land from the Iroquois that extended the boundary west along nearly the full length of the Ohio River, opening up millions of acres of new land to colonization. **[149]**

Johnson's departure from his instructions revealed his deep connections with the Iroquois; his disregard for the interests of the Ohio Indians, who also claimed this land; and his complicity with land speculators who stood to benefit from the legal redefinition of Indian land as an extension of Virginia. In addition to viewing the Fort Stanwix agreement as an early instance of dispossession by treaty that hurt indigenous interests, we can see this revision of the boundary from the perspective of Native leaders working to form national spaces that could position their societies for long-term survival. Although Indian headmen shared the Board of Trade's commitment to the boundary, their efforts to change the course of the very lines they had negotiated revealed that they did not fully accept the British prediction of what a completed boundary would mean for the frontier. The Board's notion of reconciliation was grounded on the idea of an empire composed of autonomous, self-sustaining polities, whose interests could align by participating in mutually beneficial trade. The Indians' willingness to cede land before this line was fixed betrayed doubts that colonial governors could truly control western migration, revealed fears that open trade might deepen dependency on European traders, and reflected a desire to deprive neighboring Indian states of territories that might give them a permanent strategic advantage in the new imperial relationship. "By selling distant lands, the Iroquois protected their own homeland, shifting colonial speculators into a more southern channel that victimized other Indians," particularly the Delawares and Shawnees who lived north of the Ohio River and hunted south of it. The Board of Trade's commissioners presumed that Indians feared the loss of land above all other threats and were therefore surprised when Native negotiators across the continent gave up vast tracts just as the surveys to set the boundary's position had been completed.[46]

At Fort Stanwix in 1768, the Iroquois also gave up "that piece of Land in the Forks of Susquehanna." When the Iroquois looked at Sir William Johnson's map and saw how this territory projected into Pennsylvania, where it "is or will be soon partly surrounded by Settlements," they sacrificed it to what appeared to be its inevitable colonization to keep the course of the line clear and straight. When itinerant Presbyterian minister Philip Vickers Fithian reached these "New Purchase" lands less than a decade later, this former Indian frontier seemed well advanced toward becoming a settled place. Pennsylvania's surveyor, William Scull, had so much work dividing it into tracts that he took up residence in the "rapidly growing" county seat at Sunbury, a town laid out on an urban grid in 1772. This new town at the strategic forks of the Susquehanna River overshadowed nearby Shamokin, once a Delaware Indian town that had served as the point of entry from Philadelphia into Iroquoia.[47]

As the Iroquois' national interests found affirmation at Fort Stanwix, the Ohio Indians, whose lands the Iroquois had bartered away, found themselves relegated in the eyes of the British to the status of a wandering, uncivil people. The Ohio Indians were so "exasperated on account of the boundary treaty held at Fort Stanwix," wrote General Thomas Gage, commander in chief for North America, that they urged an alliance of southern and western nations to oppose British expansion by force. Johnson denied that the Shawnees had any national claim to land south of the Ohio River. Not only were they deemed dependents of the Iroquois, but even their homeland north of the river was "more than they have any title to, having been often moved from place to place by the six Nations and never having right of soil there." The Ohio Indians, seeking their own national space along both sides of the Ohio, cited the "old agreement" made at Easton in 1758, which barred any white settlement past the last ridge of mountains. In the ten years since, however, the diplomatic process that created the surveyed boundary had changed this expectation of permanence among other, more powerful negotiators. No line that nations had the right to renegotiate could be guaranteed to last forever.[48]

From 1771, the Cherokees and Creeks competed with each other to sell additional lands to Georgia in order to pay off accumulated debts to merchants. When this southern "New Purchase" was finally surveyed in 1773, Georgia added some 1.5 million acres further upriver along the Savannah River and above the tidewater between the Altamaha and Ogeechee Rivers. As they surveyed the new course of the revised boundary in the summer of 1773, British surveyors and their Native counterparts marked designated line trees "GR. on one side for our King and the Indian Mark on the other side." When deputy surveyor Philip Yonge produced "A Map of the Lands Ceded to His Majesty by the Creek and Cherokee Indians," he represented a prospective plantation landscape in a space left blank on previous maps. **[150]** Alongside rivers and creeks drawn to scale as a navigable network of waterways, he annotated riverside lands with letters, A through F, each standing for a different soil type, from "Dark Chocolate" hardwood bottomlands to poor sandy pinelands. He marked "convenient places for mills" as well as sites of abandoned Indian fields, their fertility proven by past use and cleared for planting. Within months of the survey of the southern New Purchase, some fourteen hundred settlers had claimed these lands and brought three hundred slaves to work them. Governor James Wright annotated a blank space on the map between the headwaters of the Ogeechee River and the Oconee River, expressing his desire to "get this Line altered" to "take in about 30,000 . . . acres of extraordinary fine land" in what was, for the moment, still Creek territory. Although British officials scrutinized these "new purchases," north and south, and criticized Johnson for encouraging and accepting a cession that they had not authorized, all were approved. The very diplomatic process that Britain had initiated was predicated on defining Indian societies as nations that had every right to sell land in a formal treaty proceeding.[49]

Elsewhere in the early 1770s, Britain disallowed attempts by Indians to sell reserved land to colonists. Georgia speculator Jonathan Bryan attempted to secure an audacious ninety-nine-year lease of millions of acres of land along the Gulf Coast from the Lower Creeks. South Carolina councilman John Drayton negotiated a leasehold for lands reserved to the Catawbas. Between 1768 and 1770, the Cherokees attempted to convey three blocks of land in independent grants, each containing hundreds of thousands of acres "beyond the Established Boundary." **[151]** They earmarked two of these for the mixed-race sons of Indian trader Richard Pearis and Assistant Superintendent of Indian affairs Alexander Cameron, and they proposed another grant of 178,000 acres to Edward Wilkinson to retire an £8,000 debt. In each of these cases, Indians attempted to establish paths of connection across the boundary line by leveraging their power to sell, lease, and give away their land outside of the treaty process. They did so to open direct channels of trade in order to gain access to European goods; to secure new sources of diplomatic presents, which Indians had come to expect at every formal negotiation but which British authorities were increasingly reluctant to give; and to enhance the power and wealth of men who might serve their interests as allies and go-betweens. Before the boundary was completed, Indians and colonists conducted a flurry of secret meetings to take advantage of the final moments in which it still seemed possible to alter its course.[50]

Virginia's expanding claims on western lands, secured through negotiations with the Cherokees from 1768 to 1771, presents the most dramatic example of pre–Revolutionary War boundary erosion. More than any other colony, Virginia challenged the geographic idea behind the proclamation based on its prewar settlements west of the mountains. In the 1760s, its governors used the fact of western settlement to open western spaces that the Board of Trade and the superintendents of Indian affairs were working to close to colonization. More than any other colony, Virginia was dominated by a gentry class committed to expansion through speculation, whose members believed that they ruled a province driven by historical destiny to grow. Virginians were the last to come to terms with neighboring Indians over the location of their portion of the boundary and the most determined to undermine it. The Board of Trade denied Francis Fauquier's petition to open western lands for settlement in 1759, but his letters to London describing Virginia's settlements beyond the mountains influenced the way the Board understood and represented the proclamation's words. Not only had Virginians taken up land "to the head springs of all our rivers which run into the Atlantick Ocean," wrote surveyor Joshua Fry in 1751, but several hundred people had moved beyond to settle along the Great Kanawha River, a tributary of the Ohio. The Board drew its line across Bowen's *Accurate Map* to include these prewar grants within Virginia, departing from geographic principle to acknowledge this claim. Years before the proclamation sought to confine colonists to coastal watersheds, Virginians had already crossed this threshold.[51]

At the 1768 congress at Hard Labor Creek, South Carolina, the Cherokees at first insisted that the Kanawha River itself should be considered a "natural boundary" with Virginia. Fauquier's successor as governor, Norborne Berkeley, Baron Botetourt, rejected this, arguing that it would "so much contract [the] limits of this Colony" that it would endanger its security. Enforcing it would require a forced removal of the "people who have been encouraged to settle to the Westward" of this "proposed boundary." The Cherokees capitulated to this demand and ratified a straight north–south line that put the whole course of the river within Virginia. Even after he helped win this new territory, Lord Botetourt was not content to maintain the Hard Labor boundary as Virginia's final limit. The map he sent to London illustrated what lands could be purchased from the Cherokees out of those the Iroquois had relinquished at Fort Stanwix. **[152]** Beyond the Kanawha, Botetourt imagined a new space for colonization extending to where he thought the ridgeline of the Cumberland Mountains to be. He annotated each parcel of land on his map to show that it could be justified as a legitimate extension, citing past treaties as evidence. Botetourt described this map as showing the lands that, as he put it, "remain to be purchased of the Cherokees in order to secure peace betwixt them and us in all eternity." Such language of permanence paid respect to British efforts to use the model of international diplomacy to set boundaries that were designed to last. His map made a credible appeal that these new Virginia boundaries would legitimize settlements that were now facts on the ground. Virginia's campaign for new western lands was thus fully engaged in the British diplomatic process for acquisition described in the proclamation, by which Indian nations could alienate territory at a properly constituted "public Meeting or Assembly." If Virginians were to occupy these new lands legally, Virginia would have to take part in another congress led by the southern superintendent, to meet Britain's new standards for transparency and compensation.[52]

Botetourt's vision of a greater Virginia set the stage for a new round of negotiations with the Cherokees at the Lochaber Congress in South Carolina in 1770. The Cherokees complied with Virginia's request and granted a huge new cession of land. From the point at which the North Carolina portion of the Indian boundary met Virginia's southern border, this new line extended due west for approximately one hundred miles before angling back to the juncture of the Kanawha and Ohio Rivers. **[153]** Virginia Colonel John Donelson surveyed the Lochaber line, and the map he produced to document it reveals how wildly his survey diverged from the course agreed at Lochaber, adding some 10 million additional acres of land to Virginia. **[154]** Instead of surveying a line that pointed north toward the juncture of the Ohio and Kanawha Rivers, Donelson, assistant superintendent Alexander Cameron, and a delegation of Cherokees veered west, terminating the boundary at a new point more than fifty miles farther down the Ohio River. The Board of Trade believed that the Cherokees had diverted the line agreed to at Lochaber in favor of a termination point at the Louisa River "for the Sake of a natural Boundary." The Cherokee leader most involved in negotiations over this process, Attakullakulla (also known as the Little Carpenter), explained that he and other Cherokees who accompanied Donelson "altered the course a little" and had "given away some land by the river side as my brothers were settled upon it and I pittied them."[53]

The Cherokees agreed to a huge new cession of land that they were not required to give and to successive revisions of their boundary that shrank the size of their nation. These actions make sense by looking less at the land the Cherokees gave up and more closely at one small place that they refused to trade away. Despite pressure to do so, they would not yield Long Island, a narrow, four-mile-long strip of land in the middle of the Holston River, about 140 miles from their closest towns. In 1755, trader Richard Pearis came to Long Island to set up a store. The Cherokees confiscated his letter from Virginia governor Dinwiddie and turned him away. In 1766, Virginia's agent asked, and was again denied, permission to establish a trading post at Long Island. John Stuart asked the Indians to include Long Island with the lands that they were willing to cede to Virginia at the Lochaber Congress in 1770. Although they yielded on every other point, Stuart reported that "no Consideration would make them consent" to give up this place. The Treaty of Lochaber specifically stated that the boundary was to be run six miles to the north of the island. When it came time to survey the revised line, Long Island was chosen as the spot where the Indians and surveyors would meet to begin, perhaps to make sure that no error in the survey could grant it away.[54]

The Cherokees kept possession of this site to keep the British away. Although the image of the Cherokee Nation on Stuart's 1764 "Map of the Southern Indian District" suggested its rising prominence, the Cherokee people faced serious threats to their corporate integrity. From a population that may have approached twenty thousand earlier in the century, smallpox and war had reduced their number to fewer than seven thousand. Throughout their participation in the boundary line negotiations, the Cherokees sought to establish a defensive perimeter against encroachment that aimed at keeping at least a hundred miles between the last Cherokee town and the nearest white settlement. At the Augusta Congress in 1763, Attakullakulla made the first proposal to define the boundary in these terms, stating that the "lands towards Virginia must not be settled nearer the Cherokees than the southward of New River," another name for the Kanawha. When Virginia's House of Burgesses proposed in 1769 to extend its southern charter line into the west, this was rejected out of hand by John Stuart, because it would then "run within less than sixty miles of the Towns." During negotiations over the southern New Purchase in 1773, it became clear that they would have to give up lands that came "20 miles near[er] their Towns than they at first proposed, which they did with vast reluctancy, and Strong expressions of

Concern." The Cherokees experienced white proximity to their towns as a deeply disturbing threat. Attakullakulla told those assembled at Lochaber that he felt as if he were "Stepping out of the door, to be at the white peoples Settlements," and that it seemed "like coming in our Houses and Encroaching on us greatly." Saluy (the Young Warrior of Estatoe) reported that the Cherokees felt "much cramped in" by Virginia's expansion. Hunters had found "Paths trod[d]en by Virginia people" in the woods. It seemed as if they could "See the Smoke of the Virginians from their doors." Imagining these sights and sounds of encroachment expressed a growing apprehension that Europeans were already intruding into the heart of Cherokee territory. The Cherokees believed their very survival was at stake in defining a coherent edge to their homeland.[55]

Cherokee headmen were willing to give up lands closer to Virginia settlements because the deer, "frightened by Numberless white hunters," had already fled. The "extension of our Boundaries in the Indian Hunting grounds," wrote John Stuart, "has rendered what the Indians reserved to themselves" to the west of the "ridge of Mountains of very Little use to them." "This the Cherokees are sensible of, and therefor[e] are easily induced to complement away great Tracts." Despite the huge swathes of land they gave up, however, the Cherokees held fast to this idea of maintaining a margin of unsettled space around their towns. By creating a new corridor for Virginia's legitimate expansion, they encouraged its determined colonists to take up lands well to the north and west. Although they could not prevent settlement from advancing, they used their influence over the boundary's location to channel the expected intrusion of future colonization away from their homes. Comparing the changing boundary from 1768 to 1771 to the positions of Long Island and the Cherokee towns makes this strategy visible. The Cherokees regarded Long Island as a center point from which renegotiated lines could pivot toward the interior, opening a space for new settlement beyond a well-defined buffer of land that extended about a hundred miles from the Cherokee towns. When Virginia governor John Murray, Earl of Dunmore, led a militia force in 1774 against the Shawnees to defend new Virginia settlements along the Ohio River, he entered a space that the Cherokees had helped open for colonial expansion.[56]

From the Augusta Congress of 1763 to the Treaty of Camp Charlotte in 1774, which concluded Lord Dunmore's War, British officials fixed a frontier for North America. The new Indian boundary reflected the reality of where colonial populations already were and registered the power of Indian nations to direct where they would expand in the future. Each segment of this negotiated boundary made this relative influence visible: the Chickasaws, Choctaws, Creeks, Cherokees, and Iroquois maintained territory that helped define five large Indian states, which the superintendents scrupulously called "nations." These new Indian polities helped fill much of the space in the Board of Trade's new map of empire. A large space on the map that remained outside any jurisdiction was the area occupied by the Ohio River Indians—the Mingos, Delawares, Wyandots, Shawnees, and Miamis. Unlike the other tribes they regarded as negotiating partners, the British affirmed the Ohio Indians' status as dependencies of the Six Nations. The violent attacks they had launched across the Ohio River during the Seven Years' War had convinced British colonists as well as officials that they were a people so committed to savagery that they could not qualify for the trappings of nationhood, even as a diplomatic fiction. There would be no clearly delimited Indian state in the Ohio country, leaving an uncharted void into which Indians, colonial officials, speculators, and squatters continued competing for land and influence.

Joseph Purcell's 1775 map of the Southern Indian District displayed the final position of the boundary under British rule. **[155]** The knowledge generated by these surveys, although designed to mark the edge of an autonomous Indian country, also opened this territory to new scrutiny. After John Stuart, his assistants, and his surveyors made the overland journey from St. Augustine to Pensacola and back again to conduct their negotiations, Purcell illuminated a corridor across Creek country. **[156]** Along this three-hundred-mile route he marked the abandoned "old fields" of the Apalachees, assayed the land and soil, and constructed a well-documented itinerary for those who might make the trip again. The boundary surveys represented bands of linear space in striking new detail, recording geographic knowledge that could only whet the appetites of provincial governments for new cessions. Armed with the accounts of travelers and synthesizing the information he found on maps by the "best Authorities," Purcell revised his map of the Southern Indian District in 1781, filling in the spaces between these surveyed zones and adding a new census of the 13,162 arms-bearing men among the Indian nations. [**157**]

The Board of Trade attempted to leverage the Crown's sovereignty over territory to reform a dysfunctional colonial system, but no drawn line or surveyed boundary could prevent settlers and speculators from staking claim to land they thought was theirs for the taking. Britons of property and standing on both sides of the Atlantic regarded western migrants as social "refuse"—men and women with the "worst morals and the least discretion," prone to violence and "remote from the eye, the prudence, and the restraint of government." As open fighting with French and Indian forces wound down after 1760, settlers pushed into new riverside lands in surprising numbers. To officials and military commanders tasked with keeping the peace, their appearance on contested frontiers sharpened long-standing critiques of the colonial settlement system. General Thomas Gage, appalled by the "scandalous Disorders" caused by settler violence, concluded that the "Reins of Government [were] too loose to enforce an Obedience to the Laws." John Stuart denied people "living so remote" even the title of "Colonists," because their "Distance from Commerce" forced them to make "cloth[e]s and other stuffs for themselves." This new generation of squatters, he concluded, were "by no means the sort of people that should set[t]le those lands." He recommended that new lands on the British side of the line be closed to North Carolina and Virginia migrants, and reserved for "industrious people," particularly those worthy Protestant yeomen families recruited by the Crown from Britain, Ireland, and Germany.[57]

The logic of the line as a solution to frontier violence was the clear demarcation it promised between two distinct systems of political economy—one devoted to hunting, the other to raising stock and crops—that would not compete for the same resources. By making their living "hunting and plundering" like Indians, those "idle and disorderly vagrants" (known as "Crackers" in the southern backcountry) had forfeited any claims to land on either side of the line. By stigmatizing white frontiersmen, Britain did little to stop the flow of settlers to the new lands they desired, but it communicated a message to neighboring Indians that these people did not enjoy the protection of the law. When Georgia trader William Frazier reported to Indian commissary Roderick McIntosh the "outrage" that Upper Creek leader Emistisiquo had seized his deerskins, the official told him he was "happy" he had done so, because Frazier had traded for them outside a designated town, which the rules proscribed. When an incredulous Frazier asked him, "[W]ho were to be the Judges[,] the White people or Indians[?]" McIntosh told him: "the Indians to be sure." In another instance, John Stuart appeared to condone an Indian revenge killing. East Florida governor James Grant was blunter when he wrote that he would not "go to War with an Indian Nation for Scalping a Woods Man, who generally deserves it when it happens." A clear line allowed Britain to disown these "strag[g]ling Vagrants or Vagabonds" who "broke through Treaties and Orders." Although it was "next to impossible" to "punish such wandering People," the British used the existence of a negotiated boundary to signal that they fell beyond the pale of civilization.[58]

Colonists who had previously ventured beyond settled jurisdictions to claim unpatented land had, by and large, found these efforts rewarded by subsequent grants that authorized their possession with official titles. The boundary's threat to settlers' interests was not that it could stop migration across the frontier but that enclosing provincial jurisdictions weakened expectations that land settled outside of them would receive the post hoc legitimation of a formal grant. Without formal possession, sanctioned by the state, real property could not be mortgaged, sold, or preserved against legal challenge. Thus, creole elites pressed their governors to renegotiate the line not merely to pursue their private interests in possessing new tracts of land but also to prevent a land system taking shape that could not be organized by the land offices of their own colonial governments, over which they had influence. Colonists were not simply "land hungry"—driven by base, almost instinctual desires for possession that made them immune to appeals to reason and the rule of law. They recognized that the longer the boundary remained fixed in position, the more likely settlement systems would take shape around the limits it imposed.

Experiences of fraud, violence, and encroachment had convinced many North American Indians that Britain "projected their Ruin" in

1763. Britons denied such an overt intention but nevertheless believed that it was the demographic destiny of Native peoples to retreat into the interior with every passing generation. Speculators surveyed lands in the expectation that the line would continue to erode, allowing colonial governments to expand the bounds of their jurisdictions to incorporate them. Richard Jackson (a colonial agent, Shelburne ally, member of Parliament, and law officer to the Board of Trade) argued that the ban against interior colonization should be a "Temporary Provisio[n]" and not a "permanent one," because the "Prohibition of British Settlements beyond the bounds described in the Proclamation . . . might deprive us of a valuable Trade, in British Manufactures for Growth raised by British Subjects in America." Some imagined the interior set off in perpetuity as Native territory, but the formal language describing the line always hedged when it came to fixing it as a permanent boundary. The Board urged a ban on interior settlement in 1763 "immediately, at least," but did not rule out future development. In "Time to come," predicted Secretary of State Egremont, the land would become "Property" owned by Europeans. The king's proclamation declared an end to new grants only "for the present," not for all eternity.[59]

Senior military officials actively promoted the creation of new interior colonies. Sir Jeffrey Amherst, appointed British North America's governor-general, with military authority over Canada and the Indian country, advocated settling them around strategic positions. In 1766, he called for "taking Possession of Our Conquests to their utmost Extent" by creating four governments along the Ohio and Mississippi Rivers, "so as to have the entire Command of the Country, and to gain all the advantages which may be reaped from it." To promote his vision of interior colonization, he circulated a map among senior officials in the mid-1770s that pictured Quebec divided into two provinces and showed the domains of existing American colonies enlarged by additions of interior territory. In 1768, Gage wanted to impose government over the Illinois country in response to the disorders at the town of Vincennes, where it was reported that French residents had encouraged Wabash Indians to kill British traders. The Board of Trade had envisioned the Indian country as a space inhabited solely by indigenes; its plan had not taken into account the French Canadians, who, as Gage put it, "have always been remarkable for roving in the deserts and seating themselves amongst the Indians." The Earl of Shelburne, as secretary of state for the Southern Department in 1767, approved in principle that colonies around Detroit and Illinois could be established in response to this strategic concern.[60]

New knowledge about continental geography challenged the Board of Trade's premise that such interior colonies would be inevitably lost to empire. The army's 1766 expedition detailed every bend, island, and tributary of the Ohio River along its 1,164-mile route. Engineer Thomas Hutchins's sketches recorded critical junctures of its course in voluminous detail, documenting a continuous channel of safe navigation. Captain Harry Gordon, an engineer who built roads through the woods during the Seven Years' War, compiled these into a single chart of the "River of Ohio" at a general scale. **[158]** Policy dictated that the river mark a limit to settlement, but seeing the land in prospective terms as a new agricultural frontier was a deeply rooted presumption for those who scouted territory on behalf of Britain. Gordon assumed a future in which British farmers would till the "Rich luxuriant Soil" of trans-Appalachian lands. Great "herds of buffalo" along its banks attested to an abundance of "good pasturage" for English cattle. His running surveys of the Ohio accumulated the "proper knowledge" that "affirmed" its reputation as perhaps the "most pleasant, the most commod[io]us, and most fertile spot of earth known to European people." Astronomical observations along its course established that the Ohio flowed east to west almost entirely within the 36th parallel, an ideally temperate zone for agriculture. Regardless of how others might parse the language of the proclamation, army mapmakers could scarcely see this river and the productive land along its banks in any other way than as space to be colonized.[61]

This survey demonstrated that the Ohio River was navigable from Fort Pitt to the Mississippi, establishing a commercial route by which those who lived far from the Atlantic coast might participate in transatlantic commerce, thereby undermining the economic principle that justified the proclamation's prohibition on western settlement. In the late 1760s, the Board of Trade's idea that British North America could be held in a permanent state of equipoise between confined colonies and defined indigenous states was coming undone. In 1767 and 1768, Secretary of State Shelburne presided over the abandonment of western forts. As troops departed the interior for the increasingly restive coast, Britain scuttled the plan of 1764 for regulating Indian commerce and downgraded the authority of the Indian superintendents to oversee it, returning this power to colonial governors. Along the Ohio and Mississippi Rivers, in the spaces where no recognized Indian nation claimed territory, American land speculators exploited this moment of uncertainty to propose new interior colonies that could bring them under civil jurisdiction and complete the map of North America.[62]

Connecticut militia colonel Phineas Lyman read the proclamation as an invitation to turn decommissioned soldiers and officers into a corps of colonial entrepreneurs. He organized members of his former regiment into the Company of Military Adventurers and prepared it to settle a new British colony along the eastern banks of the Mississippi River, which he vowed to "Coloniz[e] with the same Zeal and Industry as the Coast of the Atlantic was above a Century ago." Lyman touched on the Board of Trade's most cherished ideas about empire to make a case for his new Mississippi colony. "New Colonies should be established upon better Principles than the Old," he declared, seeking to convince the commissioners that his scheme would promote rather than undermine their vision. His model colony, he pledged, would not unleash settlers into the interior to encroach on Indian land "without Regard to the King's Proclamation." Because this riverside territory had superior "Convenience for Navigation" and rich bottomland soils, he would develop it to export valuable staple crops and "purchase the Commodities of Britain," harness the energies of settlers who might otherwise be "dangerous and useless" squatters, and open a dynamic and respectful commerce with nearby Indian tribes. A well-regulated Mississippi colony established on these terms, he believed, could exert a "Powerful and Salutary Influence on the Old Colonies on the Continent of America, which were at first established with too little Foresight and precaution." In 1773, Lyman transported a group from Connecticut to settle a twenty-thousand-acre township grant at Natchez, within the jurisdiction of West Florida. When the Board of Trade finally denied his application to establish a separate colony to the north along the river, it dashed the hopes of an American who had applied to become a collaborator in Britain's project to improve the empire.[63]

The Grand Ohio Company petitioned the Board of Trade for millions of acres west of Virginia—lands ceded by the Six Nations and the Cherokee Nation—that it intended to form into a separately governed colony. This new province would be called Vandalia to honor Queen Charlotte of Mecklenburg's supposed Vandal ancestry. Colonial Secretary Hillsborough and the Board of Trade rejected the proposal because Vandalia violated its governing principle for North American empire: that colonies should "lie within the reach of the trade and commerce of this kingdom," where Britain could better keep the "colonies in a due subordination to, and dependence upon, the mother country." Settlers seeking lands along the Ohio River in the proposed province would live "above fifteen hundred miles from the sea," "separated by immense tracts of unpeopled desert," "utterly inaccessible to shipping," and far beyond the pale of British trade, culture, and law. The commissioners stood resolutely against "colonization . . . in the *remote* countries" for these reasons. "Let the savages enjoy their deserts in quiet," urged Hillsborough. Those who would take up the king's land in America should "direct their settlements along the sea-coast, where millions of acres are yet uncultivated."[64]

The Grand Ohio Company's leading "Associates" challenged the Board of Trade's reasoning, the consistency of its policy pronouncements, and the wide divergence between its general view of an imperial system and the realities of North American people and places. Hillsborough and the Board clung to the image of a line through the mountains as a permanent bar to interior colonization despite the fact that Vandalia was to be seated in a territory relinquished through the diplomatic process dictated by the Proclamation of 1763. This land now belonged to the Crown, and it was absurd to leave it as a "useless wilderness," when it could be "settled and occupied by his Majesty's subjects." Far from being beyond the reach of commerce, the region was "well watered by several navigable rivers" that would make it economical to ship goods to the east as well as the west. With "Williamsburgh, the capital of Virginia, at least 400 miles from the settlements on the Ohio," squatters already there were "out of the reach and contr[o]l of law." Without a government over them, they would only grow more ungovernable. The idea that migrants denied the chance to inhabit these lands legally would risk their lives and fortunes elsewhere defied experience as well as common sense. Those seeking land from Maryland, Virginia, and Pennsylvania had not left their temperate provinces to "remove to the scorching, unwholesome heats" of East and West Florida, nor had they ventured north to "Quebeck, Nova Scotia, and the Island of St. John's" to "settl[e] in these new provinces." They understood that a "rich, healthful, and uncultivated country" lay just over the mountains. This presumption that colonists could be forced to go where Britain wanted them to rather than choosing their own course revealed that the Board misunderstood the "very nature of colonization itself."[65]

Hillsborough resigned as Colonial Secretary when the Privy Council overruled the Board of Trade and dismissed his objections to the Vandalia proposal. His successor, the Earl of Dartmouth, ordered the Board to approve a charter for the new colony on May 6, 1773. The Grand Ohio Company offered Britain a means of regulating Virginia's otherwise uncontrollable western expansion. Its officers vowed to create a "regular and uniform System of Government," one committed from the outset to collecting quitrents and paying the salaries of Crown-appointed officials in a volatile backcountry region. Before its charter could be implemented, however, the Sons of Liberty threw crates of East India Company tea into Boston Harbor, precipitating a crisis that prevented the colony's establishment. The "Lands Intended for the New Government" of Vandalia lived on as a bounded space on Purcell's "A Map of the Southern Indian District" in 1775. Like Connecticut's Military Adventurers, the Grand Ohio Company sought the profits of large-scale land speculation along the new frontiers of empire. Their petitions for interior colonies came tantalizingly close to succeeding, because they proposed what Britain's political leaders wanted to hear in the wake of the failure of its postwar Indian policy: that Americans would partner with Great Britain to make ongoing colonization compatible with colonial dependence.[66]

Reformer John Cartwright imagined how such collaborative colonization could form the foundation of a new constitutional order in his 1775 tract, *American Independence*. He imagined Britain's Atlantic empire formed into a "Grand British League and Confederacy" and illustrated its new political geography with a map of "British America, Bounded and Divided." [**159**] Next to the existing seaboard colonies, Cartwright imagined nineteen new states, each with a "frontier accessible to shipping" that would "enable it through commerce to become a respectable member of the grand British confederacy." Although Chicasawria, Chactawria, Senekania, and other new states had Indian-derived names, all were intended as future settler societies to be governed by white Protestants. Cartwright believed that American colonies, as "opulent and respectable" societies, should operate as "free and independent states" under the general "umpireship" of the king. These new states would govern internal affairs, while the Crown exercised dominion over the "seas, lakes and great rivers of North-America." Although he acknowledged that the king should defend the "rights [and] independencies of the several tribes or nations of Indians in amity with or under the protection of the British crown," his map erases the Indian nations as sovereign powers. An acolyte of John Locke, who deemed Native lands "legally vacant" and believed they "could be appropriated without [the Indians'] consent," Cartwright saw no justification for Britain to "hinder the American states" from sending out colonists to impose civil society in this territory. No "chain of feeble forts in a wilderness, or the pronouncing [of] this wilderness to be part of the province of Quebec," he predicted, "will form a mighty barrier truly, against the swarms that will one day pour westward, from the too populous states upon the sea coast!" Uttered by a marginal voice in British politics, no ministry considered putting this "Grand Confederacy" scheme in place. Yet Cartwright's vision was the first to attempt to reconcile imperial political economy with the natural-rights theory that informed colonists' sense of their independent authority in America. It showed most clearly how this unification of interests would divest Native Americans of the power to make choices about their land. Although his map illustrated little more than a fantasy of reconciliation in 1775, Cartwright's idea that states would multiply rather than expand to govern the west proved prescient. As Virginia relinquished its vast claims on western lands to Congress in 1784, Thomas Jefferson imagined the diverse interests of a continent balanced in a stable union of settler states. As sketched by British reformer David Hartley from Jefferson's report to Congress on a "Plan for the Temporary Government of the Western Territory," this vision established a framework for settlement and dispossession in the new United States that culminated in the Northwest Ordinance of 1787.[67] **[160]**

⚜

"If we cast our eyes over an American map," Cartwright wrote, it should be obvious that the "plan we recommend is the plain suggestion of Providence." As much as he rejected the prevailing geographic vision of restricted settler colonialism that officials had attempted to impose on British America, Cartwright had more in common with his adversaries than he cared to admit. He shared their presumption that America could be radically restructured from London and that its economy, government, and population could be moved around the map at will. Overlooking the continent on his map from a fictive vantage point high above the surface of the earth, he delighted in making "alterations of boundaries" and tracing "districts of future dominion" by which the continent was to be peopled and governed.[68]

No clear proclamation line ever marked the place where colonial British America ended and Indian country began. In the course of two decades of multilateral negotiations between Britain and the Indian nations, many proposed, perceived, and conflicting lines imagined a partition of the interior. Each of these contended for authentication and acceptance as a single, absolute boundary. The line was never a promise that colonial societies would be forever halted at the peaks of the mountains but rather a legal order to stop new land grants as British authorities took stock of their vast new territories, assumed control over the process of colonization, and debated the form of government they could best impose on the continent. Britain waged war with France in North America to contest the idea that the Appalachian Mountains confined its empire to the Atlantic coast. But as diplomats charged with securing peace on the frontier put in place a new order for Indian affairs, this vast mountain range became a natural starting point for the division of the continent between Europeans and Indians.

The congresses were as important in shaping perceptions of historical time as they were in dividing geographic space. The treaty talks recognized hunting grounds as a contemporary, constituent part of indigenous national domains rather than a primitive wilderness that was free for the taking. They legitimated most western European settlements where they stood, agreeing to wipe out memories of violent dispossession and reprisal with the promise of peaceful occupation and trade. One reason why a North American boundary could never be imposed by a single geographic rule was that each society along its course sought to use it to define a corporate space that reflected its particular geographic aspirations. Differences in diplomatic clout with the British, proximity to white settlement, and histories of responding to colonization were embodied in the lines each nation was able to draw. Headmen at the congresses demanded a line that protected their towns with buffers of distance, preserved exclusive hunting grounds to enable their ongoing participation in European trade, and offered unobstructed freedom of movement between the far-flung communities—loosely joined by kinship, history, and cultural affinity, but not by coercive authority—on which the integration of their nations depended. The boundary that took shape was a composite of these differences that had to be negotiated piece by piece with each nation. As a result, the negotiated line of 1768 bore little resemblance to the proclamation line first imagined in 1763.

This process, by which each small segment was subjected to scrutiny, debate, and verification, built the line around shared landmarks and gathered firsthand geographic knowledge that gave it credibility at different scales of representation. The line could function at the scale of a cadastral map to show precisely where on the landscape the last legal grant of property could be made. As mapmakers pulled back from the details of local spaces to compile these surveys into a single continental line, this boundary helped envision Indian states as possessing coherent territories, the first criterion for nationhood. The line became more important, not less, as British officials worked to complete the map of North America by bounding old colonies, creating new ones, and affirming the political status of the largest Indian nations. The ultimate goal for this comprehensive partition of North America was as much about shaping the sensibilities of subjects as it was about organizing them into clear jurisdictions. British officials spoke consistently of the pressing need to "conciliate" the Indians. The negotiations at the congresses endorsed not only their corporate rights over the land but also their capacity to act as sensible inhabitants of a diverse empire.

Seeing the Indian boundary as a process rather than a location helps clarify Britain's vision for North America and explain why it failed. Not only did the congresses provide a forum in which Indians' wants and fears could have influence, but through the participation of the colonial governors, provincial elites found an institutional channel for expressing geographic goals that gave them a stake in remaining engaged in the negotiations. Seeing the line change from proclamation to revolution—bowing inward into the continent until the rupture of independence erased it as a legal barrier to expansion altogether—traces how the agents of settler colonialism in British North America pursued progress through expansion. For American creoles, colonization began in the mythologized pasts of seventeenth-century foundings, at which the first settlers, their ancestors and predecessors, risked their lives to establish outposts of civility at the edges of a savage wilderness. It looked ahead to ongoing demographic increase, economic growth, and cultural refinement. This historical understanding of colonization as the common project of autonomous colonial societies pointed it from coasts into adjacent interiors, aligning the interests of settlers, speculators, and provincial officials by uniting them in a common endeavor of adding territories to and enlarging the jurisdictions of their particular colonies. In the Board of Trade's vision of American empire, epitomized by the idea of a general Indian boundary line, this colonial drive for expansion could be cut off at the mountains, broken into individual units of settler desire, detached from home societies, and directed toward the underpopulated frontiers it had earmarked for new settlement, where land awaited individuals in presurveyed townships and tracts. As colonists pushed against the limits the boundary attempted to set, forcing new negotiations over its location, each resurveyed line revealed their opposition to the diminishment of their societies as they sought to reclaim western lands.

So vast was the trans-Appalachian west that Britain could only attempt to take command of it by establishing fortified points of space along its frontiers and measuring a single line across it over the span of a dozen years. The Indian boundary represented the Board of Trade's attempt to reconcile the opposing interests brought together in the same sovereign space at the least possible cost. The commissioners understood that rapid intrusions into indigenous homelands and hunting grounds would provoke violent retaliation, and given the staggering war debt accumulated to finance the Seven Years' War, they saw that it would be impossible to defend a frontier into which settlers were allowed to advance. The proclamation made a virtue of necessity by declaring the continental interior off-limits to new settlement on the basis of the shared humanity of all the king's subjects and envisioning that a natural balance of interests would emerge through the power of mutually beneficial commerce. The lines that resulted masked this incapacity to enforce order and police movement across a vast territory that remained largely unsurveyed and unmapped.

Just as Britain completed the final surveys of the boundary, the violence and political disruption of the War of Independence made moot the promise of a negotiated frontier. Colonists objected to the elevation of peoples they regarded as rude and uncivilized to the status of honored nations, and they saw the rights they believed they had earned over a long history of turning wilderness into property compromised by the loss of land that fell within their colonial charters. The new powers assumed by the Board of Trade and the superintendents to create this boundary seemed to overturn the expectations for indirect rule from the metropolis, negotiated authority, and due regard for provincial interests. As George Washington wrote in 1767, he never "look[ed] upon that Proclamation in any other light . . . than as a temporary expedient to quiet the Minds of the Indians." Inevitably, he predicted, it "must fall of course in a few years especially when those Indians are consenting to our Occupying the Lands." Although he wished his own schemes to patent trans-Appalachian territory to be kept secret for the moment, he refused to sit on his hands while the Indian boundary was marked, reasoning that "any Person . . . who neglects the present opp[o]rtunity of hunting out good Lands & in some measure Marking & distinguishing them for their own . . . will never regain it." In 1770, Washington and two companions departed Fort Pitt by canoe and headed up the Great Kanawha River, locating 64,071 riverside acres that they divided into eight great tracts of land, about a third of which Washington claimed for himself. [**161**] Colonists never relinquished the hope that whatever British officials negotiated with the Indians would be undone in the fullness of time to open the land for occupation.[69]

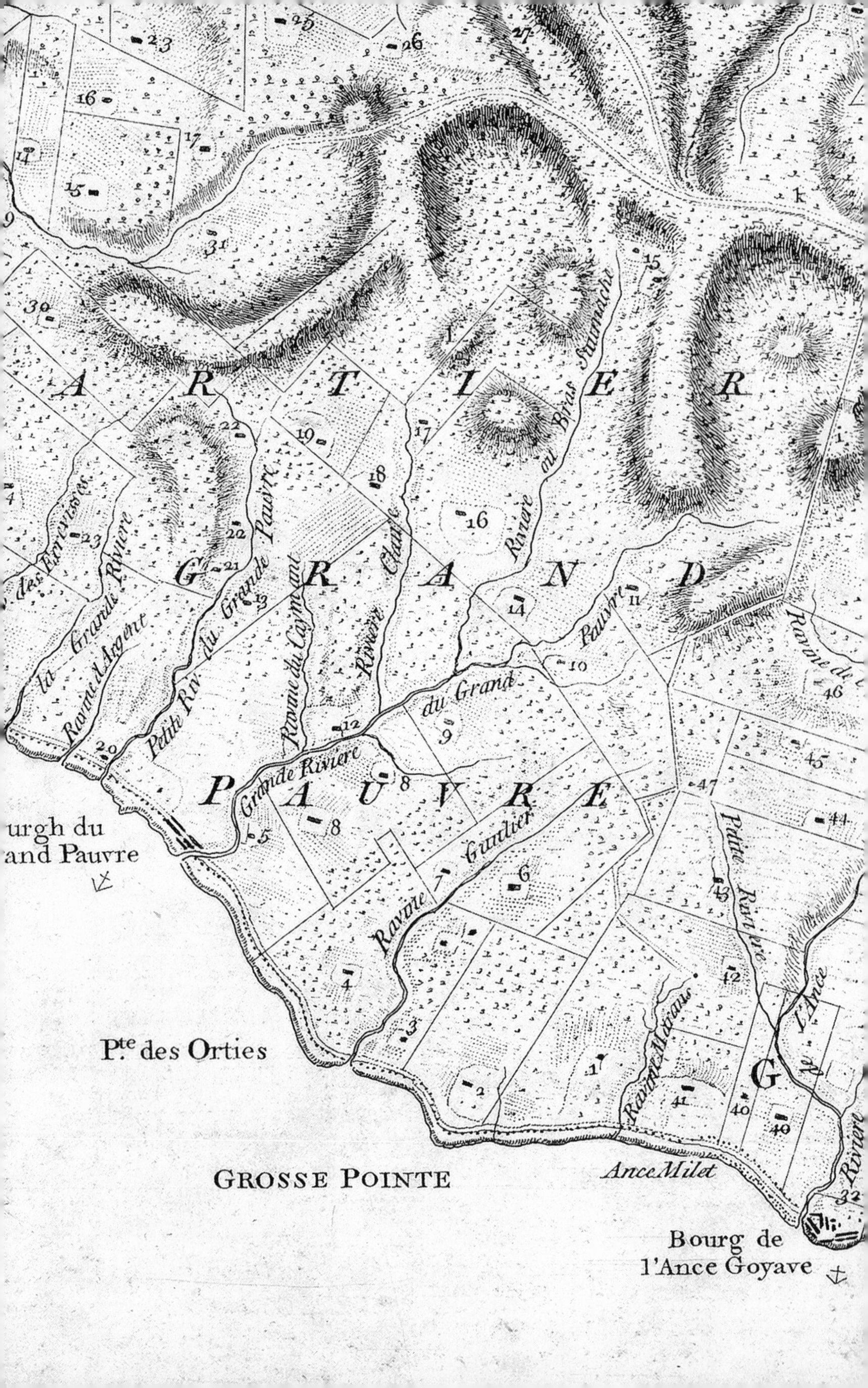
A R T I L E R
G R A N D
P A U V R E
Grande Riviere du Grand Pauvre
Petite Riv. du Grande Pauvre
Riviere Claire
Riviere ou Bras Salamache
Ravine du Cayman
Ravine d'Argent
la Grande Riviere
Ravine Gautier
Ravine Moreau
Petite Ravine
Riviere de l'Ance
urgh du
and Pauvre
Pte des Orties
GROSSE POINTE
Ance Milet
Bourg de
l'Ance Goyave

{ FIVE }

Charting Contested Caribbean Space

In 1763, the Lords Commissioners of Trade and Plantations drafted a master plan for America illustrated by an annotated map. The novel pattern of shapes, lines, and colors they marked across the continent, however, should not obscure the importance of a much simpler annotation in the lower right-hand corner. With an ink wash of pale yellow, they targeted the recently ceded islands of Dominica, St. Vincent, Grenada, and Tobago for colonization. The reformed imperial system the Board of Trade envisioned would involve new tropical islands as well as new mainland colonies. Dominica sat uneasily between restored French Guadeloupe and Martinique, some two hundred miles from the Ceded Islands' capital of St. George's, Grenada. At the southern end of the Lesser Antilles, the tiny Grenadines connected two larger islands, Grenada and St. Vincent, like links in a chain. Tobago, the southernmost Ceded Island, lay some eighty miles to the southeast, within sight of Trinidad at the edge of South America's continental shelf.

The Board of Trade identified these islands as places where "planting, perpetual Settlement & Cultivation ought to be Encouraged." "The Chief Object of the New Acquired Islands in the W. Indies," it explained to the king, was to "Exten[d] W. Indian Products of all kinds, as quickly as Possible, to the Benefit of the Trade of Your Majesty's Kingdoms." Although the commissioners could not yet know what Florida would become or how Canada might change under British rule, they knew exactly what these islands were for: planting sugar, purchasing slaves, and making money. While vast sums would have to be spent peopling the mainland, the settlement of these islands could begin at once to produce revenue for the Crown. With the exception of Grenada, with its sizable population of French planters and African slaves, the other islands appeared to be "almost Entirely Uncleared and Uncultivated." To put this program of colonization into practice, the Board designed a new system for distributing land that depended on gathering and representing information about each of these islands. It anticipated a "very large Capital immediately to be laid out by Settlers in the purchase of Negroes and [the] Erection of Buildings." To attract planters to these new islands, it called for surveys to describe their lands and divide them for purchase. Unlike other overseas provinces, which had developed over time through the independent activities of settlers, proprietors, and companies, the British state had conquered the Ceded Islands and taken command of their development directly. Thus unrestrained by the constitutional limitations of initiating charters that bound their hands elsewhere in British America, imperial reformers viewed these islands as laboratories for experimentation in improved metropolitan governance.[1]

Britain compiled statistics on the islands' commerce and mapped their lands and coasts. This new standard for knowledge reflected the wider-scale and longer-term conceptions of empire, which saw colonies less as individual territories and more as contributing societies within a larger British Atlantic system, each with particular natural endowments, economic capacities, and population profiles. The developmental trajectories of existing tropical plantation colonies had long ago been set in motion and could not easily be changed. The Board of Trade took advantage of the free hand offered by the chance to settle new colonies without the interfering precedents of long practice. Although the story of Britain's Caribbean past could not be rewritten, its future could be narrated in new terms by making the Ceded Islands into ideal island colonies.

Dominica, St. Vincent, Grenada, and Tobago came into focus as Caribbean places over the course of two centuries of European cartography. They sat in the midst of the long archipelago of the Lesser Antilles, a well-trafficked place for Atlantic commerce by the mid-seventeenth century. Sometimes called the Windwards (because, compared to the Leewards to the north, the trade winds reached them first), these islands remained hidden in plain sight within a region that plantation development and ongoing European military conflict had otherwise opened to view. To better claim the economic spoils of the sugar revolution, Britain initiated its own surveys of territories and sea-lanes in the region in the early eighteenth century. From the decks of passing ships, English navigators viewed the looming interior mountains of these mysterious islands, sights that tempered the promise of fertile tropical soils with the suspicion that "Dragons; Vipers and other most venomous and dreadful Creatures," not to mention human cannibals, made them savage places.[2]

After 1713, new British surveys determined locations, clarified routes of navigation, and scouted new colonies. Yet charts of the Windward Islands produced by naval surveyors reveal how incomplete this knowledge was. British navigators coming from Europe or Africa aimed for Barbados as a first port of call in the West Indies before sailing between these islands and into the Caribbean Sea. In 1716, the Admiralty gave Captain Bartholomew Candler command of the *Winchelsea* and the task of making new surveys in the West Indies to "Amend Our Sea Charts." When Candler approached Barbados, he sketched a view of Britain's most valuable colony, whose denuded landscape bristled with visible windmills for grinding cane. Rum was plentiful here, but Candler sailed to remote Prince Rupert Bay, Dominica, "for wood and water, for there is no wood at Barbadoes." He found an island "full of Tall Trees" and "Inhabited onl[y] by Indians." Candler sounded this "very good bay" and sketched the coastline of the island before departing. A 1726 expedition to prepare St. Lucia and St. Vincent for British occupation met with hostility from Carib Indians, who told Captain John Braithwaite that they would oppose any attempt of "settling amongst 'em." Although the voyage failed to yield a new colony, the published account that documented it included a newly detailed map of the Lesser Antilles, annotated with comments on the numbers of its Native inhabitants, as well as the extent of French incursions into the southeastern Caribbean.[3] **[162]**

During the first half of the eighteenth century, French settlers had taken possession of sugar lands up and down this chain of islands leeward of Barbados. French officers oversaw the construction of forts and cultivated an alliance with the so-called Black Caribs of St. Vincent. Confirming a diplomatic agreement first forged in 1730, Britain, France, and Spain agreed in 1748's Treaty of Aix-la-Chapelle to leave Dominica, St. Lucia, St. Vincent, and Tobago as "neutral islands," to be possessed by no European power. Such a resolution appealed to each state because it prevented its rivals from establishing positions near its own West Indian colonies, from which trade could be disrupted and invasions launched. Neutrality catered to the economic interests of West Indian planters, all of whom faced the same glutted European sugar market that new sites of production would only make worse. These islands sat astride the lanes by which ships typically entered the Caribbean, and this diplomatic agreement, by setting them outside any imperial domain, advanced the international law principle that mariners from civilized states should enjoy open access to the sea.

As Britons sounded the alarm about French forts in the Ohio River valley, they also noted the ongoing presence of French plantations in the Windward Islands in violation of this agreement. French officials had posted the order for "evacuating the four neutral islands" in the "most publick places," but they refused, if colonists chose to ignore the edict, to "hunt them out like wild boars." The goal of decolonizing the neutral islands by "obliging all the English and French settlers" to "withdraw themselves and their effects" proved impossible to enforce. By 1753, the "disputed neutral island" of Dominica, far from showing signs of reverting to a Carib homeland, was inhabited by some four thousand French subjects, whose slaves had cleared "great quantities of cultivated land," and was governed by officials who bore commissions from the governor of Martinique. As violence flared in the woods of North America, these "islands still swarm[ed] with French inhabitants."[4]

The French possessed "very fine Plantations" in St. Lucia, Dominica, and St. Vincent—and, with them, the capacity to muster men and outfit attacking vessels to "Invade & take all our Leeward Islands," an easy sail downwind from their ports. With their "decisive advantage in the fertility of their islands," the French at mid-century had assumed a commanding position in the Atlantic sugar economy. North American vessels were known to have sold their lumber and grain in Jamaica, departed in ballast, and stopped at St. Domingue to buy cheaper French sugars, many of which were reintroduced into the Atlantic economy as British produce. This "drain of Money from the Sugar Islands" thus came to "enrich [Britain's] Rivals in Trade." Commerce, the connective tissue of empire, expected to bind the islands and the continent together as a well-functioning body, had become a malignant force that threatened to starve it. Not only were gold and silver coins being siphoned into the French Atlantic economy, but this outgoing flow of money shifted the balance of colonization in the region. Such a trade "Discouraged the improvement of sugar works and Retarded and Obstructed the further settling" of Jamaica, while spurring a "Surprising Encrease" of plantation settlement in French Martinique, Guadeloupe, and St. Domingue. If left unchallenged, such a trend "must in the end ruin" the "Sugar Colonies and Translate the Sugar Trade to the French." Some planters had already uprooted their slaves from Barbados, Jamaica, and the Leeward Islands; deserted their plantations; and resettled in "foreign sugar settlements" such as Dutch Guiana and Danish St. Croix. Although aggrieved British planters railed against the unscrupulous merchants who profited so handsomely from this damaging trade, it was hard to fault the "fair Trader," who in following the plain logic of his own self-interest refused to buy expensive British sugars, which could not be sold at a profit in a market supplied by lower-priced French produce.[5]

By adding the Ceded Islands to the British West Indies in 1763, the Board of Trade sought to restructure this dysfunctional political economy. Prime Minister John Stuart, Earl of Bute, commissioned John Campbell to explain the logic of settling new sugar islands to the British public. Those who would wash their hands of American empire were willfully blind to the "prodigious value of our sugar colonies," he contended. They failed to see how transatlantic trade in slaves, sugar, cloth, and goods of every description, from knives to silk handkerchiefs, propped up the nation's naval supremacy as well as its surging manufacturing sector. Those who unfolded Campbell's "Map of the Caribbee islands" could see a new frontier open to British enterprise in the many small islands stretching from the Virgins to Trinidad. [**163**] It sheared them off from familiar landmarks in the West Indies so that they floated at the western edge of an almost trackless Atlantic Ocean, connected only to Guiana, a long-standing object of English colonizers' schemes on the South American mainland. At this scale, these new Caribbean possessions came into focus for map viewers as distinctive places capable of awakening national desires for tropical fortunes. By breaking cartographic convention in this way, the map made islands sighted by Columbus appear new and exotic. A printed key for colorists assigned a hue for each empire with a foothold in the Lesser Antilles: yellow for French Guadeloupe, Martinique, and St. Lucia; blue for Spanish Trinidad; brown for the Danish Virgins; and green for Dutch St. Eustatius. As polychromatic as this chain of islands appeared, the spread of British red across this strategic archipelago signaled the rise of Great Britain as a hegemonic West Indian power, positioned to command the whole.[6]

Inscribing a Plantation Landscape

On October 17, 1763—just ten days after the king issued his proclamation—the commissioners of the Board of Trade assembled in their Whitehall chambers and "took into consideration the method of disposing of lands in the Islands of Grenada, Dominica, St. Vincent's and Tobago." They drafted a plan, which the Privy Council approved on March 26, 1764. It called for surveying the land, dividing it into tracts "proper for Plantations," and auctioning them off in London to the highest bidder—a process of sale and settlement to be overseen by an independent land commission. Those willing to pay a premium for these new tropical lands found their rights as freeholders restricted by rules that both limited the number of acres they could acquire and made their possession of them contingent on peopling and planting them. Outside New England, colonists claimed real estate by headright and purchase, and selected the locations of the tracts they were granted. In its reformed program for New World colonization, British officials controlled the process of taking possession of American land. In place of colonists—whose independent choices about where to take up land, aggregated together, had previously determined how colonies were occupied—surveyors selected sites for planting, divided the islands' arable acres into tracts, and drew a grid of preformatted property lines across new maps. These surveyors reported to the newly created Commission for the Sale of Lands in the Ceded Islands, whose members were appointed to act outside the entanglements of provincial political and social networks and directly accountable to the Treasury. Because these islands had so few British inhabitants, their provincial governments began without councils or assemblies in the first years, effectively circumventing the influence of local elites on the distribution of land by granting it all before representative government was instituted.[7]

The Lords of Trade recalled their predecessors' attempt to mold Caribbean development on the island of St. Christopher in the 1710s and 1720s. After France ceded its parts of this jointly settled island at the Treaty of Utrecht, the Board redistributed this land, setting conditions on its ownership and setting aside small tracts for poor settlers. A half-century later, to prevent the "acquisition, by all-grasping individuals, of large quantities of unsettled and uncultivated land" in the new islands, the land commissioners likewise predetermined how much land to distribute, how much to reserve, rules for tract size, tract location, and the conditions of sale. To circumvent speculators, no prospective planter could purchase more than five hundred acres on each island. In this regulated system, little was left to chance transactions, the vagaries of the marketplace, and local influences and decisions. As envisioned by the Board of Trade and implemented by the Treasury, this plan molded land into building blocks, out of which perfected island societies could be assembled. Each parish was to be populated by between fifty and a hundred planters, their thousands of slaves, and a few score of "poor settlers" (to work as provisions farmers, tradesmen, and overseers, and to serve in the militia) for whom one-fifth of the land was reserved to give away in ten- to thirty-acre tracts. A prohibition on the sale of these small tracts would prevent the "great Planters" from buying out "all the poor Settlers, as they ha[d] done in the other Islands, to the great Diminution of their Numbers."[8]

But before Britain's new tropical territories could be put up for auction, purchased, and inhabited, they had to be mapped. The first order of business for bringing the Board's "Propositions into Execution" on each island was to "appoint able Surveyors" to make a "Map of the Island, in which the Parishes, the several Reservations, the Allotments of Lands, together with all Roads, Rivers and Bays, shall be particularly and accurately described." In the spring of 1764, even before the newly appointed land commissioners left Britain to take up their posts, the "Chief Surveyor and his assistants . . . set out immediately" to begin this work. As they fanned out across the new islands, they scouted likely locations for parish towns and island capitals, which they had been instructed to "mark out upon the Map." The first of the Treasury's sixteen separate instructions to the Land Commission was to "cause as exact a Survey to be made (with all possible dispatch)" of each of the islands. The surveyors were to characterize every "cultivable" acre, because no "persons of moderate Fortune would risque a considerable part of that fortune in the purchase of Lands, the Nature and Quality of which can be but little known, and the Advantage of Settlement very precarious." An "exact survey" would show "all the Lands belonging to the Crown" "divided into lots, numbered, and laid down on paper." Only after the surveyors were "sent there, to examine . . . the Lands," would it be possible to "put a Value on them."

Surveyors were to complement their maps with narrative descriptions of the quality of each newly defined tract before the commissioners could fix a minimum price per acre. Enabling wealthy investors from Britain to create sugar estates promised to accelerate the pace of building viable societies and thriving economies. Charging high prices for land would not only raise revenue for the Crown, to help defray the costs of settling the islands in those first years, but also vet the financial resources of prospective planters, who would each need approximately £10,000 to "stock" their lands with "Negroes[,] Cattle, Buildings, Mills, Stills, and other Plantation Utensils." After seeing the auction advertised in the *London Gazette,* prospective buyers who might be "induced to undertake Settlements in these Islands" could hire West Indian agents to inspect the lands or bid on them sight unseen after examining representations of the land in texts and images that captured their appearance and described their worth to new standards of specificity.[9]

The prospective planter could buy a tract by delivering up 20 percent of the purchase price and then making installment payments of 10 percent of the total cost every year. This initiating transaction was to begin a paper trail with a documentary record of the purchase and the terms of its sale. To apprise London of the "fresh Grants" made in the interim, updated property maps were to be "transmitted every Six months, to Our Commissioners for Trade and Plantations," with copies to the Treasury. As the Land Commission granted new tracts, its chief surveyor was to revise an official record of the changing cadastral landscape, drafting new maps in which each numbered tract corresponded to a landowner's name and narrative description of it. These new records of real property would aspire to the level of resolution met by the maps that the British and American gentry commissioned of their own estates. Such a record promised a far richer description of private land than the perfunctory sketches that filled the plat books in other colonies. These "exact and authentick Surveys" were to be deposited at the provincial secretary's office in Grenada, where they could be consulted to verify grants in case of dispute. From the Ceded Islands, however, a copy of the file of documents for each tract (which was to include detailed estate maps, agricultural assessments of soil quality and terrain, inventories of wood and water resources, and natural history observations about the tract's place within the island's landscape) was to be sent to London, where it was to fill a new metropolitan archive of colonial landholding. As these files joined the official papers of the Treasury and Board of Trade, this imagined repository of documents would enable these agencies of empire to exercise oversight that went well beyond adjudicating property ownership. Such records opened views onto colonial society at the scale of the individual colonist developing a tract of land. Since every "Grantee must be tyed down to some Terms of Cultivation," those that did not "conform to the Terms upon which the Grants were made, as to the number of Negroes and Whites to be kept on the said Lands & the degree of Cultivation to be carried on," would lose their lands and be warned off them by provincial authorities.[10]

By 1776, Britons could see this plan realized in a series of finely etched maps, drawn by the Land Commission's chief surveyor, John Byres, and published in London. [**164–166**] Just thirteen years after the Treaty of Paris brought these enigmatic islands into the realm, the maps showed them to be fully occupied colonies and comprehensively known places. The maps' common style emphasized that these islands were fundamentally similar places, each suited to the Board of Trade's program to develop them into productive sugar islands. Byres pictured each of the islands as divided natural spaces, one part fertile clearings and another part rugged wilderness. In a broad ring around each island, rivers flowed to the seacoast across unobstructed plains and property lines carved up these promising lands for sugar agriculture. At the center of every island, densely etched hachure lines cast dark shadows across forested interiors to signify impassible mountains, uncultivable spaces that therefore remained unmarked and unclaimed. The comparatively small size of these islands permitted Byres to merge cadastral and jurisdictional mapping in a single image to show even more clearly than was possible in other regions how private land ownership structured colonization. On these maps, one could take in a view of the whole island and yet still see the boundaries of every tract, no matter how small. Drawn at a scale of a half mile to the inch, they pictured colonial space at roughly ten times the resolution of comparable North American maps, demonstrating, tract by tract, how settler occupation established sovereignty across whole colonies.[11]

Published "references" sold with the maps made individual colonists visible as legal presences, alphabetically by name and then cross-indexed by the numbered tract on the map that he (or, more rarely, she) owned or leased. On Dominica, these ranged in size from lot number nine—the six-acre tract of one Baptist, a "free Negro"—to lots seven through eighteen, brought together to form the jointly owned 1,981-acre Rosalie Estate scheme. Rosalie's unique arrangement aside, the property grid of the island revealed a populous community of planters, none of whom had been allowed to speculate in huge tracts at the expense of Dominica's rapid cultivation. The social innovations designed to steer a sustainable course for the development of new sugar islands appeared throughout the maps, from the "reservations of wood-lands" across the mountain peaks of Tobago; the grid of "Garden Lots" that promised to feed the inhabitants of newly fortified port cities; glebe lands set aside for Protestant ministers; lots designed to pay the salaries of schoolmasters; lands reserved for poor settlers; and those set aside for batteries, barracks, and fortifications. To these goals were added the creation of botanical gardens and exercising grounds for island militias that the Treasury authorized the Commission to provide for by "appropriating small parcels of land." The Board of Trade's plan for America was to reconcile the particular interests of colonists with the general interests of the empire. In his maps of the Ceded Islands, Byres showed that they were not only surveyed and settled but also fully realized societies, in which enlightened land policies had shaped private interest toward the goals of social balance, productivity, and defense. For the first time, Britons could behold the Board's vision for colonial reform as an image, one that could be framed and mounted for display.[12]

As Land Commission surveyors began tracing boundaries around prospective plantations, provincial surveyors working for the new government of the Ceded Islands colony planned new cities and redesigned those that Britain had inherited from France. Two years after naval surveyors certified Prince Rupert Bay, Dominica, as a prospective port, James Simpson laid out a plan of the new town of Portsmouth that imposed a grid of 206 lots, laid out on a line established by the compass across 105 acres of sandy coastal plain between the North and Indian Rivers. **[167, 168]** Where nature disrupted this rectilinear order in the form of a "small Rivulet," Simpson sketched how it could be redirected by canal to conform to his urban grid. Other sites of state power—a riverside jail and a customhouse overlooking the harbor—were relegated to its edges, along with tracts reserved for a "Market Place" and "Bleaching Ground" that anticipated the bustle of trade and domestic life in which future inhabitants of this empty site would take part. Simpson shaded the portion of this site already cleared in green, revealing the footprint of a seized French plantation that had protected its soils from the salty spray of the surf with a stand of trees. Slaves laboring under British masters would have to clear all the remaining woodland (marked in red) to prepare a space for the new port town of Portsmouth to emerge.

Wills Hill, Earl of Hillsborough and Britain's secretary of state for the colonies, sought to make Dominica, flanked north and south by French colonies, into a "place of Strength & defense." As he approved plans to fortify Portsmouth, Hillsborough stressed that to keep Dominica British, it would have to be "well and speedily peopled with white inhabitants." Simpson's plan of lands "about the Town of Portsmouth" pictured this undeveloped site as it appeared from the harbor: beyond the few houses along the shore, a plain opened to reveal the clearing into which Britain would plant its new colonial capital and establish a hub for Caribbean trade. **[169]** This plan extended the reach of the town beyond its urban core to organize a rural hinterland. Ownership of a town lot came with the rights to a two-acre garden lot marked off north of town, providing townspeople with land to graze livestock and supply themselves with "necessary Vegetables." This innovation associated Portsmouth with other towns built to occupy volatile frontiers that also granted owners a garden lot, such as Savannah, Georgia (1733), and Lunenburg, Nova Scotia (1753). In keeping with the Board of Trade's plan to grant land to poor white settlers, Simpson plotted 409 acres in three tracts on a parcel of wet and wooded land outside Portsmouth's boundaries, a mile upriver from the "Free Negro Town" at the shore. **[170]** Although the Board of Trade made the regulated granting of plantation tracts a priority for the development of the Ceded Islands, its approach to urban design also used cartography to set a process of settlement in motion. Before "one House is allowed to be built" in any new Ceded Island port, it demanded that a "regular Plan of a Town" (like the ones James Simpson drafted) fit the properties of urban landowners into a predetermined grid.[13]

These plans for Portsmouth expressed an official intention to build the economic and social infrastructure that could make Dominica into a self-reliant colony. The smaller Leeward Islands colony had been malformed by centering the administration for an entire archipelago in Antigua, where this distant "Seat of Government" became a "Gulf which draws towards it not only the Riches but the Minds of the Provinces and People around it." Dependent Montserrat, Nevis, and St. Christopher, each lacking a capital town of its own, produced admirable quantities of sugar but also exemplified everything that the Board of Trade's commissioners thought was wrong with the British West Indies. From their reliance on imported provisions to their sharply skewed populations, which contained twelve to fifteen slaves for every white colonist, these tiny islands were perpetual targets for attack. As plans for Portsmouth made clear, each of the Ceded Islands was designed to be a freestanding society organized from a principal capital city. Although Dominica and the other Ceded Islands were first administered from Grenada, they were to develop into separately governed colonies. Each was to have its own customhouse, as Portsmouth did. A tract of potentially valuable land adjacent to the town was reserved for the use of its governor, providing a source of income that would not be dependent on external subsidies or grants from a future assembly. To maintain army and navy outposts on Dominica, this plan preempted private ownership of strategic resources near the town by appropriating sizable tracts for forts, batteries, barracks, an "exercising ground," hospitals, storehouses, provisions grounds, dockyards, and firewood. Marshaling the labor required to create this tropical metropolis from scratch entered into imperial planners' calculations. Land commissioner Hugh Graeme suggested that those who languished in British jails "condemned for petty Crimes" might be made newly useful to the empire after being "put to hard labour . . . cutting down woods to clear the Country about [P]rince Ruperts Bay or such necessary places." Simpson set three acres aside in his plan for a "burying Ground for Slaves," a provision that anticipated the high human costs of carving out a new city from a thinly populated coast of an underdeveloped Caribbean island.[14]

By 1773, this place intended as the "seat of Government and the capital of the Island" remained largely undeveloped. Portsmouth's location "to the leeward of a morass and partly in it" had rendered it "so unhealthy that almost all the people who attempted to settle there had been forced to abandon it and go to Roseau," the established French port town to the south. Although "situated in so excellent a bay and declared a free port" to legally export slaves and manufactured goods to the French islands, Portsmouth now "looked more like a ruinous deserted village than a new place of trade." Despite its disadvantages for shipping, Roseau absorbed the infusion of new settlers, grew along with Britain's military capacity and commercial traffic, and was becoming "one of the most flourishing towns in the West Indies." A 1785 "Plan of Prince Rupert's & Douglas bays" revealed how the vision for Portsmouth persisted as an imperial abstraction and how little progress had been made toward achieving it.[15] **[171]**

The British quest to anglicize Grenada left its mark on the maps of the island's capital and largest port, St. George's. Rear Admiral Richard Tyrell undertook a series of "running" surveys of Ceded Island harbors in the early 1760s, sketching what he could see from the deck of his ship. His hasty "Draft of the Harbour of Fort Royal" measured its depth with a smattering of soundings and located anchorages, hazards, hills, and structures. **[172]** Without a rigorous plane-table survey, its depicted shape was little more than an impression of the coastline. Published French maps provided more polished and precise views of this harbor, but Tyrell's crude sketch demonstrated an independent British command of geographic information about the island. Such "actual Surveys reported to the Admiralty" established the harbor's "Depth of Water and Spaciousness" and affirmed that it was "so well formed by nature" that it provided "perfect security from the tempestuous weather" during hurricanes and put residents "out of the power of an Enemy by sea" in times of war.[16]

After a furious defense of Martinique in January and February 1762, French forces had capitulated quietly in Grenada, where the British inherited a residential, maritime, and military complex that included a fifty-year-old masonry fort that overlooked the harbor from its four bastions. Will Cockburn mapped the town and charted its harbor in 1763, following a year under British occupation. **[173]** After erecting a battery on the grounds of the hospital to occupy the highest ground and flanking the harbor with an earthen redoubt to command its entrance with artillery, the British began transforming the small French town near the shore. In name, law, and fact, the British took possession of a place they now called Georgetown. A 1765 surveyor's map revealed the division of its handful of urban blocks into 160 town lots, each numbered to correspond to an owner named in an attached list. **[174]** Those with French names predominated, their ongoing ownership made visible because the mapmaker wrote their lot numbers in black ink to signify their grant from the "French Governors, before the surrender." Roughly one-third of the lot numbers were inked in red, which denoted a post-conquest grant from the Crown to men and women with English and Scottish surnames. These included all the valuable water lots along the shore and the harbor. French townsmen now walked familiar streets that bore new English names. They might see Governor Robert Melvill (one of Britain's conquering generals) make his way from the Government House on lot 45 to the seashore along Melville Street. To circumnavigate the town's central square and return to his chambers, he would have followed streets renamed to honor Prime Minister George Grenville, war hero John Manners (Marquess of Granby), Secretary of State George Montagu Dunk (Earl of Halifax), and First Lord of Trade Hillsborough. By 1765, these distant imperial patrons could see their surnames across the new body of cartography that arrived in London, signifying streets, rivers, and bays on St. John Island in the Gulf of St. Lawrence, the mosquito-infested inlets of East Florida, as well as here, in the new townscape of St. George's, Grenada, once known by the French as Fort Royale. A new ordinance gave English names to the parishes and other towns of the island and forbade the use of the French names by which they had previously been known.[17]

During intervals of peace between Europe's eighteenth-century wars in the West Indies, a few intrepid colonists had attempted to settle Dominica, but none remained long until a bold push by the French at mid-century brought some fourteen hundred white planters and more than five thousand slaves to inhabit this supposedly "neutral" island. Unlike the other Ceded Islands, clustered together in a contiguous band far to the south, Dominica was an island apart, its peaks visible from two neighboring French sugar islands, Guadeloupe and Martinique. British officials found that more than one-third of its available land had been claimed but less than one-tenth had been planted by the existing French settlers, who numbered 1,718 and lived in 475 houses scattered across ten districts. They used its eighty-three small rivers to "Water their Cattle," but the British dreamed of hundreds of new sugar works that could be powered by their currents. They believed that of 90,374.5 "uncultivated" French acres, "near ninety thousand are capable of Culture." Simply to "Cultivate properly" the small portion of the island that had already been cleared, they estimated that an additional 6,743 "Working Negroes" would need to be added to the current population of 3,145. By this measure, the most wild-eyed projection expected new planters to import some 300,000 slaves as they turned a sparsely settled, diversely planted island into a fully exploited plantation colony, utterly devoted to sugar production. Aside from preserving the outlines of French administrative districts, the British took possession of Dominica intent on changing it.[18]

Just as the British designed a perfect paper capital city for Dominica, their plans for the island's countryside used maps to accelerate the creation of large sugar estates. Under the Board of Trade's rules, individuals could not amass huge tracts of undeveloped land, but in one instance, individual landowners joined their grants in partnership to better promote and develop them. Pooling twelve separate grants to form the Rosalie Estate on Dominica's windward side, Lieutenant Governor Charles O' Harra, William Stuart, James Clark, and Robert and Philip Brown exhaustively mapped their collective portion of Dominica, remeasured to contain 2,219 acres. After spending more than £2,300 on the land alone and more than £54,000 on slaves, overseers, buildings, and other start-up costs, they were eager to recoup a huge investment. The partners hired provincial surveyor Isaac Werden to survey every acre, perch, and rod of Rosalie land in 1776. His "Plan of the Rosalij Comp[an]y Estates" presented this early modern real estate development scheme in detail, capturing a small part of Dominica's transformation as the British colonized it.[19] **[175]**

In little more than a decade, the partners had established a large and well-ordered sugar plantation at the mouth of the Rosalie River and named it Sea-Side Estate. Slaves had cleared and planted about a third of this tract's 548 acres in cane across twenty-three fields. Where the river met the shores of Rosalie Bay, they had constructed a showplace plantation settlement that featured scores of buildings clustered together to grind, boil, and pack sugar and house, feed, and monitor its few white employees and its many enslaved workers. Across a small stream from the "Mansion," slaves slept in a compound of fifteen "Negro Houses" at the juncture of three of the many roads that connected cane fields, pastures, and woodland lots to the central complex. A large formal garden, intersected with walking paths, filled the space between white and black living spaces. The Rosalie Company spared no expense to present Sea-Side Estate as a rationally ordered place for work and life. Along with the regularity of the slave quarter's structures were other signs that the partners intended to maintain the plantation's population rather than following the usual course of business in the British West Indies by working slaves to death. It devoted fifty-two acres to provision grounds and another thirty-three to grazing land, part of which was stocked with imported Guinea grass to provide livestock with rich feed. A large hospital, overseen by a resident nurse, and a "Yaw House," isolated near the shore to prevent the spread of the infectious bacterial disease yaws, provided facilities to care for sick slaves. The layout of the settlement put a premium on efficiency and visibility in the process of turning cane into sugar at a profit. Watch houses overlooked the fields to guard against the first signs of fire. A dung heap provided the means of enriching the fields after the fashion of the most improved English farmer, and three "Trash Houses" stored pressed cane husks to provide a renewable source of fuel for the boiling coppers. A bridge built over the Rosalie River carried cane by wagon from fields located at the south of the estate, while those further inland were served by a canal that passed beneath the "Clarks Hall," from which bookkeepers might look up from their ledgers to watch rafts of harvested cane arrive at the grinding mill.

Where the river emptied into the sea, Isaac Werden drew two schooners and a ship flying the red ensign of the merchant marine plying the waters around Rosalie's moorings. This maritime scene echoed the signature image used to illustrate early modern maps of British America: a tropical port crowded with vessels waiting to transport staple commodities across the Atlantic Ocean. The mapmaker added a pictorial view of Sea-Side's settlement in a large watercolor inset at the corner of his "Plan." Next to his bird's-eye overview of its structures and layout, this image captured the poetry of the place for the eyes of prospective British planters. It showed how Rosalie's partners had put a stamp of civility onto Dominica's wild landscape. The dense forest that covered the highlands terminated in an abrupt edge halfway up the mountainside, where slaves had stopped cutting down trees. A three-story sugar works loomed above the tree-lined banks of the Rosalie River at the center of a complex of workshops and farm buildings. Behind the great house, slave cabins clustered together in a tidy village. Even the necessary house, separated from the mansion by a neat fence line, bore a steeply hipped, ornamented roof, whose style was repeated in the manager's residence.

Nothing underscored the idea of harmonious order in this image of plantation settlement more than the two figures pictured in the foreground. Along the road shaded by the leaves of growing canes that bordered field 22, a white man wearing a tricorne hat and a blue coat gestures with his left hand, as if speaking words of greeting or instruction, to a faceless black man dressed only in undyed breeches. Both hold walking sticks—suggesting that they had just encountered each other along the path—but neither threatened nor feared violence from the other in this unguarded corner of the estate. Rosalie was a place, this image said, in which an enterprising Briton could assume the stature of an improving gentleman on an impressive scale and become rich by helping to create a society characterized by beauty, productivity, and utility.[20]

These images of Sea-Side Estate held out the promise that four other tracts pictured in "A Plan of the Rosalij Comp[an]y Estates" stood ready to be transformed by slave labor into full-fledged plantations. The partners carved the rest of their property into four portions, each between three hundred and five hundred acres, which they named the Retreat, the Grand Fonds, Rosalÿ Valley, and Newfoundland, each of which converged on the Rosalie River to take advantage of Sea-Side's access to the sea. They commissioned Werden to create separate plans for these properties that fit like pieces of a puzzle into the master plan, to show prospective investors where they could claim a share of the venture. **[176–179]** Sober West Indian hands had warned that these islands were "not the promised land" but something more like the "garden of Hesperides," whose "fruits cannot be gathered without Herculean labour." The "expense and toil of clearing away woods, and erecting habitations," should give unwary newcomers pause, they cautioned, as should the "extreme sickliness and mortality of the country," which would make these rugged islands, dense with vegetation, "melancholy mansions of disease and death." At Rosalie, company slaves had already built roads, cleared fields, planted provisions, and erected structures to form the hearts of new plantation settlements on these tracts, lowering the risks for purchasers of establishing plantations and increasing the premium that could be charged for them. At Grand Fonds, slaves had already planted thirty-eight acres in indigo, and a nearby "Woodcutters House" sheltered those tasked with clearing a new cotton field on steep ground. Only Newfoundland's 442 acres remained entirely in "Wood." Across the acres of these nascent plantations, Werden denoted forests with repeating icons of widely scattered trees and bushes, making uncultivated Rosalie seem as unthreatening as a park, easy to traverse and clear. His two-tone scheme for topographic shading used dark color for declivities and steep grades and light colors for plains and plateaus, which made the terrain appear less rugged than it in fact was. The Rosalie Company used maps to accumulate a stock of working knowledge about its lands and represent their potential for plantation agriculture in the most favorable light.[21]

This surface image of prosperity and potential belied Rosalie's tumultuous history. The Rosalie partners had purchased the property virtually sight unseen. No road existed by which they could easily reach their tracts to inspect them, so they bought them "in the dark," as they put it, based on reports of surveyors who described the "Surface and quality of the Land." The "Settlement of a Sugar Plantation" on the Sea-Side Estate held out a model to which new investors could aspire, but Werden's bucolic "Plan" concealed the "great loss of Slaves" who died felling the trees and cane-holing the fields in preparation for the first crop. The surveyor pictured a placid sea beyond the estate's wharf, but it was in fact located on the notoriously rough windward side of the island. Before new installment payments to the Crown came due in 1777, the partners commissioned another map to petition for relief from the burden of debts under which their enterprise struggled. [**180**] Werden's second plan presented a different picture of the land from the one he had so artfully composed to promote Rosalie. At the same time as he drafted picturesque views of its busy harbor, showplace plantation, and rolling hills, he created this utilitarian image as a legal document—and vouched for it as a "Sworn Colony Surveyor"—to attest that much of the property purchased was a wasteland that should not have to be paid for. To the partners' "great astonishment," almost a third of the land, more than 628 acres, had to be written off as "impracticable and wholly unfit for Cultivation and Improvement being broken Mountains, Deep Vall[ey]s, steep Ridges and inaccessible Precipices." In its darkly shaded mountainside and obstructed plains, we can see a mirror image of West Indian landed affluence, a landscape rendered to show that some spaces were too wild to bring under control.[22]

The quality of land in Britain's new territories appeared differently at different scales of representation, and this resulted less from mapmakers' attempts to mislead investors than it did from the way variation in scale changes the way salient details can be depicted cartographically. At a quarter mile to the inch, John Byres's *Dominica* incorporated topographic details from Werden's surveys, but the lightly shaded outlines by which he represented the company's lands suggest no insurmountable obstacle to their development. Byres's simple dichotomy of barren mountains and arable plains broke down when Isaac Werden mapped the Rosalie Valley at approximately eight times this resolution. Other purchasers who had selected pre-surveyed properties from Land Commission maps could be forgiven for thinking that their lands were unobstructed by ravines and crevasses, rendered as they were without topographic relief. When purchasers took up their tracts, they found a world of variation where they had expected flat and fertile terrain. As their payments came due and they struggled to make them, many sought exemptions from the high pound-per-acre charges on those parts of their estates that proved, on closer inspection, to be "impracticable."

In contrast to Dominica, where limited prior colonization left it exposed to far-reaching British schemes, French planters had settled Grenada extensively. More than half of its 67,425 acres were "under actual Cultivation and Improvement," and on them, slaves produced sugar, coffee, cocoa, rum, and molasses worth more than £200,000 per year. Even the least valuable and most mountainous lands had already been "claimed as private Property under Grants from the French King." Beyond the work of surveying those few parts of the island that had not yet been granted, little remained to be done to discover the island's attributes. Instead, the Land Commission remapped Grenada to redefine this French island as a British place. When the British made it the center of a provisional government for the Ceded Islands, they found a place that was already divided, occupied, and developed by French planters. As reported by military governor George Scott, although the French had occupied Grenada since the mid-seventeenth century, they had failed to produce a "compleat Map of this Island." To demonstrate Britain's superior command of this long-contested place, Scott commissioned a resident French mapmaker, one Monsieur Pinel, to create a new image that described Grenada's natural and human geography. **[181]** Pinel's map presented *L'Isle de la Grenade* as a composite of eighty-seven separate sugar and coffee plantations, with each planter's *habitation* located within a bounded tract, which, when joined together, spanned the island and subdivided almost all of its territory up to the edges of the mountainsides. Every tract on the map had a place within the jurisdiction of one of seven named *quartiers*. Every *riviere* bore a name, and only a few stretches of Grenada's coastline appeared without annotations that designated anchorages, towns, bays, and points.[23]

Although it was drafted at the command of a British governor, Pinel's map of Grenada represented the island as a French place. Although we know nothing about "M. Pinel" beyond his map, the image he created was at the intersection of French cartographic style and British imperial desire. Several British precursors for this comprehensive settlement map—including William Mayo's *A New & Exact Map of the Island of Barbados* (1722), Robert Baker's *A New and exact map of the island of Antigua* (1749), and Patrick Browne's *A new map of Jamaica* (1755)—identified specific plantations with icons of multistory houses and cane-grinding windmills and located them within a well-rendered landscape divided into administrative districts. Decades before Britain launched its Ordnance Survey, the first sections of the *Carte de France* (1750–1815) used advanced geodetic surveying methods to present the kingdom as a network of human settlement imposed on a detailed natural topography. *L'Isle de la Grenade* represented the island at a comparable scale and with similar precision to this emerging and monumental work of state cartography. It enabled its new British masters to behold, at a glance, the location of its "Harbours, and the proportion of cultivated Lands" over which Britain now asserted sovereignty. In his first official dispatch to the Board of Trade after the conquest, Scott included a manuscript version of Pinel's map to illustrate his lengthy assessment of the "true advantages" of the new colony, which was then published in London.[24]

As a French island, Grenada had followed a diversified approach to tropical agriculture. Before it was ceded, slaves cultivated coffee, cocoa, cotton, indigo, and spice crops, along with sugar. When the British arrived, diversification gave way to intensive sugar monoculture. Planters who owned played-out fields in Barbados and the Leeward Islands arrived in Grenada to claim its undercultivated sugar lands. Although Pinel's map presented the island as a place in which the arable land had already been claimed, within this cadastral grid remained an abundance of woods covering highly productive virgin soil. Grenada had languished as a dependency of the larger and more profitable island of Martinique and seemed a "rough Diamond" that the French had "neither Skill nor Implements to polish." Transformed by British enterprise, government, and trade, Scott predicted it would become a "most brilliant Colony" and the empire's "most valuable Island" (save Jamaica) in a decade. The map that James Casey drew for Governor George Macartney in 1778 pulled back from St. George's urban grid to picture an emerging plantation countryside that had taken shape in its hinterland after sixteen years of British rule. **[182]** Signs of capital-intensive sugar production abounded. Thomas Townsend's plantation occupied a site just east of the Lagoon, on which Jacques-Nicolas Bellin had located three lone structures in his "Port et Fort Royal de la Grenade." **[183]** It now featured rows of slave houses and the signature footprint of the structures (mill, boiling house, curing house, distillery, cane house, and workshops) that composed a large-scale sugar works. West of town, a canal connected another plantation to a network of roads and bridges that terminated at the new quay in the harbor.[25]

As assistant to the army's quartermaster general in London, Lieutenant Daniel Paterson was well positioned to see the stream of official manuscript maps sent from America. In 1767, he redrew the map of troop deployments in North America as the army pulled back from Indian country and consolidated its forts and garrisons. From this metropolitan position of access, he also created *A new plan of the island of Grenada* in 1780. **[184]** Pinel's *L'Isle de la Grenade* had created a template for the representation of the Ceded Islands that John Byres followed closely in his maps of Dominica, St. Vincent, Bequia, and Tobago. By updating this French map with "English Names" and "other Improvements," Paterson completed the set of images that represented these colonies as British places. He altered Grenada's appearance so that its "irregular

Chain of vast Mountains" seemed more rugged and formidable than had Pinel's. This sharper topographic distinction between highlands and lowlands focused attention on the locations of ideal sites for large-scale sugar planting. His accompanying pamphlet assessed the land in each of the island's six parishes in terms of two qualities: its amount of unobstructed land that could be planted in cane, and its access to mountain-borne rivers.

The best land "rises in gentle undulations from the Sea" in "large tracts easy of Cultivation" that terminated at the feet of looming mountains. A number of "fine Rivers" descended from these formidable heights to provide a ready source of power to turn "Mills for Grinding the Canes." French settlers had planted chiefly on the leeward side of the island, where the land was "exceedingly broken," and their diversified plantations featured smaller crops of sugar, supplemented by the coffee and cocoa that slaves cultivated in the uplands. By 1780, the British had downgraded the commercial value of this uneven coast and focused interest on the plains of the windward side of the island, which by these terms seemed "one of the finest Sugar Countri[es] in the West Indies." Here and throughout the island, as Paterson's map showed, the arrival of water-powered sugar mills signaled the redevelopment of Grenada as a place of intensive sugar monoculture. After nearly two decades as a British colony, these planters operated ninety-five water-powered grinding mills, which enabled them to process sugar on a larger scale compared with the thirty still remaining that operated by wind or livestock. The borders of the six French quarters that divided the island remained, but each of these had been renamed as an English parish. The shape of the cadastral grid by which the territory had been divided into properties likewise persisted, but most of Grenada's original colonists had been displaced by British newcomers. Grenada's engineer, Harry Gordon—the same army officer whose map of the Ohio River established its navigability—now owned a 240-acre plantation marked number 42 on Paterson's map, once the property of Dupleix Montaigu. The British completed what the French had begun: by the late 1770s, the island had "but little Wood remaining excepting on the inaccessible tops of the Mountains." Grenada epitomized the British vision for the Ceded Islands as a lucrative commercial outpost. By the end of the 1770s, it had surpassed

Barbados and was second only to Jamaica, producing sugars worth over £800,000 per year.[26]

In 1762, George Scott had judged the French inhabitants of Grenada wanting as colonial subjects. Like patriotic Britons everywhere, he feared the French as a rival for American power and, at the same time, denigrated their capacity to colonize New World lands. After examining the customs ledgers, Scott calculated how many thousands of hogsheads of sugar had been exported and the value of goods Grenada had consumed. Confident in the "superior Skill of an English Planter" to extract twice as much sugar from an acre of land, Scott anticipated Grenada's rapid development into a major exporter of sugar, rum, and molasses. The French planters were "frivolous," "vain & ostentatious," and "fond of military Trifles." Although "naturally indolent & very much given to Luxury," they were also a malleable people who if "mixed with English Planters" for a time might "learn Frugality & Industry, & many of their Follies & Vices will in a great Measurement wear off."[27]

Anticipating that Britain might retain conquered Grenada at the peace, migrants had arrived from Barbados, St. Christopher, and Antigua to "purchase Plantations conditionally" from French planters during the interlude between the conquest and the Treaty of Paris. Such rapid migrations validated the Board of Trade's expectations that settlers would move from established colonies to populate newly acquired territories, but Scott ordered this practice stopped to prevent departing French planters from converting their property into cash and then leaving with it for another part of the French dominions. He nevertheless admired the spirit of enterprise that had created a market for Granada estates even before the military government had time to put in place a system for regulating the sale of land. He did not want to discourage such planters, on whom the "speedy Settlement of this Island" would depend, and proposed that they be granted lands on generous terms so that they could "soon bring them to Perfection." In addition to collecting all the maps and customs records left behind by the French, Scott perused the "Archives of the Island" to review its laws of inheritance and property that would have to be revised to make this French island British. Like others who ascended to prominent positions in the new empire after publicizing their expertise in American affairs with reports and maps, George Scott, who also commissioned Pinel's *L'Isle de la Grenade,* was named lieutenant governor of Dominica in 1764.[28]

As British commentators never tired of pointing out, they fully expected conquered French Grenadians to learn to appreciate life under a British system of government. In the new political economy of empire, religion, loyalty, and culture—the very factors that the conduct of the war forged into a new sensibility of Britishness—meant less after the peace, "because it is people that are mostly wanted in these Islands." Although prudent laws prohibited Frenchmen from purchasing coastal land (from which they might take part in smuggling with nearby French islands), every foreigner who populated a British colony was in fact an "advantage to the State; more so than [a] drain on the mother Country already too much exhausted . . . from so long and expensive a War; carried on in all the Corners of the World." More than 3,500 French colonists—masters to more than 20,000 slaves—lived on Grenada, Dominica, and St. Vincent in 1763, places the British were determined to possess through occupation. Because so many British colonists "only th[ought] of making a rapid fortune to enable them to return to Europe to spend it there, leaving only servants on their Estates," French colonists in the British Caribbean might have even been preferable to Protestant Britons. They were more likely to form a "Yeomanry of the West Indies," remaining on the islands to maintain white rule over so many subjugated Africans.[29]

The Land Commission recognized the property rights of French Grenadians. Although the official British view was that the French had squatted illegally as "Encroachers upon the Neutral Islands," new rules about French landholding sought to entice them to remain for the contributions they could make by keeping the slaves at work as well as for their "knowledge in the Culture peculiar to these Islands." The fourth and ninth articles of the Treaty of Paris provided the new French subjects of the king of Great Britain with eighteen months to sell their property and emigrate with the wealth they had accumulated, a flight of capital and people that officials determined to prevent. Those who decided to remain would become the king's "new subjects," with comparable rights to his "old subjects." In place of outright grants to their land, they received long-term leases and the promise that these might be converted into permanent free and common socage grants once they had demonstrated their loyalty. If the British failed to honor French land rights, it was conceivable that Grenada's twelve thousand French-owned slaves could be "instigated by their Masters and the Priests, to commi[t] great Outrages and may become Rebellious," destabilizing British rule. The "religious and national Attachments" of the French would soften, it was argued, under a legal regime that respected their interests in lands that they had "cleared by their own Industry." Should the buildings or land they owned be taken for some public or military purpose, they should receive "full and ample Compensation." Officials should show "indulgence and encouragement" to French leaseholders by promising that their leases would be renewed on the "most advantageous & favorable terms." The British were scrupulous to avoid the appearance that French property rights were open to predatory seizures by the provincial government. As "experience shews," reasoned the Board of Trade's commissioners, the "possession of property is the best Security for a due obedience and submission to Government." Defined during the war in moral terms as an insidious enemy, peacetime reformers recalculated the value of French planters as colonists.[30]

As British forces took possession of Grenada in 1762, George Scott promised to send out "Draughtsman to take the Map of" the uncharted islands between Grenada and St. Vincent the moment he could spare an "armed Vessel." Until then, he could report only that Grenadines were a scattered archipelago of islets, many with only a few acres of "Surface," "Besides a Multitude of Rocks with their Heads above Water." These scraps of tropical land could be "turned to Use of some kind," he speculated, perhaps as mule pens to supply animal power for the cane mills or as diversified producers of minor commodities like coffee, cocoa, and cotton. Whatever might be built there would be exposed to the attacks of marauding privateers in wartime, who could retreat behind "Multitude of Shoals & Rocks" to evade pursuing men of war. Who would invest the £10–£20,000 required to set up a large-scale sugar works, when seaborne raiders might approach the coast under the cover of darkness and "ruin four or five Plantations in a Night's time, & carry off all their Slaves"? By early 1763, the Board of Trade could report that the "Grenadillas" were made up of "about Thirty" small islands and keys. A few French planters had settled on the two largest, Carriacou and Bequia, before the war, and the Land Commission dispatched a surveying expedition to Bequia in the early 1770s to pin down the locations of their lands. In place of vague written "concessions" that documented these claims, chief surveyor John Byres divided the island into thirty-six visible tracts owned by French planters as well as recent British purchasers. **[185]** What remained amounted to seven hundred acres that he laid out in allotments "for the use of poor Settlers." This image of tiny Bequia, seven square miles of mountainous Caribbean land owned by just seventeen people, was revealed to a wider British audience when it was published alongside Byres's plans of Dominica, St. Vincent, Bequia, and Tobago in 1776.[31]

By the mid-1760s, investors eyeing other unclaimed "Waste Lands" in the southeastern Caribbean arrived in Carriacou, the "principal of the Grenadines," where some two hundred people had lived in 1750. Drawing on a ready pool of available investment capital, British colonists took up practically all of its eight thousand acres, settled slaves on their lands, and began producing cotton. Between 1776 and 1790, Carriacou increased cotton production by 30 percent, abandoning other crops and producing this staple commodity on estates that rivaled sugar plantations in size and profitability. Walter Fenner's *A New and Accurate Map of the Island of Carriacou in the West Indies* (1784) delineated a plantation landscape that exported 1 million pounds of cotton per annum, accounting for roughly one-eighth of all British West Indian cotton in the 1780s. **[186]** By 1790, some four hundred British colonists commanded the labor of more than three thousand African slaves on the island. Owned entirely by absentee investors, these cotton estates exhibited strong economies of scale and high rates of productivity per slave worker. The heyday of Carriacou ended after the invention of the cotton gin opened up the vast southwestern cotton lands of the United States. For two decades, however, the island "prospered more uniformly and generally than any other of our West Indian islands."[32]

Geographies of the Carib War

John Byres surveyed St. Vincent for the Land Commission in 1764 and calculated that the island contained 84,286 acres. His 1776 *Plan of the Island of St. Vincent* overlaid a cadastral grid of approximately 175 individual tracts on a natural topography defined by mountains and plains, representing it in the same visual style as Pinel's and Paterson's maps of

Grenada and his own maps of Dominica, Tobago, and Bequia. On the whole, this series of maps celebrated Britain's rapid occupation of the southeastern Caribbean, but Byres's map of St. Vincent also registered a dissonant note of thwarted imperial ambition. Although the Treasury took in £162,854 from sales of 20,538 acres, much of the island's potential plantation land remained unsurveyed and unsold in 1776. Before the Board of Trade could act to prevent it, the Crown had granted thousands of acres to General Robert Monckton and Lieutenant Colonel George Etherington, turning two war heroes into land speculators who stood to pocket profits from the sale of these lands. More damaging for the Board's intention to accelerate sugar planting in the islands, resistance from the so-called Black Caribs forced Britain to hold off on surveying much of St. Vincent. Byres labeled the entire northern portion of the island as "Mountainous Lands Granted to the Charibs." The "most extensive and finest" part of the island on its windward side, he wrote, was likewise reserved as a Carib homeland. As long as they possessed it, this excellent land would remain "in wood, useless and unoccupied."[33]

A dozen years before Byres's map appeared in print, the Land Commission began selling lots in St. Vincent, before any of the land had been surveyed. It took for granted that its orders for dividing parishes, creating towns, and granting moderately sized plantation tracts could proceed unimpeded. However, the Board's general plan for tropical colonization faced a unique challenge in St. Vincent, where a large indigenous population remained "very jealous of any settlement by Europeans upon this Island." Compared with previously planted Grenada, thinly settled Dominica, and practically unoccupied Tobago, St. Vincent's Indian problem challenged the presumption that West Indian islands were open to transformation into sugar colonies and invited comparisons to the program of conciliation underway along the frontiers of settlement in North America. A 1764 "State of the Island" report revealed that fewer than two thousand French colonists along with some seven thousand slaves and free blacks occupied the island. Only thirty-two men merited a designation as "Credible Planters." They specialized in growing coffee and produced smaller quantities of cocoa, tobacco, and maize for export. This population lived in six quarters clustered along the leeward and southern coasts, the largest of which, Routtia, boasted 107 houses. Among them lived seven Frenchmen identified as "Men of Ill Conduct," who had "frequently endeavoured to raise Insurrections." The whole colonial population inhabited a portion of the coastline from the point of Ribichi Bay in southeastern St. Vincent to the Chateau-Belair Bay, about midway up the leeward coast, a stretch of scattered settlement that laid claim to less than half of the potentially arable coastal plain. The "far greater part was possessed, tho' little cultivated, by the Charibbs," whose estimated numbers ranged from fifteen hundred to five thousand. This partial occupation of St. Vincent verified British prejudices about French colonial indolence and suggested the work ahead: a bold campaign to intensify colonization where the French had attempted it and extend it across the island.[34]

The head of France's hydrographic office, Jacques-Nicolas Bellin, represented St. Vincent as a place lacking in civil jurisdictions or formal settlements. **[187]** Across a stylized landscape of widely spaced peaks, he imagined three "chemins," or routes, connecting leeward harbors to windward lands, depictions that discounted the famous ruggedness of the island's interior as well as the difficulties in traversing it. Unlike Bellin's map of Grenada, with its clearly bounded parishes, port towns, and coastal plains littered with plantation houses, his map of St. Vincent portrayed a wild place. When the British took possession of the island in 1764, they discovered that Bellin's pictured roads did not actually exist. For the next three years, the Land Commission awaited permission to build a road broad and level enough to carry wagons loaded with hogsheads of sugar from the windward coastal plain to the leeward port towns of Richmond and Kingston. Rough winds churned up punishing surfs on the windward sides of the islands of the Lesser Antilles, focusing the search for secure ports within the best natural harbors and bays on calmer, leeward coasts. The Board of Trade's plan for the islands included an intention to build "great roads" through the mountains that could connect these leeward ports to windward lands earmarked for new sugar plantations. Since sugar that could not be exported was worth next to nothing, securing routes across the new islands was a prerequisite for purchasers of windward-side properties to realize the value of these expensive lands. Although the combination of Caribbean winds, currents, and topography made this task especially important for the settlement of the Ceded Islands, road-building schemes preoccupied the governors of other new British territories as well. In East Florida, Governor James Grant planned an ambitious program aimed at "opening a Communication between the different parts" of his new colony overland. His counterpart in Nova Scotia, Michael Francklin, likewise put a priority on establishing roadways: "Roads of Communication from One part of the Province to another will give life and vigour to the different Settlements, for without them the Farmer is unable to transport his produce, and consequently the country must remain in a languid state." In these remote windward reaches of St. Vincent, far from safe anchorages, the land, although "mostly covered with Wood," was "more level than the Leeward side," which the French had planted so tentatively. Although it was not yet accessible to prospective British planters, Richard Tyrrell had inspected the "good, Rich Soil" of Carib country during his coastal survey and pronounced it "fit to bear Sugar Canes."[35]

The British had long known that a large indigenous population lived on St. Vincent, but the Board of Trade's commissioners had not accounted for their presence in their conceptualization of the Ceded Islands. They counted on revenue from land sales and duties on sugar exports to offset the costs of establishing places like East Florida and Nova Scotia, where it was predicted that colonies capable of contributing to Britain's Atlantic commerce would take longer to develop. Nathaniel Uring's 1726 map marked the leeward coast as a place of "Indian hutts" and the windward coast as a place of "Negro Habitations." Despite these documented observations, the Board classified St. Vincent, with its large indigenous population (and Dominica, with its small one), as a territory open to intensive settlement. Early pamphlets promoting the tropical colonization scheme understood the Black Caribs (said to have descended from the native "Yellow Caribs" and the survivors of a wrecked slave ship) alongside North American Indians. In the same year that Superintendent of Indian Affairs John Stuart assured the southern Indians gathered at Augusta of their corporate land rights under British rule, Chief Commissioner for the Sale of Lands William Young echoed this language of conciliation to suggest that the Caribs would be secure in the "possession of their little dwellings" and would be counted, along with the French who remained in the islands, as "new and useful subjects." Treaties with the Maroons of Jamaica set a precedent for how autonomous, non-English subjects might be counted as "part of the strength" of St. Vincent. Although Britain recognized the rights of Creeks, Cherokees, and Chickasaws to vast tracts of hunting lands after 1763, no

British commentator (at least none who wrote about the islands before the humanitarian controversy sparked by the Carib War of 1772) extended this expectation of extensive territorial rights to the Black Caribs. Instead, they imagined them as proprietors of "little cottages, and spots of provision ground," which would leave the bulk of the windward coast open to plantation development. The land commissioners vowed to treat St. Vincent's indigenous population with "Justice, humanity, and good usage" as they conceived of schemes to displace them from the desirable potential sugar lands they occupied.[36]

But "how best to dispose of them consistently with their own happiness, the peace and safety of the present British subjects settled in the island along with them, and the future population and culture of the colony"? Young proposed a compensated process of removal by which the Caribs could claim a share of uncultivated woodland for every parcel of cleared, windward-side plantation land they gave up. This "idle, ignorant, and savage people" spoke an unintelligible patois and, armed with cutlasses and fowling pieces and encamped in "thatched huts" on some of the island's best sugar ground, they lived in a squalor that fit the primitive character of their culture, society, and economy. Despite the remarkable fertility of their lands, they eked out no more than a mean subsistence with a few hardscrabble plots and "what fish they catch and game they kill." Their negotiated dispossession would acknowledge the Caribs as "adopted subjects" whose need for land could be met by granting them uncultivated forests too steep and rocky for sugarcane fields. A full five years was to be allowed for the Caribs to gather crops, build new huts, and clear new provision grounds at another spot picked out for them on the island. As they left, the planters who owned adjoining tracts would absorb parcels of the former Carib homeland into their plantations. Young's scheme used ideas of exchange to imagine how a society of diverse imperial subjects could coexist in the same space. His plan for reconciling the Caribs to the transformation of St. Vincent resembled nothing so much as planned *marronage,* the process by which runaway slaves lived off the land beyond the bounds of plantations.[37]

Another proposal suggested the islet of Bequia in the Grenadines as a place to which the Black Caribs might be transplanted and where, by leaving St. Vincent, they might become productive subjects. "It would not probably be very difficult to persuade the Indians to leave St. Vincent," reasoned merchant John Campbell in 1763, "for an island at least equal in extent to all that they can possess there, with which they are perfectly well acquainted, and where they might live in safety, after their own manner and undisturbed by strangers." Once resettled on Bequia, they could grow provisions for their "British neighbours," whose hungry slaves would soon labor in the sugar cane fields across the cleared woodlands of their former haunts. Like the poor whites expected to inhabit the small, sandy plots reserved for them on the seacoast, the Caribs had a role to play in St. Vincent's accelerated development from a tropical wasteland into an expanding plantation colony.[38]

These schemes of voluntary removal fit the Board of Trade's presumption that the state should reshape the demography of colonial populations to keep the peace, establish sovereignty over recently acquired territories, and ensure that inhabitants remained enmeshed within the British Atlantic economy as producers and consumers. The proponents of Carib removal conceived it to be an enlightened process of matching peoples to environments in which they might make an optimal contribution to the larger imperial economy and society. "Naked, barbarous, despicable, as they are, they are still human creatures" who would bind themselves to British interests so that they could live and prosper, unmolested, on their new, more suitable, lands. In this respect, they considered the Caribs at first like other British subjects in the post-1763 order: they could claim rights based on their private, material interests, but not the liberty to live wherever or however they chose within the king's dominions to pursue them. On the new map of empire, the British saw American inhabitants as movable subjects.[39]

Just as the Proclamation of 1763 halted new land grants in North America's Indian country, Britain exempted Carib lands at first from surveys and auctions. The Board's commissioners acknowledged that the "wild Caraibes and Negroes . . . who consider themselves to be, and really are, an independent people" were "very jealous of their property and sufficiently numerous to defeat any settlements attempted to be made without their Consent." They affirmed that "no survey can be safely made, or Settlement undertaken, until that consent is obtained, and the affections and good Will of these people conciliated." A 1764 instruction from the Treasury confirmed that "no survey should be made of lands occupied or claimed by the Charaibs" until their status and compensation could be negotiated. From 1764 to 1771, the Land Commission modeled its diplomacy on the practices of the North American Indian superintendents, disbursing some £1,500 worth of presents to Caribs who came to the port city of Kingston to negotiate. Nevertheless, the Caribs refused to acknowledge British sovereignty over their part of the island, denied permission for the building of a road into Carib country, and rejected offers to purchase land or permit its private sale by individuals. As talks stalled, the Land Commission took up its long-delayed mission to "trace out Great Roads of Communication between Town and Town in each Island." It dispatched surveyors and a gang of hired slaves to make a "Communication from Leeward to Windward" across St. Vincent in late 1768.[40]

A Carib band led by Joseph Chatoyer confronted surveyor Levi Porter and his party at the banks of the Yambou River, "which they chose to consider as their boundary." As the Caribs watched the "progress that was making in the admeasurement of the lands" and observed the growing familiarity with which colonists passed through this disputed zone, they made opposition to the windward highway a focus of their resistance. Porter returned in the spring of 1769 with an enslaved road crew and a forty-man military detachment assigned to protect them. Some two hundred armed Caribs "unroofed" a thatched house on the Massarico River that served as a barracks and held the soldiers they had captured hostage until their demands that "no roads, or further surveys, should be made . . . in the wood lands" were met. They "insulted and obstructed" the "Surveyors and slaves" sent to mark out the road's path, and declared that they would not "permit any road to be made through their country." They maintained that they "owned no sovereignty to any prince," and that since they had never ceded lands on the windward side of the island to the king of France, he had no right to "cede them to the King of Great Britain." They put the British barracks hut to the torch, "broke up the roads lately made in several places," and "felled more trees in some parts of the roads, in order to obstruct the passage." Urged to approach the Caribs "with the gentlest hand, and in the mildest manner," Chief Commissioner William Young forbade the surveyors from "proceed[ing] any longer in tracing the road through the Indian country." As the surveyors departed that "fine cream part of the island," it became clear that the Caribs "denied any subjection to his Majesty, and were determined to preserve their independence."[41]

After this incident, Britain's definition of the Black Caribs as a people changed in ways that eroded their stature as legitimate inhabitants and denied their rights as subjects. For over a century, the British justified the annihilation of the Africans who resisted them in the West Indies because they were "beyond reason or persuasion." Commentators now elaborated on stories of the Black Caribs' presumed origins as the "descendents of a cargo of Guinea slaves" who washed up on St. Vincent as castaways and wrested control of the island from its true inhabitants, the so-called red or "yellow Charibbs, an inoffensive quiet people." With their numbers augmented by "runaway negroes from Barbadoes" and the "disbanded Men from the Privateers of all Nations," the Black Caribs were a particularly "Savage and Barbaro[u]s" "Mixture of native Charribeans, Negroes, French," and "Molatoes." This motley society of outcasts colluded openly with the French at Martinique to secure arms that they intended to use to overthrow British rule. Unlike Jamaican maroons, they refused to come to terms. Lieutenant governor of the Ceded Islands Ulysses Fitzmaurice concluded that as long as "these Charibbs are permitted to occupy a large extent of country, without any mixture of white inhabitants, they will retain their fierce intractable nature, continue uncivilized, lawless, disaffected, and of no use." No longer the population of primitive, improvable hunters and farmers destined for a productive role within an expanding plantation society, they now appeared as a people predisposed to wage a permanent war of rebellion against legitimate authority.[42]

Even as British opinion on St. Vincent turned strongly against the Black Caribs, officials made a final bid for conciliation. As with the North American Indians, who felt threatened by settler encroachments, they learned that the uncertainty as to "what will be acknowledged theirs" was behind the Caribs' obstruction of the road survey. The Caribs believed that the king himself had ordered a prohibition on settling their lands—a credible interpretation that might have derived from their own understanding of the Proclamation of 1763—which aggressive settlers sought to evade. The governor of Martinique reflected the tenor of these fears when he reported that the "English have begun a road to cross the part of St. Vincent which belongs to the Caribs, with the aim of making themselves masters of it and then hunting the Caribs." Yet the British did not legitimate the indigenous possession of woodlands in the way they had on the continent by acknowledging hunting grounds as belonging to the territories of the Indian nations. In St. Vincent, where good land was scarce, the British were prepared to "allow them good, proper, and sufficient lands for their support, maintenance, and comfort," but not to recognize their right to possess the lands they claimed but that were "little cultivated by them."[43]

In January 1771, the Land Commission met with Black Carib leaders at Morne Garou, the volcanic mountain in the north that marked the "Borders of the Country which they Claim." The commissioners conveyed the king's "Gracious Intention of preserving and defending them in their Liberties, and of Confirming them in the Possession of certain Lands necessary and Convenient for their Subsistence and also making them presents in Money for such Lands as they should Relinquish." They worked to dispel the idea that the road survey was the first step in a "fixed design" for "enslaving them." As a final resolution of the standoff, the commissioners proposed that the windward coast be partitioned. Britain would purchase four thousand acres of land adjacent to already-platted tracts in the southeast, where no Caribs currently lived, and guarantee that the "Whole of the Land inhabited by them and Woods adjacent" on the central windward side of the island would remain intact as an inviolable homeland. To the commissioners' "astonishment," the Black Caribs rejected this offer, maintaining a "fixed resolution not to consent to our settling any part of the country claimed by them."[44]

After years picturing St. Vincent as a place that could be shared by a mixed settler class of British Protestants and French Catholics, their many slaves, and the inconvenient population of indigenous people the Crown inherited when it gained title to the island, the commissioners declared the Black Caribs beyond the pale of civility. "We conceive it to be impossible that so small an island can long continue divided between a civilized people and savages, who are bound by no ties of law or religion; and who, from their situation among woods, are even exempted from fear of punishment," they wrote. "Every day produces some great inconvenience to the civilized inhabitants, by their slaves being enticed away, and harboured by these savages; and a declaration of war with France threatens almost inevitable ruin to the colony, as experience teaches us, that the efforts of the French would not be wanting in stirring up such an enemy." By rejecting every opportunity to secure their wants within British society, they argued, the Black Caribs had opted out of a system of political economy by which their actions might make sense. Such a challenge to British sovereignty and security meant that the "sale of the lands is no longer the most important object, but the honour of the Crown now becomes concerned for the protection of its subjects, against a race of lawless people, who, when prompted by liquour, or ill-designing persons, may commit any kind of violence without being subject to controul." The best way of "reducing them to obedience," they proposed, "will be to carry a road through their country, under the protection of a sufficient military force; and after allotting them proper lands for their comfortable subsistence, to sell the remainder; which will very amply repay the expences incurred by the arrangement, and contribute to keep them in order by mixing white inhabitants amongst them."[45]

Richard Maitland, the colony's agent in London, calculated the cost of Carib intransigence. As a consequence of retaining "possession of [a] full two thirds of the cultivable and richest land in the island, which they have frequently declared their resolution not to quit one foot of," they had "put a stop to the cultivation and settlement of the colony for the last two years," a point confirmed by statistics drawn from the muster rolls and the customs records. Thus, the burden for the island's defense and government fell entirely on the planters who controlled the remaining third of the arable land on the leeward side, the "most broken parts of the country" for which they had paid "no inconsiderable prices" in the expectation that each new plantation tract would entitle them to a share of the prosperity envisioned for the Ceded Islands as a whole. Plagued by the ever-present threat of attack and unable to bring to bear the full resources of the island to defray its costs, St. Vincent faced inevitable economic decline if it continued to tolerate the Caribs' claims. Their "absolute and immediate removal from the island" was the only remaining measure that could reverse this trajectory. Maitland proposed what was perhaps the first African colonization scheme to achieve this end, suggesting that the conquered Caribs could be transplanted to a ten-thousand-acre tract somewhere on the "coast of Africa."[46]

Britain invaded Carib country to "complete the settlement of that very important and valuable island." By the end of 1772, two army regiments arrived to force them to terms modeled on the 1738 treaty with Jamaica's Maroons, designed to make them "useful subjects instead of dangerous enemies." If they declined, instructed colonial secretary Hillsborough, troops were authorized to forcibly transport them off the island, perhaps to "some unfrequented part of the coast of Africa." In preparation for the attack, army engineer Harry Gordon and chief surveyor John Byres reconnoitered the windward coast in search of safe anchorages and landing places. Hostile islanders fired at them when their vessel came into view of the shore. The campaign against the Caribs met with "many obstacles; those savages are desperate and determined[,] and the issue of our war against them is become doubtfull," wrote John Pownall, former secretary to the Board of Trade and now joint colonial undersecretary. With their intimate knowledge of the rugged country, the Caribs inflicted casualties on troops and destruction on plantations with impunity, refusing to meet in open engagement with British forces and evading pursuit by escaping to the security of the interior. After the army established fortified posts along the windward coast, its troops completed the long-delayed coastal road, from which they laid waste to Carib provision crops. When news reached London that British troops had slaughtered a "parcel of innocent savages in cold blood," parliamentary critics began to question the whole strategy of financing empire by expanding plantation slavery. Public outrage at the war launched the first sustained critique of slavery as well as a new negative assessment of the worth of West Indian islands in the British Atlantic world. Many Britons questioned the moral authority of an empire that would use violence to attack "an independent nation, the original and rightful possessors of the island of St. Vincent's."[47]

A "Peace is made with the Charibs, and they are suffered to continue in the possession of part of their Land," wrote Dominica's attorney general in March 1773. "I hope they may prove better subjects than formerly, tho' I am much afraid they are a Treacherous lott." John Byres's 1776 *Plan of the Island of St. Vincent* captured the negotiated settlement that ended the Carib War. The moment of capitulation was painted by Augustin Brunius, who depicted scenes of British West Indian life with the patronage of Chief Land Commissioner William Young. In his image, versions of which were later engraved and reproduced, the Caribs "lay down their arms" at the feet of General William Dalrymple at the army's encampment at Macaricau, as the treaty required. An officer on the far right of the image holds a map of the island on which the territorial divisions crucial to forging a peace would be drawn. The treaty that ended the war fell short of the goal of removing the Black Caribs from St. Vincent, but it did offer the British much of what they had previously sought. The Caribs acknowledged British sovereignty over the whole of the island, and they ceded four thousand acres of prime windward sugar land to the Crown. The treaty defined their diminished homeland as all of the territory between the Analibou River on the northern leeward coast and the Byera River at the center of the windward coast. In place of a vernacular understanding of the landmarks between the colony and Carib country, a new boundary would be signified by "lines to be drawn by his Majesty's surveyors, from the sources of the rivers to the tops of the mountains," lines that Byres helped measure in the field and then drew across the surface of his 1776 map. As in North America, private land sales were prohibited to preserve the integrity of this reservation and prevent its piecemeal alienation. Like the Jamaican Maroons, the Caribs were now obliged by law to return runaway slaves to their masters. The "Roads, ports, batteries, and communications" that the army constructed along the length of the windward coast during the war would become permanent, granting the British command of the island's full coastline. Despite the Caribs' effective use of irregular military tactics and the leverage this gave them to negotiate for a measure of corporate autonomy on their own lands, their quest to prevent a road through their territory came to an end with the treaty of 1773. At the conclusion of this violent campaign to take possession of St. Vincent, in a treaty solemnized a decade after the formal cession of the island in the Peace of Paris, the Black Caribs formally entered into the bonds of subjecthood.[48]

Metropolitan readers who sympathized with the Caribs' plight did so not only because of their own increasingly broad humanitarian sensibilities but also because they admired, despite the islanders' savage appearance, the primitive virtue driving their uncompromising quest for absolute liberty. The road meant more than access to lands that might someday be added to the cadastral grid that Byres had begun to trace. It was also the means by which an expanding imperial state had the ability to count them and describe where and how they lived. They were slotted into the role that the Board of Trade had envisioned for them: as adjuncts to a plantation colony, who occasionally ventured across the new boundary to sell fruit at Kingstown's market. As his maps of the Ceded Island appeared for sale in London, Byres petitioned for a commission to survey new roads across St. Vincent's mountainous core. After years of occupation, the British still lacked a "thorough knowledge of all the interior part of this Island, which serves at present only as retreats to dangerously numerous bands of runaway Negro[e]s." By March 1779, Governor Valentine Morris reported that a path had been traced across the island to connect its leeward and windward military posts.[49]

Colonizing Tobago

Although "formerly a well settled Colony," Tobago had become, by 1763, "almost an entire waste of Woodland, there being no Settlement or Habitation upon it, except a few Huts belonging to the Car[ibs]" and the temporary encampments of a few French turtle fishermen. A century earlier, as English planters cultivated sugar in Barbados, St. Christopher, and Antigua, Tobago had attracted the attention of the Duchy of Courland, a small trading state on the Baltic Sea. Thomas Spencer and Nicholas Benoist documented the Courlanders' efforts on an undated seventeenth-century manuscript map. [**188**] After 1642, when they settled three hundred colonists at Courland Bay on the island's north shore, Spain blockaded Tobago and fomented a Carib uprising, forcing the settlement's evacuation. The Courlanders returned and were joined by the Dutch—who planted some twelve hundred settlers to grow tobacco, sugar, and cocoa at a colony they named New Walcheren—who were, in turn, countered by the English. Johannes Vingboons's view of the Dutch harbor reveals the limited scale of the outposts that the Dutch and the Courlanders had established by 1665. [**189**] To promote the island to prospective English colonists, John Seller printed a new version of Spencer and Benoist's "The Island of Tobago" in the early 1680s, a copy of which was included in the Blathwayt Atlas, the one-volume compendium of maps, plans, and charts assembled for the Committee on Trade and Foreign Plantations, predecessor to the Board of Trade. **[190]** The English, Dutch, and French contended for possession of the island until 1748, when it joined Dominica, St. Lucia, and St. Vincent as one of the "neutral islands," defined as out of bounds for European settlement by the Treaty of Aix-la-Chapelle. France's minister to Whitehall admitted that Tobago was not "very essential to their commerce" in the Caribbean but had value primarily as a strategic hedge against British settlements and shipping. The French fort on the island, in compliance with the ninth article of the treaty, was abandoned and burnt to the ground.[50]

Islands and coastlines left unsettled and undefended by European states attracted small-scale traders, who arrived to plunder shipwrecks, dive for pearls, and cut valuable stands of mahogany and logwood. On his hand-painted vellum nautical chart from 1722, mariner Robert Egerton pictured dilated coves, treacherous reefs, and an interior whose forbidding mountains were visible from the sea. **[191]** Neutral Tobago became a haven for "Turtlers & Woodcutters" whose temporary habitations, along with those of the native Carib Indians, were marked by tent-shaped icons on a 1749 British manuscript map. **[192]** Even as this image of Tobago's interior demonstrated how itinerant inhabitants made a living from the natural resources that could be extracted from its woods and waters, the information it provided about the navigation around the coast anticipated how Britain might retake the island. It located eight established anchorages and offered a detailed inset of an ideal natural harbor, Man of War Bay. Coastal views pictured how the island's profile appeared from the deck of a ship approaching from the northeast (as it would arrive if borne by the trade winds) and clarified the tricky navigation between the "Rocks of St. Giles" and its southeastern most point, a passage that linked Tobago's northern and southern coasts. As Britain prepared to claim the island in the early 1760s, military surveyors dusted off this early chart to make new copies that could guide British vessels into its harbors. Naval captain John Stott reconnoitered the northern coast again in 1764 to confirm its safe navigation. **[193]** His chart also included a rough appraisal of the interior, locating three encampments inhabited by fewer than 120 Caribs, who spoke "broken French," in addition to "several strag[g]ling parties of Turtlers, both English & French[,] who have no fixt habitation but live where they find the best fishing." A few enterprising timber cutters had also set up temporary camps to "cut wood or large hard Timber for the Barbados Windmills." Beyond these insignificant pockets of occupation, Stott's chart pictured Tobago as devoid of inhabitants and open to settlement.[51]

James Simpson, appointed the first "Chief Surveyor of the Southern Charibbee Islands," came to Tobago in 1764 with three assistants and a force of seventy slaves and found the southernmost of the Ceded Islands to be a place without "Inhabitants or Habitation." His was a provincial office that laid out the lots of new cities and assisted the Land Commission in preparing the islands' plantation tracts for sale. After slaves built a village of "temporary huts" to provide the expedition with shelter during the downpours of the rainy season, Simpson began his mission to impose a grid of prospective property lines across an island that was "totally covered in Wood." Alone in the immensity of this tropical wilderness, he feared that his party would succumb to the malignant vapors that he believed its dense foliage trapped. Although two assistant surveyors died before the land commissioners arrived, Simpson managed to plat four thousand acres for the first land sale in May 1765. Experienced colonists cautioned that the prospect of so much unsettled territory should inspire clearheaded calculations of the human costs of its initial colonization. Because of the "dampness of the Air in Lands cover'd with woods, making the Negroes extremely Sickly," perhaps a third of them would die every year, a "consideration [that] must depre[c]iate the value of the Land in this uncultivated Island, at first."[52]

Bringing Tobago's wild landscape into order was a task, Simpson believed, on par with the "Reduction of the Charibbs in St. Vincent" and the mapping of the "uncultivated parts of Dominica." Because neither the Dutch nor the French had attempted a formal survey, no preexisting land grants existed that British colonists could appropriate. Here, the Board of Trade's formula for the creation of parishes, towns, and roads could be imposed as a de novo creation. Within the outlines of a well-rendered coastline, Simpson began representing the natural and legal geography of the interior. He measured tracts of new plantation land from the coast and the banks of the Great Courland River into an interior that his "Map of part of the Island of Tobago" began to cast in topographic relief. [**194**] He fit smaller, one-hundred-acre parcels, reserved for poor settlers, along sandy strips near the shore at the junctures left behind by the location of larger tracts. He set aside roughly one thousand acres behind Barbados Bay as the future site of the new George Town, for which he produced a model town plan resembling those he had designed for Portsmouth, Dominica, and Kingstown, St. Vincent. [**195**] In unsettled Tobago, an "Exact Map made of whole Island, & the Spott for the Capital fixed upon" would, he believed, be important for promoting land sales, because buyers would clamor for tracts close to the island's center for commerce and government.[53]

As his assistants and hired slaves marked off these lines in the woods with compass and chain, Simpson left a few untenable strips of land—a mile or more long and a few hundred feet across—unnumbered and unmarked. Tobago's lack of previous plantation settlement offered an opportunity to correct another historic failing of the British West Indies. According to the prevailing theory of terrestrial desiccation, the excessive clear-cutting of forest cover triggered the region's periodic and punishing droughts. Believing that the "fertility of these Islands depends upon those refreshing Showers, which are produced by the Preservation of a sufficient quantity of wood," the Board's plan for the Ceded Islands sought to mitigate the "entire Waste of Wood" that had prevailed in the other sugar islands. Left to their own devices, self-aggrandizing planters might convert every arable scrap of land to sugarcane fields and cut down every stand of woods to fuel their boiling houses. In addition to reserving land for roads and towns, the new governments aimed to "reserve all the Lands upon the Summit of the highest Hills to be preserved in Woods; that the Colony may enjoy fruitful Rains[,] which experience has taught us are wanting in those Islands where the Highlands are cleared of wood."[54]

For every intensively developed section of the coastal plain, reserves of hilly forest land, whose topography made them "least adapted for Cultivation," were thus set aside to trap moisture that would produce reliable rains. Britain's tropical forest preservation policy in Tobago, one of the first of its kind anywhere, demonstrated the imperial state's concern for the long-term viability of overseas colonies as well as the problems that resulted from allowing colonists to exploit the natural resources of their islands in pursuit of short-term profits without restraint. Although planters had indisputably created the denuded, depleted, and eroded landscapes of Barbados and the Leeward Islands, the theory that this heedless cultivation was to blame for the punishing droughts these islands had faced in recent years proved to be false. Just as the Proclamation of 1763 demonstrated the Crown's intention to halt the expansion of colonization on the mainland for the good of the empire, the gaps in Tobago's cadastral maps, which remained unmarked by any surveyor's line and bore the caption "Reserved in Wood for Rains," expressed the intention of establishing a centralized *dominium* over American territory that aimed at serving a greater imperial interest. Cartography was the means by which officials were able to intervene in the process by which the king's territory became, or was prevented from becoming, private property. Maps were the working documents by which the Ceded Island land commissioners laid out tracts for sale. For their superiors at the Treasury, these maps documented how much had been paid and how much was owed; and these images, once published, disseminated evidence to a broad public of Britons, some of whom had begun openly questioning the value of colonies, that the Board of Trade's vision for enlightened expansion was being realized.[55]

In 1774, John Fowler published *A Summary Account of the Present Flourishing State of the Respectable Colony of Tobago,* which praised the island's "universally rich and inexhaustibly deep" soil, which produced "amazingly luxuriant" vegetation year round. The bulk of his pamphlet cataloged the sales by which purchasers had claimed 57,408 acres, for which they pledged to pay a total of £154,058, plus nineteen shillings. Fowler listed these buyers by name under the heading of each division, populating so many yet-to-be-settled neighborhoods with a cohort of gentlemen planters whose landholdings could be located by number on the map. Those who wondered about the wisdom of acquiring unsettled Tobago at the Peace of Paris could see how quickly these lands sold, how much revenue they had already generated, and who would inhabit them. The statistics Fowler gathered showed that by 1768, the "speedy settlement of Tobago" seemed "beyond all measure of dispute," as investors drove up the price of land at auction to an average of more than four pounds sterling per acre. When John Byres published his *Plan of the Island of Tobago* in 1776, completing his set of published cadastral maps for the Ceded Islands, this image celebrated a triumph of careful imperial planning. Britain had acquired a virtually uninhabited place that lacked even the rudiments of civil order and had set in motion a process of colonization by which it could be populated and become prosperous.[56]

Hidden in these impressive accounts of lands sold were a number of distressed and dissatisfied adventurers who complained to the Treasury about the high costs of the land they had purchased and the poor quality of the land they received. To pay the £100,000 balance due on their lands, planters in Tobago "will have their Negroes and Estates Sold, after having endured uncommon hardships and fatigues to defend them" following "unfor[e]seen Disappointments, Losses and Misfortunes" that had "Reduced this Infant Colony to the lowest Ebb." During the first years of Tobago's settlement, surveyors produced maps so that prospective "Purchasers might view the premises before the Sales." Yet as the purchasers compared images of their tracts with the realities of Tobago's irregular interior, they charged the mapmakers with failing to venture beyond the coastline. The geometry of property boundaries they drew appeared, after actual inspection, to be nothing more than "imaginary Lines, drawn upon paper only." Had the buyers seen these uncultivable lands with their own eyes, disgruntled purchasers believed, the Crown would never have sold an acre. Robert Walton paid £1,507 for two hundred acres of Tobago land in the Great River Division, marked as lots 30 and 31 on the Byres map. Nestled between two rivers that joined to form the Great River, his prospective plantation was by all appearances a fertile river valley with easy communication to the coast. Actual inspection by surveyor Hector Fraser, who visited the site to "obtain a perfect knowledge of the Lands," proved them to be "impracticable." As installment payments came due, the proprietors entangled one another in "endless Law Suits" to disentangle their confused boundary lines. They looked to London to remedy this failure to make Tobago's landscape live up to the promise of its cartographic representation. The whole scheme to demand high prices and full payment in the short term for Tobago land was founded on the "fallacy of those Sanguine Expectations" that had discounted the "innumerable difficulties and Expenses which attend the Settlement of New Estates in desert Islands."[57]

Because it lay "near Trinidado[,] a Spanish Settlement" from which Britain's long-standing adversary could "easily annoy an infant colony," Tobago remained vulnerable. The British also feared the "Reception and Concealment of fugitive Negroes by a Convention with Spain." Just as the Spanish had worked for years to undermine slave society in the Carolina Lowcountry by offering runaways safe haven in Florida, they might likewise destabilize the Ceded Islands. A few years after Fowler urged colonists to invest in the expensive purchase of Tobago land, the Spanish governor opened Trinidad to foreign settlement on far more favorable terms. Don Emanuel Falques invited "all persons," from disgruntled French leaseholders in Grenada to British planters seeking to settle their slaves on rich new soil, to become naturalized subjects of the king of Spain. Those willing to venture an additional nineteen miles from Tobago across Galleons Passage could forgo the privilege of paying four pounds sterling per acre for one of the small predetermined plantation tracts mapped so deliberately by James Simpson and John Byres. When they reached Trinidad, "Land will be given out as much as one desires up to One-Thousand *Quarrés*," or 3,200 English acres, "in whatever part of the Island that one likes." The "first ones to arrive here will have lands in the flat Country, in the middle of which run rivers, & on the Oceanfront & near the Port," which had been sown to great success in indigo, sugar cane, coffee, cotton, and tobacco.[58]

Admiralty captain Edward Columbine reconnoitered Trinidad in 1762, producing a manuscript map that sketched in its mountain ranges, harbors, and ports, and the smattering of geographic details that he could glean about Britain's menacing new neighbor in the southeast Caribbean. [**196**] A visitor described the island to Governor Macartney in 1777, confirming its potential for plantation agriculture. Among the mountains of the northern range were lands suitable for the "Cultivation of Coffee, & Cacao," and "between these Mountains" was "flat Terrain, & rather vast, where one can grow Sugar & chiefly Cotton & Indigo." Altogether, it contained "about 57,000 Acres of plains or of nearly flat Land, suited for forming the most excellent settlements, & where the carriages will be able to go in all directions." Such natural bounty made the threat posed by the king of Spain's offer of free land and access to trade on the Spanish Main credible. These early surveys and assessments of Trinidad's worth also identified the island as a possible new frontier for British expansion, and a number of French planters had already taken up the Spanish call to settle it. Just as it took command of nascent plantation colonies begun by the French in Grenada, Dominica, and St. Vincent, Britain captured Trinidad in 1797 and added it to its empire at the Treaty of Amiens in 1802.[59]

In 1763, the Board of Trade put forward an ambitious vision of planting improved Caribbean colonies in the Ceded Islands. By taking control of the process by which colonists acquired land, the Board of Trade sought to mitigate the most destabilizing social features of tropical plantation slavery. These islands were to be populated chiefly by African slaves, but with enough poor and middling whites present to prevent uprisings, make them less tempting targets for attack, and provide the social and cultural infrastructure to turn them into recognizably British societies. Although every Ceded Island parish was to reserve eight hundred acres of its least favorable sugar land for the "Accommodation of poor Settlers," the Byres maps show few signs that these small plots survived the islands' development. Only on thinly settled Tobago did the Land Commission set aside substantial tracts for poor settlers.[60]

Britain demonstrated its capacity to populate new sugar islands in emphatic terms. By the early nineteenth century, Dominica's enslaved population of 5,872 had increased by nearly four-fold to 22,083, while St. Vincent's 3,430 had increased by more than seven-fold to 24,920; Tobago, practically uninhabited in 1763, now had more than 17,000 slaves. In the space of two years during the mid-1760s, British slavers brought 10,432 Africans to Grenada. The island's black population more than doubled under British rule in less than a decade, from about twelve thousand to over twenty-six thousand despite high mortality rates. This was not demographic evidence of the society that the Board of Trade had sought to implant but rather the product of a robust, well-capitalized slave trade. As a result of these surging slave imports, the islands soon featured the most skewed ratios of enslaved blacks to free whites anywhere in the Caribbean. When the Board's commissioners conceived of their settlement plan, they noted the Leeward Islands' notoriously imbalanced populations—with ten or more slaves for every free British colonist—as a demographic problem that demanded mitigation. Just prior to the abolition of slavery in the British West Indies, Tobago's ratio of black to white stood at 32, Grenada's at 30, Dominica's at 18, and St. Vincent's at 14. Britain mobilized settlers, labor, and capital as effective instruments of colonization, but the societies that developed around an unrelenting focus on plantation agriculture bent to the social pattern dictated by large-scale sugar monoculture in ways that socially conscious land regulations did not—and perhaps could not—divert.[61]

This program for colonization attempted to prevent the growth of large plantations, but after they had acquired land at auction, planters throughout the islands strained against the rules designed to restrict the size of their estates. In Dominica, planters pressed surveyor Isaac Werden for the chance to purchase "Detached" parcels of land near their plantations in private sales, rather than bidding for them at public auctions. One planter had drafted his own "Dyagrams" to show that the sixty acres he desired was "broaken & bad." He insisted that he only wished to acquire it to "keep of[f] bad Neighbours." He might have known, as Werden did, that some free blacks on the island had sought such small tracts and were willing to pay cash to claim them. Whatever his true reasons, he behaved like other sugar planters in the British West Indies in seeking to expand the scale and profitability of his plantation rather than accepting the modest allotments provided by the Land Commission's policies.[62]

As soon as the Privy Council granted Grenada's governor permission to hold elections to constitute the colony's first assembly in 1767, the Board of Trade's vision of a polyglot empire came into direct conflict with settler expectations of how British polities should be constituted overseas. After they were permitted to vote in 1766, French residents of Grenada next sought the right to sit in the assembly. To allow them to do so, British officials exempted new subjects in the colonies from the terms of the Test Act, which forbade Catholics from holding office, and established set-asides to guarantee French membership in the assembly, in the council, and among parish justices of the peace. Making inclusion manifest in these ways roused British settlers to resistance. Protestant assemblymen passed a law that sought to bar Catholics from serving; Protestant councilmen withdrew, denying that body a quorum, rather than serve alongside Catholics. British colonists refused to pay their taxes and wrote pamphlets and memorials that expressed their outrage that their rights as a self-constituting settler class could be disregarded by fiat. They drew a bright line of customary British liberty across the Board of Trade's general scheme for colonial development. Policies of toleration served a general British "Interest" by helping to promote the "Speedy Settlement and farther Cultivation" of the Ceded Islands by increasing "his Majesty[']s Revenue and the Number of His faithful Subjects." But this demographic vision ran counter to a settler conception of political liberty that was exclusionary at its core, reserving the right to serve on juries and in legislatures—the two defining institutions that guaranteed English political rights—to a privileged class of Protestant, landowning subjects.[63]

This controversy escalated into a constitutional crisis because it exposed the Board's assumption that directives from the metropolis, if calculated to serve Britain's general interests, should override the partial privileges of any particular class of Britons, even if these rode roughshod over expectations of settler autonomy. Rigging the laws of Granada to integrate its French inhabitants into this ruling class, argued these "old subjects," made a mockery of the principle that Britons abroad bore the same rights as Englishmen at home. When bound together with the whole host of land regulations imposed to organize Grenada's development as a British colony, this program of directed colonization disrupted the traditional process by which colonists stood to affirm those rights by transforming territory into productive and civil spaces through their own initiative. When the Board of Trade directed Governor Robert Melvill to "constitute an Assembly for the Island of Grenada" in early 1766, it dictated that the government would lack a critical power other colonial polities enjoyed: the right to grant land. Only by the king's prior assent could it pass any bill that confirmed "any Titles or Claims to Land whatever."[64]

An anonymous proponent of the Board's policies dismissed such customary privileges as irrational when compared with the promise of adding the wealth and productivity of new subjects to the commerce of Great Britain, and wrote off its opponents as members of a self-interested faction. Seen through the lens of this controversy, however, the Board's vision for America appeared as a premeditated "oppressive plan" that attempted to diminish the contributions, stature, and rights of British colonists within the empire. Although the pains taken to redefine the French as British subjects alienated colonists everywhere in British America, such policies of inclusion did little to retain them as productive inhabitants. By 1765, the French on Dominica and St. Vincent had "almost entirely deserted these Islands." By 1778, Grenada's preconquest population of 1,262 French men capable of bearing arms had declined to 324, a remnant community that controlled less than 10 percent of the island's real estate.[65]

Britain's attempt to tax Grenada's planters via a direct order of the king triggered a second constitutional controversy. Before its assembly convened in 1767, the Crown imposed the traditional West Indian 4.5 percent duty on the value of the island's exports. After merchant James Campbell challenged tax collector William Hall's right to extract this money, the Court of King's Bench ruled in *Campbell v. Hall* (1774) that this assumed power was illegitimate. This direct tax, ordered without any legislative approval, contradicted the expectation that "settlers and subjects" who ventured to the southeastern Caribbean to purchase land at the king's invitation would possess the traditional rights of English subjects in protecting their property. Together, this British effort to tax the colony without legislative approval and the ongoing effort to integrate French Catholics into its government gave British colonists "just cause of apprehension" that they were to inhabit not a properly constituted British colony but rather a conquered territory exposed to the "arbitrary and tyrannical" actions of the king.[66]

Britain's vision of controlled American colonization depended on obtaining a comprehensive body of geographic information and both disseminating and managing this data effectively. After a decade spent "laying out & surveying his Majesty[']s Lands in the New Ceded Islands," John Byres possessed an archive of manuscript materials as well as a "thorough knowledge" of the islands to which his published maps attested. Yet these exacting images concealed much that remained uncertain. No one knew the rugged topography of the interior of St. Vincent as well as Byres. The shapes of properties seemed self-evident from the bird's-eye perspective of his maps, but closer to ground level, the "Irregularity of the surface" of the land combined with the "multiplicity of very intricate boundarys" between land already sold and that which was "still remaining to be disposed." To anyone but him, "who ha[d] attended the laying out of these Lands from the beginning," the true location of these lines would be "totally inexplicable." As many landowners realized, to their dismay, the maps' pretensions to comprehensive geographic knowledge fell short in practice. By 1775, some Dominica purchasers were ready to abandon tracts that the Land Commission's manuscript maps had presented in glowing terms. Modern attempts to find evidence of Byres's property lines in Dominica have struggled to locate these boundaries within its present-day landscape.[67]

Since the "first great object of Attention should be the speedy Settlement of these Islands," the auctions did not "wait the general Survey of each Island." The actual work involved in traveling between the islands and marking off every prospective property line across rough and sometimes contested terrain swelled the expenditures of the Land Commission, which sold thousands of acres "subject to future surveys" that were never completed. Despite devoting unprecedented resources to collecting information about geography and property so that the islands could be colonized from a rational master plan, the British built no bureaucratic system that could put this data to use. After the land was sold and most of the payments collected, the Treasury dissolved the Commission for the Sale of Lands in the Ceded Islands in October 1774. Acting chief surveyor Alexander Forbes did not know what to do with the surveyor's office he maintained in Roseau, Dominica, and the sizable collection of maps, plans, and survey notes it contained. He showed a copy of the "general Plan" of Dominica to the new governor, Thomas Shirley, and explained the "Importance of the Plans to the Island." Although he argued that "some steps ought to be taken for their preservation," Forbes was told to keep the maps himself, which he did, although as a former employee of the defunct commission, he was no longer paid. He maintained his office unofficially and made its maps available for public view before depositing them with the colonial secretary's office at the end of 1775 and petitioning the Treasury for back pay. In 1776, the Treasury ordered the former commissioners to send "all the Surveys, Plans, &c" they had created since 1764 to London and dispatch copies of cartographic materials relating to each of the islands to their respective governors. The Board of Trade envisioned building a comprehensive, continuously updated imperial archive of land and property in the Ceded Islands that would give it the capacity to monitor colonization from across the Atlantic; however, Britain's early modern state lacked the resources and governing capacity to manage the data it generated.[68]

From the outset, the Board of Trade's commissioners understood that their general plan could not be imposed in the same way for each of the islands, which "differ[ed] from each other in many Circumstances of Advantage and Disadvantage, arising from Situation, natural Soil, Cultivation and Inhabitants." The geographic surveys were empirical investigations designed to uncover the particular conditions of Dominica, St. Vincent, Grenada, the Grenadines, and Tobago, and the information they returned was to shape distinctive policies tailored to these realities. "What is the Size & Extent of the Island? What Number of English or French Acres is it computed to contain? What is the Nature of the Soil and Climate? and if it differs in these Circumstances from other Islands in the West Indies, in what does that Difference consist?" they asked. "What Rivers are there, and of what extent and convenience to the Planters? What are the principal Harbours . . . what is the depth of Water and nature of the Anchorage in each of them? . . . What Forts and Fortifications . . . if any further Fortification is necessary, what shou'd it be & how situated?" Despite this openness to local variation, the British plan itself was wedded to a general idea of the islands that proved difficult to modify—a point made with violence during the Carib War of 1772. It began from the principle that each of these islands was fundamentally the same kind of place, and the common scale and graphic style of representation on maps by Pinel, Byres, and Paterson made this presumption visible.[69]

At a general, Atlantic level, sugar islands fit into a colonial category defined by the high value of their exports and their capacity to consume large numbers of African slaves. In this respect, Britain's colonization of the Ceded Islands was remarkably effective. From 1765 to 1773, the Crown sold 165,502.5 acres of land on Dominica, St. Vincent, and Tobago, netting a staggering total of £687,582.18 in installment payments by 1778. By the early 1770s, the Ceded Islands produced more than 13 million tons of sugar, more than twice as much as venerable Barbados. So well entrenched was plantation society in the Ceded Islands after a decade that it survived two crises that might have otherwise led to their abandonment. Some investors found themselves too deep in debt to withstand the dual shocks of the transatlantic credit crisis of 1772–1773 and the long disruption of the American War of Independence of 1775–1783. Unable to borrow against the value of their new lands as credit vanished, many purchasers defaulted on their payments. Former land commissioner William Hewitt returned to the southeastern Caribbean in 1777, empowered to investigate and resolve outstanding claims, and found two thousand granted acres abandoned in Dominica. The war unleashed privateering attacks that virtually cut off the islands from the Atlantic economy, and the French conquered Dominica (1778), Grenada (1779), St. Vincent (1779), and Tobago (1781), returning all but Tobago at the Peace of 1783. Despite the spate of bankruptcies, the planters who survived to rebuild their fortunes did so quickly and to great effect, improving the commercial system that made sugar planting so profitable. Just before Parliament abolished slavery in Britain's overseas dominions in 1833, the Ceded Islands (including Trinidad, added to the empire in 1802) accounted for just under one-third of the total sugars produced in the British West Indies. After a tumultuous half century, the Board of Trade's plan to expand the British West Indies by developing new sugar islands had become a reality.[70]

The Main tranchin
Cotton Indigo &
Soft muddy bottom
Cape River
White Water
Mangrove Swamp
Mangrove
Gorge
Entrance in white water

{ SIX }

Defining East Florida

The British pinned their greatest hopes for recolonizing America on Florida, a place that had long kindled European desires for tropical wealth. Of all the territories added to the empire in 1763, it was the one they knew the least about. As troops occupied posts in Mobile and Pensacola along Florida's Gulf Coast, Governor James Grant arrived in St. Augustine with a mission to create a new plantation society out of the terra incognita of the peninsula. Surveyors in the pay of the army, Admiralty, and Board of Trade spent years charting the contours of Florida's coastline in an attempt to describe it as a place for settlement. Their expeditions converged on the Keys, a remote archipelago that they struggled to define and defend as part of the mainland. As these surveyors generated the first rigorous cartographic record of the region, however, their maps revealed geographic realities that conflicted with the Board's designs. By the time the War of Independence ended these British experiments in state-sponsored social creation, a new version of a plantation society, modeled on the Carolina Lowcountry and accelerated by the Board's streamlined land system, had taken root in East Florida.

Europeans did not know Florida in a way that could be easily represented on a map. The Board of Trade's vision of a coastal, commercial empire had to be refined when settlers began occupying distinct lands. The Board planned Florida's occupation at the scale of the continent and, as settlers and soldiers arrived to take possession, charged surveyors with the task of reconciling general expectations with particular islands, coastlines, inlets, rivers, and terrains. The maps and charts they produced in the 1760s and 1770s were instruments of empire that suggested how Florida could—and could not—become a southern continental capstone for British North America. Florida in the seventeenth century was an "external, strategic, maritime periphery" with which Spain defended the routes that linked its colonies to the Atlantic world and from which it strove to disrupt the southward expansion of British colonial societies on the mainland. By colonizing Florida, Britain sought to reverse this geopolitical relationship, extending its command of land and water within and around the Caribbean.[1]

The British idea of Florida changed after 1763. Beginning with an open, undetermined view of the province, colonizers worked to pin down its terrestrial shape, specify the contours of its coastlines, clarify the relation of its islands to the mainland, and delineate the navigation of its waterways. The Board promoted East Florida as one of several officially sanctioned new frontiers for settlement in the 1760s. It presumed that colonists would farm and fish in the maritime northeast and purchase African slaves to grow sugar in the Caribbean, but precisely what those who came to Florida would do with the land they were granted remained open to innovation.

Florida seemed to defy categorization. Was it a collection of islands at the northern edge of the Caribbean, or an extension of the subtropical mainland into which the signature staple commodities of the Carolina and Georgia Lowcountry—rice and indigo—might be transplanted? The very map on which the Board's commissioners reorganized the empire, Emanuel Bowen's *An Accurate Map of North America,* pictured Florida as an indeterminate place, part islands and part mainland. Britain's lack of knowledge dictated a strategy of simultaneous discovery and development that began from the uncertain edges of an immense coastline and proceeded into a largely unknown interior. The search for good harbors identified places from which maritime power could be projected into the Atlantic Ocean and Caribbean Sea; the search for inlets opened sites on which settlers might take up lands with good soil and access to river navigation; and the establishment of settled borders with the Creeks defined East Florida as a bounded domain.

Imagining Florida

Spain claimed the southeast by virtue of its explorers' wide-ranging sixteenth-century *entradas*. Europeans at first considered all of North

America south of its cold-water fishing grounds as the "land called Florida," named for the superabundant growth of its lush natural landscapes. Later images of North America, such as Abraham Ortelius's "La Florida," pictured an expansive Spanish region that spread well beyond the peninsula and the panhandle. [**197**] From the first settlement of St. Augustine in 1565 through the end of the seventeenth century, Florida's territorial bounds extended at least two hundred miles from this military outpost, encompassing Indian missions from Apalachee Bay along the Gulf of Mexico to the Savannah River on the Atlantic coast.[2]

During Queen Anne's War (1702–1713), forces from Carolina laid siege to St. Augustine and joined Creek and Yamasee allies in enslaving Christian Indians and laying waste to mission towns. Spanish dominion fragmented under the intensity of these wartime raiding parties, whose reconnaissance reports led mapmakers to picture the peninsula in a striking new way—as a landmass crumbling into the Caribbean Sea. Edward Crisp's *A compleat description of the province of Carolina* (1711) was the first published map to redefine southern Florida as an aggregation of islands, an image sustained by Guillaume Delisle's *Carte de la Louisiane* (1718). [**198, 199**] General James Oglethorpe, advancing from the recently planted Georgia colony, led a second failed siege of St. Augustine in 1740. In 1742, Antonio de Arredondo, engineer of the Havana garrison, drafted a record of this destructive military adventurism. [**200**] His "Descripcion geographica" marked the stages by which Britain had reduced Florida from one of North America's major regions to a marginal space at the continent's edge, a circumscribed enclave of bays and islands secured by a handful of forts. Its image of the peninsula as a broken remnant had, by mid-century, become a cartographic commonplace. Such maps made sense as a reflection of what St. Augustine had become to Spain: a besieged outpost at the edge of a maritime zone centered in Havana. This image of a broken peninsula appealed to British expansionists because it undermined Spain's claims to North America. Seeing St. Augustine clinging to the edge of the mainland, surrounded by marginal coastal islands, affirmed their historical understanding of Spain's decline. Such maps showed that Spain had withdrawn from all the "space conquered or ravaged by Soto," which now reverted to the status of "a large desert," open for occupation by a more capable imperial power.[3]

Florida's disintegration on eighteenth-century maps revealed how, in the absence of systematic surveys, the arrival of new information, however incomplete, could lead mapmakers to revise their conceptions of basic geographic forms. Admiralty surveyor George Gauld understood that this conception of the peninsula as a cluster of islands was an error born of empirical observation and reasoned synthesis. The "Masters of little Vessels who traded from St. Augustin[e] to the Havana," as well as the fishermen who sailed from Cuba to the Keys, had "made sketches" of the landscape that reflected their "own practical knowledge of the coast." From the deck of a ship, the low-lying terrain separated by inlets along the southern peninsula appeared as so many islands, a geographic presumption confirmed by the extension of the Keys from the Cape of Florida into the Gulf of Mexico. Respected cartographers such as Jacques-Nicolas Bellin used these sketches—"with the assistance of a little verbal information, and a great deal of imagination or conjecture," thought Gauld—to see Florida as a detached archipelago rather than part of the mainland. **[201]** Spanish surveyors had in fact been hard at work charting Florida's coasts, but the images they produced reflected the same maritime orientation of Spain's eighteenth-century engagement with the province. Spanish colonists came to Florida to fish and harvest dyewoods, as they would in any other Atlantic commons in the maritime spaces of the greater Caribbean; they inhabited its only significant European settlement largely as soldiers manning a military outpost to which surrounding territories appeared as an Indian-occupied borderland into which they ventured, from time to time, to evict interlopers by force. Maritime experiences, seaborne communication with Havana, and a discrete pattern of defensive occupation affirmed the Spanish idea that Florida was a collection of islands at the edge of a continent.[4]

Once Britain took possession, the idea of a peninsula "intersected with arms of the sea," which divided the "land into a great number of islands," ran counter to the Board of Trade's vision for reformed colonization. The Board had made the "extension of Settlement" into Florida—a place targeted along with Canada and the Ceded Islands for "planting, perpetual Settlement & Cultivation"—an essential element of its overall plan to reshape American empire. A search of the Plantation Office's library of maps and manuscripts, however, yielded no "certain Information" about the "Coast, Harbours and Rivers of Florida" or the "Variety of Produce" that "may be raised in that extended Country." The army's first manuscript map of the province, compiled by engineer Thomas Wright "from the latest and most Accurate Surveys," retained these supposed southern islands. **[202]** Thomas Jefferys likewise presented British readers of the *London Magazine* with an image, compiled from "original *Spanish* and *French* charts," of a Florida in pieces. **[203]** John Gibson's picture of a similarly fractured peninsula appeared in the pages of the *Gentleman's Magazine*. **[204]** Although the very map on which the Board's commissioners imagined North America followed this convention of a fragmented Florida, the geography they envisioned for it was one of a contiguous mainland, crossed by rivers that penetrated into the interior. Without any certain "idea of the nature and geography of this country," the Board divided the "great Tract of Sea coast from St. Augustine . . . to the mouth of the Missis[s]ippi" into "two distinct gov[ernment]s": peninsular East Florida and the Gulf Coast colony of West Florida. Just as it had reduced a sprawling New France to a compact Quebec, the Board's new boundaries dismantled the once-vast claims of Spanish Florida and shaped two prospective settler colonies into modest, bounded domains. This definition of East Florida allowed for the territorial expansion of Georgia; confined since its founding in 1733 to "narrow limits" by its proximity to hostile Spanish forces, it could now take on the proportions of a full-fledged mainland colony and become a "flourishing Province."[5]

In Florida, more than any other territory acquired in 1763, the Board made cartography an imperial priority. Although the Spanish had claimed Florida for "two centuries, or more," they had undertaken "no actual survey of the country." The "Country is by no Means known," realized East Florida governor James Grant when he arrived in St. Augustine to assume his post in 1764. He thought his collection of "exceedingly erroneous" Spanish charts and maps, some seized from Havana during the British occupation in 1762–1763, to be "not worth a farthing." They revealed that the "Spaniards," confined to St. Augustine for much of the century, "knew very little of the Interior parts of the Country." As a major commanding troops in the 77th Regiment of Foot (Montgomerie's Highlanders) during the Seven Years' War, Grant was captured at Fort Duquesne, burned Indian towns during the Cherokee War, helped take

Martinique from the French, and battled the Spanish during the siege of Havana. He shared the general view of British military and civil officers that imperial indolence was a Spanish character trait, one verified by sloppy, error-riddled maps. Given the "mangled condition" of East Florida on these maps, as surveyor Bernard Romans put it, Britain would need to reconstitute it from scratch, amassing new geographic knowledge through firsthand surveys and drafting new maps, plans, and charts.[6]

William Stork's 1766 *Account of East-Florida* advanced the broadly shared ideal that the province was an ideal hybrid of continental mainland and Caribbean islands. The "great Luxuriency of all West India Weeds" found throughout the countryside revealed a fecund natural world that vitalized everything that sprang from its soil. The Board of Trade anticipated that Florida's "Commercial Advantages" would include the production of "Indigo, Silk[,] Cotton and Many of the Commodities now, found in the West Indies only." Its mild climate, "in which the productions of the northern and southern latitudes seem to flourish together," promised perfect conditions for subsistence as well as market agriculture. To illustrate his text, Stork produced *A new Map of East Florida*, giving this vision of tropical agriculture in North America a geographic form. [**205**] Stork's map drew together the scattered islands "exhibited in old maps" until the spaces between them shrank to the size of broad rivers. He dedicated this new image of East Florida to the First Lord of Trade, the Earl of Hillsborough, and presented a copy of the map to its commissioners personally in 1766. Stork's text and map pictured Florida as the Board wished to see it: a fertile, uninhabited, contiguous body of open land. Thomas Jefferys's version of this map, published in the third edition of Stork's pamphlet in 1769, fused its territory together even more securely. [**206**] When mapmaker Emanuel Bowen produced a new edition of his *Accurate Map of North America,* a revised state of the map on which the Board first visualized its imperial plan, he reengraved his copperplate to affirm Florida's continental integrity and reflect the new British consensus about its form.[7] [**207**]

In 1763, Sir William Johnson diverted his attention from managing northern Indian affairs to take stock of the problem of geographic ignorance along the new southern frontier, frame it in the language of imperial political economy, and send his thoughts to Secretary of State Henry

Conway. The British lacked "accurate knowledge both of the Country and the people" and knew nothing about the "course & capacity of the Rivers," information vital for diplomacy, defense, and commerce. They were "totally ignorant of the Coast," largely "unexplored" along some fifteen hundred miles from New Orleans to St. Augustine. The passage around the Cape of Florida that linked the Atlantic Ocean to the Gulf of Mexico was "so much frequented & yet so dangerous by being so little known." Beyond three identified port sites—St. Augustine, Pensacola, and Mobile—many other, better "places may be found whose Names we have never heard of." There was peril as well as promise in this lack of knowledge about Florida's ports, because those sites capable of concentrating inland trade and commanding water routes, if left unclaimed, might fall into the hands of the Spanish as they sought to erode Britain's sovereignty in the region. Such a threat demanded that "every Port which is capable of carrying on the slightest Commerce should be settled." The "great object in every Colony," Johnson concluded, "is the encouraging [of] Population wither by fo[r]eign Settlers or . . . retaining the old or Civilizing the wild or rearing the Young, or by the introduction of Slaves." Florida was "so little known," he noted, that it "will require some judicious person to explore it before the Government can determine positively what projects" to begin there to put such industrious hands to work. Before a single step could be taken in the direction of Florida's improvement, Britain had to employ "Engineers for the shore" and "Draftsmen for the Sea" and complete an "accurate survey of the Country and the Coast."[8]

With interest piqued by the prospect of a British Florida, prolific London mapmaker Thomas Jefferys published images of Spanish St. Augustine as it had appeared before the Peace of Paris. [**208**] He presented the town as a grid of twelve blocks, flanked by two Indian settlements outside its fortified perimeter but otherwise isolated in a landscape of wetlands and forests surrounding the harbor. Another cartographer, John de Solis, copied a Spanish surveyor's manuscript to preserve the image of a thriving town, with scores of houses scattered across broad lots filled with orchards and gardens. [**209**] Emblems of church and army, whose troops drilled across its broad Plaza de Armes and manned the Castillo de San Marcos, dominated this townscape. The fort, begun in 1672, had given refuge to some fifteen hundred residents when South

Carolina forces burned the town in 1702. When they departed, the Spanish rebuilt St. Augustine, entrenching it within a defensive earthworks, and expanded San Marcos in the face of persistent British and Creek threats. Army engineer Philip Pittman's plan of San Marcos revealed the square shape of a "Fort Quarrè," with four corner bastions fit to mount more than seventy cannons. **[210]** His sectional drawings brought this flat geometry into three dimensions, seating it above the low-lying swamps, which had made the fort so difficult for English forces to capture at the sieges of 1702 and 1740. Built of "congealed shells dug out in the shape of stones" from a quarry on nearby Anastasia Island, the fort's tabby walls shone blindingly white in the sun.[9]

As the British arrived to take possession of an all-but-vacant town, they began destroying signs of material culture that marked it as Spanish. The army planned imposing new barracks for the men and officers of the regiment. Senior officials claimed the "best houses" and drove chimneys through the "tops of [th]e house roofs." They glazed open windows so that the "sun beg[an] to shine thro glass" and smashed through solid north-facing walls, installing doors and more windows to encourage the circulation of air. Governor James Grant hired a mason and began planning the construction of "Council & Assembly rooms" and a "Hall for the Courts," dreaming of transforming St. Augustine's plaza into a central square flanked by these core institutions of English civil government. He spent liberally to refurbish the governor's residence, making a "very habitable English house, out of a very bad [S]panish one, which was as dark as a Dungeon, without Windows or Chimney." To English eyes, nothing stood out more as a sign of the town's strangeness than the absence of proper hearths, without which houses could not provide temperate bodies with a modicum of domestic comfort during Florida's mild winters. English maps of America had long used the icon of a house with a chimney to indicate sites of colonial occupation. In St. Augustine, the first wave of troops, slaves, and colonists retrofitted actual buildings to make the town their own. For naturalist John Bartram, who toured East Florida in 1765–1766 to vouch for its promise, this rough alteration of St. Augustine sacrificed beauty for expediency, turning "well cultivated gardens" into weedy cow pastures. The town's new occupants grubbed up orange groves for firewood and cut down trees bearing figs and pomegranates, fruits that did not agree with conventional English tastes. Such destructive occupation illustrated a principle of British colonization in action. It was not enough to inhabit St. Augustine; it had to be transformed, improved, and fitted to its designated role as the provisional capital of a rapidly expanding plantation society.[10]

The British believed that it was because of the "insuperable laziness of the Spaniards" that they had, after two centuries of possession, occupied so little of the province beyond St. Augustine. Bound by the Peace of Paris to offer Spanish residents the opportunity to become new subjects of the British king, the Board of Trade's commissioners had hoped that merchants, pilots, and farmers would remain to impart their knowledge of the place. Major Francis Ogilvie was pleased to report that almost all Spanish residents had left by the beginning of 1764, as he considered them the "least Industrious of any People [he] ever saw." Engaged in "Constant War with the Indians" and "helm'd in within the Lines of St. Augustine," they had imported the very provisions they needed to survive. Beneath the reflexive nationalism of this charge stood an assertion of Florida's untapped natural bounty. Only indolence could explain why the Spanish had failed to cultivate and profit from such an abundance of semitropical land, so well positioned for commerce.[11]

Before the virtuous transformation of this wilderness could begin, however, the rapaciousness of British speculators threatened to derail it. Under the terms of the Treaty of Paris, Spanish landowners had the right to sell their property to incoming British colonists. Sensing an opportunity to take possession of virtually all of St. Augustine's immediate hinterland through this clause, merchants John Gordon and Jesse Fish negotiated with the last remaining Spanish agent in St. Augustine to purchase hundreds of thousands—perhaps millions—of acres purportedly sold by senior army officers before the handover. James Grant railed against the "exorbitant Quantity of Land" claimed by "imaginary Titles of a few Spanish Military People (shut up within their lines) to an extensive Country." The map that Gordon produced to locate these tracts subdivided the countryside around the capital into a grid of rectangular fiefdoms. **[211]** These encompassed tens of thousands of acres each and monopolized the "best Land" across a thousand leagues of territory between St. Augustine and the St. Johns River.[12]

The map was pure illusion, Grant charged. No records showed that these lands were ever "Surveyed or Granted" or "divided amongst the Inhabitants." In his sketches of this landscape, East Florida's deputy provincial surveyor, Samuel Roworth, recorded no material signs of Gordon's boundaries or the Spaniards' occupation of their tracts in this "uncultivated, and uninhabited Part of the Country." [**212**] Roworth did draw small clearings and "Spanish old fields Over Grown with Scrubs" at "Fort Mossy" (Gracia Real de Santa Teresa de Mose). [**213**] The free black soldiers of Fort Mose had escaped slavery in South Carolina and Georgia for freedom in Florida, and their encampment to the north of St. Augustine had served as its first line of defense against British attack. The community's forty-eight remaining residents sailed for relocation in Cuba on August 7, 1763, where their abject poverty suggests that they were never paid for their land. Gordon's map show the fields of Fort Mose falling within the holdings of the heirs of Captain Augstin Peres de Vellareux. When the Privy Council disallowed the speculators' claims, they quashed Gordon and Fish's attempt to profit from the unregulated sale of so much of East Florida's land. As the British arrived to take possession of a new plantation colony in 1763, they transformed the built environment of the capital, undermined the credibility of preexisting property claims, and rewrote Spain's long history as Florida's colonial ruler by characterizing its record of occupation as defensive, limited, and unsustainable.[13]

Grant ordered army engineer James Moncrief to remap St. Augustine and its environs after undertaking the first independent British surveys of East Florida. In March 1765, Grant sent the Board of Trade a portfolio of hand-colored maps. This dispatch included a new image of the reoccupied town that showed Spanish "Plantations abandoned" beyond St. Augustine's line of entrenchment and English names affixed to town lots. [**214, 215**] As Moncrief's maps widened in scale, they located isolated military outposts that marked the limits of Spanish-occupied Florida (at Matanzas, fifteen miles to the south, and at Picolata, twenty miles to the west) and pictured them as inconsequential footholds within an encompassing, uncultivated wilderness colored in green. [**216, 217**] It was clear that colonists would take up land first along the St. Johns River, the waterway that encircled St. Augustine. For East Florida to demonstrate its viability as a colony that could be settled as extensively as the Board of

Trade envisioned, Grant believed he would need to bring planters some seventy miles to the south of the capital to cultivate the place where departing Spaniards hinted there was a "fine Harbour called the Mosquettos." Over the course of his ambitious governorship, nothing was more important to James Grant than establishing a thriving plantation neighborhood behind the first major inlet south of St. Augustine. After "two unsuccessful attempts to send an Engineer to the Mosquittoes," Grant hired two Spaniards—"who say they know that part of the Country perfectly well, and that the Soil there is Rich & Good"—to accompany Moncrief on a voyage that arrived at this elusive spot in January 1765.[14]

The engineer sounded the bar at low tide and found ample clearance for small vessels. His "Plan of the Harbour of Musquitos" followed this inlet as it opened onto two rivers, the Hillsborough (named for the First Lord of Trade) and the Halifax (named for the secretary of state for the Southern Department). **[218]** Along one shore, Moncrief traced the shape of an eight-pointed star, outlining the defensive walls of a prospective garrison post. Within it, he inked in red an L-shaped footprint of a barracks that could shelter "Three Officers & Fifty Men." On the map, these imagined soldiers stood guard in their invisible fort over what was, at the moment, an uninhabited space used by Creek Indians as a seasonal hunting ground. To Moncrief, it had the lush "Appearance of a West India Country." His "favourable account of the Country," with its seemingly tropical climate and "deep Soil," convinced Grant that it "must produce any thing" planted in it.[15]

To fail to settle the Mosquitos would be to confine East Florida to the environs of St. Augustine, the space in which the Spanish had been so long sequestered. The initial expeditions to the Mosquitos produced images of a verdant wilderness, open to planting, into which the governor was prepared to locate large grants for metropolitan investors. Moncrief provided a final chart, derived from an original "Spanish Report," that located these sites within a seventy-five mile stretch of coastline, extending to the former frontiers of Spanish Florida. **[219]** Although the region's coastal rivers defied expectation by running in the same direction as the shore rather than striking out into the interior, they allowed the Board of Trade's commissioners to envision the scope of new settlement beyond St. Augustine as they evaluated petitions by wealthy British applicants to claim tens of thousands of acres of land. Moncrief's "Sketch" and his chart of Mosquito Inlet served as a kind of battle plan in Governor James Grant's quest to extend the range of East Florida's domain.[16]

The Search for Atlantic Inlets

Nothing in Lord Adam Gordon's arduous, nineteen-day voyage from Pensacola terrified him as much as the passage over St. Augustine's notorious sandbar. Although "Augustine has all the appearance of a place that will thrive," he concluded, the "Bar is an insurmountable obstacle to its being a place of Exportation." The Board of Trade initiated exploratory surveys to identify inlets that opened onto potential rivers of settlement. Lacking information about "places fit for Harbours," few ship captains would dare to venture to East Florida for lack of a place to safely anchor behind its shores. Without reliable knowledge about the "Internal Country and Rivers," it was impossible to defend against attack or know where to grant land to aspiring planters. Since "Nature has rendered it almost fatal for any" large ship "to come into the Port," plans for the development of the colony hinged on the idea that better harbors might be found south of St. Augustine. James Grant believed that East Florida was full of hitherto undiscovered rivers that flowed into the Atlantic, each of which might anchor a new plantation district, and that these settled riversides would join to form a populous mainland province that would extend beyond the small river system that Moncrief charted between the St. Marys and Mosquito Inlets. Grant dreamed of opening "a Water Communication all over the Province"; projected easy travel between an expansive plantation countryside and new coastal ports, each nestled in a safe harbor; and anticipated finding safe passage around the peninsula to connect East Florida's Atlantic and Gulf coasts.[17]

The Board of Trade directed the southern branch of its General Survey of North America to uncover this hidden plantation landscape. Before he arrived in St. Augustine in 1764 to lead it, William De Brahm had mastered the techniques of surveying as a captain engineer to Charles VII's imperial army as it fought across Europe in the 1740s during the War of the Austrian Succession. De Brahm resigned his commission at the war's end and arrived in Georgia in 1751 along with a party of Salzburgers, one of several groups of persecuted German Protestants the Board of Trade had encouraged to populate strategic townships in South Carolina and Georgia since the 1730s. De Brahm's distinctive expertise gained him appointments as South Carolina's chief surveyor and as one of Georgia's co-surveyors. During his fourteen-year tenure as a provincial official, De Brahm designed elaborate new defenses for Charlestown as well as Fort Loudoun in Cherokee country, but the bulk of his work was organizing colonization as an everyday legal process. Those claiming land by headright or purchase came to his office, warrants in hand, to request a formal survey of the specific tracts they desired. De Brahm and his assistants traveled to spaces at the edges and interstices of the cadastral landscape with their surveyor's compasses, measuring chains, and sketch books to mark off their claims on the land before the colonial council formally granted it. In a masterpiece of mid-century cartography, De Brahm represented the knowledge he had acquired about the qualities of the region's soils, its potential for commercial agriculture, and the shape of the collective matrix of property lines laid out under his authority.[18]

William De Brahm dedicated *A map of South Carolina and a part of Georgia* (1757) to Halifax and his fellow Lords of Trade. [**220**] The map pictured these two colonies in the throes of ongoing territorial expansion. It charted the uneven spread of parish and township jurisdictions that organized the region's coastal riversides into settled places. In one small portion of this territory, De Brahm traced the plat boundaries of individual properties in the rapidly settling plantation district between South Carolina's Combahee River and Georgia's Medway River. As no map had shown before, it took stock of the progress of settlement and pictured it comprehensively—at a scale of roughly five miles to the inch. When displayed as a four-foot-square wall map, this image was expansive enough to show the contour of the southeastern coast of the continent but also detailed enough to reveal the land's articulation into individually owned tracts. Its graphic cartouche shows slaves making indigo dye, a recent addition to the Carolina Lowcountry's list of transatlantic exports, which included deerskins, naval stores, and rice, the signature staple commodity that drove its economic growth. De Brahm's map described how colonists took possession of American land, how their slaves made it productive, and how plantation enterprise secured Britain's *dominium* in America. The Board's scheme for American empire in 1763 depended on generating precisely this sort of knowledge of place across the varied landscapes of the new territories. Like Northern District surveyor general Samuel Holland's "Map . . . of the inhabited parts of Canada," *A map of South Carolina and a part of Georgia* was a credential that displayed deep knowledge of the land and its possibilities. De Brahm's masterwork helped him gain an appointment as one of Britain's most important American agents.[19]

Board of Trade secretary John Pownall laid out the goals of the General Survey for the general surveyor in a 1764 letter of instruction. Although the Southern District included all of eastern North America south of the Potomac River, De Brahm was to focus his efforts on an "accurate Survey of that part of the province of East Florida, which lyes to the south of St. Augustine," and make assessments of the "Lands lying near the Sea Coast of the great promontory." Understanding these lands was the "most pressing Expediency, in order to accelerate the different Establishments[,] which have been proposed to be made in that part of the Country." As the Board entertained proposals by entrepreneurial colonists for "making Settlements upon these Coasts," such maps were "to be in great Measure the Guide" by which petitions for new twenty-thousand-acre township grants would be judged. Beyond documenting "Latitudes and Longitudes" and "Depths of the Water," the General Survey aimed at analyzing and interpreting every observable fact about East Florida that might be useful to the task of colonizing it. De Brahm should "convey a clear & precise Knowledge of the actual State of the Country." He should gauge its extent, calculate its "Quantity of Acres," identify its "principal Rivers and Harbours," and assess the "Nature of it[s] Soil and produce, and in what Points" it was "capable of Improvement." The product of the General Survey was to be a collection of manuscript maps annotated with descriptive comments as well as a written report illustrated with maps. Together, these texts and images were to describe the land's capacity for settlement both in general terms and in reference to specific, locatable spaces, providing the Board of Trade with the geographic information it needed to form a "true Judgement of the State of this important Part of His Majesty's Dominions."[20]

For nearly six years, from 1765 to 1771, De Brahm surveyed East Florida's Atlantic coast "to its southernmost Extent, by both Land and Sea." Less than a week after his arrival, he fixed St. Augustine's latitude by astronomical observation and sounded the shallow bar that obstructed access to its harbor. He chartered a coastal schooner, the *Augustine Packet*; outfitted it with rowboats and an array of surveying tools; and hired a crew of sixteen, including two assistant "surveyors well versed in trigonometry." De Brahm and his party set out from St. Augustine on February 11, 1765, to discover East Florida's coastal river inlets. Following these initial reconnaissance voyages, he drafted fifteen manuscript charts that sketched the colony's Atlantic coast, revealed its accessible rivers, and rendered judgments of the quality of the lands visible along their banks. He did not linger long enough in any one place during this four-month voyage to perform a rigorous plane-table survey, which would have aligned the shapes on these charts precisely to the contours of the coastline. His purpose was to lay down basic geographic forms in order to identify viable spaces for settlement. Six of the eight surviving charts form an interlocking image of the peninsula south of Cape Canaveral, a place most contemporaries understood as a loose aggregation of islands. **[221]** Two additional "special charts" examined critical places at high resolution: "Cape Florida" (the southeasternmost point that marked the beginning of a treacherous navigation into the Gulf of Mexico) and "Muskito Inlet" (designated as the colony's most promising prospective plantation district).[21] **[222, 223]**

As the governor waited impatiently in St. Augustine for De Brahm's report of his "Discovery of Harbours," what the surveyor observed from the deck of his schooner confirmed that the peninsula was indeed part of the mainland. In addition to the three previously charted by Moncrief, De Brahm identified four more inlets, each up for consideration as a point of entry into new districts of settlement. These charts provide little more than a glimpse of the coast from these navigable waterways. Wherever the surveyor general found inlets, he entered the rivers they served, all of which continued to follow the peculiar pattern of running "parallel to the sea," just inshore, rather than flowing toward the coast from a mountainous interior, as they did elsewhere in North America. Beyond the point where he could discern these lands through a telescope from a rowboat, they remained unmapped. De Brahm filled the blank terrestrial spaces on these charts with his own handwritten sailing instructions, observations on natural history, and appraisals of the land's value for commercial agriculture. A trompe l'oeil ornamentation gave the impression that these notes were written on a separate sheet of paper, whose torn edges had caused it to curl up in scrolls, revealing a portion of the chart underneath. This embellishment demonstrated De Brahm's proficiency with the drafting pen, but it also emphasized how small a portion of Florida's vast territory he had managed to represent on his first expedition, and how much remained to be examined. De Brahm's sectional charts invited the viewer to peel back more layers of text in the future, at which time these concealed spaces would be filled with new images derived from as-yet-to-be-completed surveys. They suggested a time when these provisional observations might be torn away completely to reveal a fully rendered map of East Florida.[22]

The four inlets he charted, sounded, and named prefigured a landscape of uncertain promise for plantation development. Entering Little River Inlet proved complicated because of its breakers and sandbars. Beyond it lay a brackish stretch of the Hillsborough River, where the "plantable land [wa]s Scar[c]e," and what little there was of it was "indifferent" in quality. Aside from a few patches of corn and cotton land, the best that could be said of it was that its abundant mangrove swamps might someday support the culture of barilla—a salt-tolerant plant used to make soda ash (sodium carbonate), a minor commodity used in dying and glassmaking. De Brahm's discovery of the New Inlet opened access to two coastal rivers that approached each other from opposite directions before merging in a small cove and emptying into the ocean. He found comparatively small pockets of planting land along each. The Shark's Head was an "intirely fresh water River, running in fresh water marsh ground" from the north. Although its course came within a half mile of the seashore, sand dunes separated it from the salty waters of the Atlantic. De Brahm estimated that 23,400 acres on the inland side of these marshes was "all good plantable land fit either for rice, Indigo[, or] Sugar canes." This amounted, in total, to slightly more land than was to be granted in a single tract to one large-scale investor. Much of the rest of the land he viewed consisted of "barren Sandy hills on which are Scatted oak Shrubs & other bushes" that "may do in time for the cultivation of the Opuntia," or prickly pear "plant, which will content it Self with a Soil fit for nothing else known at present."

From the south, the Shark's Tail River snaked through a larger expanse of freshwater marshland that contained tens of thousands of acres fit for planting rice, indigo, and corn, as well as a few knolls in the seaside marshes that Native farmers had once cleared. The forests of yellow pine he spied west of the river would yield good lumber but poor crops. He saw evidence that this New Inlet was in fact a recent breach, opened by the combined forces of torrential rains, strong contrary winds, and the current of the Florida Stream—a point of rupture between coastline and riverbank that had previously sealed this river system off from salt water inundation. In just a few days, De Brahm noted that both rivers, "at a considerable distance from their mouths by this New Inlet[, had] become Saltish." The same violent forces that opened New Inlet threatened to close it again by filling its mouth with sand and silt, a fate that seemed to have befallen Middle Inlet, which was "bar'd up," cutting off maritime access to Middle River.[23]

De Brahm concluded his initial running survey of the coast with a chart of Cape Florida, the southeasternmost point of Atlantic Florida at Key Biscayne, which mariners used as a landmark as they navigated the treacherous waters around the peninsula. De Brahm sailed the *Augustine Packet* through White Inlet, a waterway just below the thin finger of land—now Miami Beach—that divides the end of the mainland from the beginning of the Florida Keys. Along the Cape River (modern-day Biscayne Bay), the highlands sprouted "luxuriant plants," bearing fruits and blossoms, and sixty-foot-high mangroves grew in nearby marshes. The land appeared good for cotton, indigo, and corn. A visible western river, four miles inland, held out hope of untold acres of rich uplands, as well as the promise of a sheltered passage to the Gulf of Mexico. There was also the possibility that it connected, via some remote interior course, to the other rivers he had explored, providing a reliable alternative to risky open-sea voyages along the coast.

The first expedition of the southern branch of the General Survey failed to identify places that combined both of the indispensable natural qualities for plantation settlement: large tracts of fertile land along navigable rivers, and a port site featuring a sheltered harbor and clear passage to the sea. As the British arrived in East Florida, promotional writers trumpeted the idea that its tropical fertility would ease the work of settlement and deliver outsized rewards to colonists who came to claim their portions of its generative power. As the governor proclaimed the terms by which they might receive grants of land, he asserted this environmental cant of colonization, promising temperate winters, productive soils, and long growing seasons that would produce "Vegetables of every kind" without "any Art" and yield multiple crops of whatever was planted with "little labour." The inlets and rivers De Brahm surveyed in 1765, however, suggested no place at which it would be easy to clear land, plant crops, and export commodities so close to the volatile, marshy, and sandy Atlantic coast. These new representations of a previously unknown place arrived at the Plantation Office in Whitehall on October 16, 1765, complicating the Board of Trade's general plan for East Florida with newly specific geographic information.[24]

A few months after James Moncrief sailed back to St. Augustine, De Brahm arrived at Mosquito Inlet. His "Special Chart of Muskito Inlet" showed the depths of its waterways and divided the highland from the swamp and marsh at one mile to the inch, twice the resolution of his sectional charts. He sent this chart—the product of three weeks of surveying—to the Board of Trade in 1765, along with a thirteen-page discourse that acknowledged the region's promise for agriculture, but noted that other sites, where the mangrove trees grew "Straighter, higher & thicker than in Muskito Bay," might be better for settlement. He was unequivocal, however, in his negative opinion of its harbor. There was some cypress swamp for rice and bands of hardwood soil for almost any dry-soil crop, but like all the inlets along this coast, the problem was getting to these lands from the sea. His chart revealed the maze of narrow channels constricted by shifting sand banks at the inlet's mouth, leaving as little at ten feet of clearance at high tide.[25]

Grant had "laid out at least thirty Guineas" of his own money to fund Moncrief's survey and "get the State of that part of the Country and Harbor ascertained with some Degree of certainty." De Brahm's report contradicted his belief, breathlessly related to the Board of Trade, that this was the "best Harbor which has yet been found in this Province."

The fact that the two surveyors "differ[ed] very much" over Mosquito Harbor troubled Grant, who downplayed De Brahm's concerns with the inlet's navigability. This conflicting view of the Musquito Inlet marked the first fissure in the relationship between the governor and the surveyor general, who answered directly to the Board of Trade but, as the colony's provincial surveyor, reported to Grant. One measure of how much these images mattered can be seen in the official response from Whitehall to Grant's requests to fund his Mosquito colonization schemes. The Board had not yet received De Brahm's chart of the harbor, and without it, "no Judgment can be formed here for want of the Plan."[26]

De Brahm returned to these inlets five years later and found them transformed. By 1770, Shark's Tail River no longer emptied into the New Inlet, and two new inlets had breached the shore, exposing the river's southern end to the Atlantic. Commanded to discover and chart the inlets of East Florida as if these were permanent fixtures of the coastline, De Brahm revealed a shifting topography at the volatile edge of land and sea that changed soon after it was mapped. Despite their slippery nature, De Brahm built his "Report of the General Survey in the Southern District of North America" around the idea of inlets. Presented personally to George III in 1772, it pictured East Florida beyond the environs of St. Augustine as a place the British were still in the process of discovering. Along with the handwritten text of his "Observations and Remarks," De Brahm gave the king fourteen manuscript maps and charts of East Florida, all of which became part of his personal map library, the King's Topographic Collection. Eight of these described the colony's Atlantic inlets. [**224**] These images abandoned the premise of his 1765 sectional charts: that, given enough time, money, men, and paper, he would someday comprehensively map all of the colony's territory. Instead, these final charts presented a discontinuous picture of the coast, each centered on its particular inlet and each representing a body of codified geographic knowledge designed to open the territory to colonization. Their style was clean and abstract, dividing water into clear sounded channels and dark-shaded shoals—punctuated by rocks and other hazards—and marking off the land as high ground, wetlands, or sand dunes. Rivers and streams flowed into oblivion past the margins of the page, and between these shaded bands of observed terrain, De Brahm left unmarked white space.[27]

Governor James Grant's first major colonization scheme was to plant a group of Bermudians on the "River which runs South from the Mosquettos." The Board of Trade's program to "settle the whole Coast" of North America sought to mobilize populations that were misaligned with their environments to inhabit new territories where they might occupy and develop them for the benefit of the British Atlantic economy. After Bermuda's seventeenth-century tobacco planters exhausted its soil, the islanders learned to exploit a central maritime position in the western Atlantic, setting enslaved workers to the tasks of building and sailing ships and exploiting the resources of a vast "Atlantic commons" open to long-distance seafaring. Grant took pains to prepare a place for them in East Florida, scouting sites, drawing up plans, securing a five-pound-sterling-per-head bounty, offering a ten-thousand-acre tract of land, and promising to build a "small Stoc[k]ade fort in the Center of the Bermudian Settlement," defended by a military "Detachment there for their Protection." The prospect of an initial migration of two hundred colonists, whom Grant regarded as an "immense acquisition to this Province," promised to "soon make it a flourishing Colony, if they succeed." Like his superiors on the Board, Grant envisioned a self-righting empire in which enterprising colonists would not hesitate to move away from poverty and toward opportunities made more attractive by the inducements of the imperial state. The Bermudians wanted land in Florida with stands of "exceeding fine" ship timber and a safe harbor with deep inlets for oceangoing vessels. Although "Ship Building appears at present their first Object," Grant encouraged them to take the long view and "look upon Planting" and its possibilities "as a matter of much moment to their Settlement." The governor wanted to turn these artisans and traders into planters, whose example at the Mosquitoes would become the "very Foundation of this Infant Colony."[28]

A scouting party from Bermuda came to "inspect the Inlets and Harbours, to see if there [were] any navigable Inlets" that would make the "Country suitable for Trade." They did "not seem to like our Inletts much." They agreed with De Brahm that the sandbar at its entrance unfitted the Mosquitoes for shipbuilding. Grant and De Brahm attempted to entice them to a site along the St. Marys River by laying out lots for "New Bermuda," which the scheme's promoters hoped would be inhabited by "Two thousand People, at least," once the "First Bermudian Families" were "well settled." [225] This town plan featured wharf lots to give settlers access to the river, meeting all of their stated needs for a maritime outpost. If "I can bring this plan to bear of raising a Town at once in a Wilderness, upon a Navigable River, with an Inlett equal to that of Charles Town, I think it must be a means of settling the Neighbouring Country," Grant declared to the Board. This Bermudian beachhead for the British colonization of East Florida never materialized. The migrants disappointed Grant by avoiding the risks of venturing to an untested province and selected a site for their new lumbering and shipbuilding town in Sunbury, Georgia, instead.[29]

The departure of the Bermudians reflected an emerging tension between the general conceptualization of East Florida and the particular knowledge generated by the surveys. The Board of Trade assigned the colony a role within a wished-for system of political economy in which it was to take form as a substantial plantation society capable of producing high-value tropical commodities for export. This vision came with particular geographic beliefs about East Florida as a place, including an insistence on its territorial integrity as part of the mainland and the expectation of discoveries of navigable inlets, sheltered harbors, and broad rivers along its extensive Atlantic coastline. As De Brahm's surveys cast doubt on these spatial presumptions—suspicions confirmed by the opinions of Bermuda's land scouts—Grant and the Board used the Crown's power to grant land in order to draw new colonists to East Florida and, by doing so, remake it in the image of a model North American colony.

Planting East Florida

Britain built East Florida to fail through its misapplication of geographic knowledge. From its promotion of an impossibly productive natural world to its outsized township grants to metropolitan investors who knew little about their lands, the Board of Trade encouraged a mode of colonization that was "fevered, dilated," and "overreaching." Hundreds brought to work in East Florida endured extremes of want and violence on its malarial shores as delusional visions of subtropical fiefdoms collapsed.

Interspersed among the ruins of disastrous experiments in reformed plantation settlement, which made the short-lived British colony little more than a "failure in Xanadu," some colonists adapted long-established planting practices to develop tracts along the St. Johns River. Their modest successes revealed the robust power of unfettered settler colonialism to transform land for profit, especially when the state opened up new districts for development with thorough surveys, and even when their efforts were constrained by new regulations for granting land.[30]

The Board of Trade created a tiered land system that drew two types of colonists to East Florida. Wealthy metropolitan investors proposed schemes for settling twenty-thousand-acre townships "at their own expense with Protestant Inhabitants." These adjuncts to empire were to invest capital and experiment with the production of valuable commodities to exhibit the colony's commercial promise. From 1764 to 1774, Britain granted 242 petitioners to the Board of Trade the right to claim more than three million acres of East Florida land in tracts of 5,000, 10,000, and 20,000 acres. Of these, 114 claimants actually arrived in or dispatched agents to St. Augustine to obtain grants, patenting a total of 1,443,000 acres. Sixteen of these large-scale landowners made credible attempts to settle plantations within the 222,000 acres of land transferred to them under the terms of these grants before 1776. Whether measured by the number of acres approved, awarded, or developed, the scale of this distribution surpassed comparable claims of all the township recipients in Nova Scotia, New York, Quebec, and West Florida combined.[31]

The list of large grantees included government officials, military officers, bankers, members of Parliament, royal courtiers, aristocrats, and scores of merchants. The First Lord of Trade (the Earl of Dartmouth) and his sons received 100,000 acres; Prime Minister Charles Townshend received 20,000. Those of lesser stature parlayed connections with such prominent men to become East Florida grantees as well. Thomas Astle, a state papers archivist and antiquarian book collector, had been "particularly employed in Methodizing, digesting, binding, and securing under proper Covers" the Board of Trade's papers relating to the colonies. He requested, and received, five thousand acres. Many of these petitioners met together monthly at a Covent Garden tavern, the Shakespeare's Head, to discuss their plans as members of the East Florida

Society, where they began the proceedings by drinking a toast to the health of Governor James Grant.[32]

As these "English Grantees" gained a "great part of [East Florida] in their hands," others claimed a few hundred acres each by headright and purchase. In addition to the 220,000 acres actually taken up by Britain's metropolitan colonizers, East Florida's council issued grants directly to those who arrived in the colony. From 1765 to 1775, it granted 210,673 acres to 576 petitioners. Settlers claimed one hundred acres as a personal headright, claimed another fifty acres for each free or enslaved dependent, and could purchase up to one thousand acres more at the price of five shillings sterling for fifty acres. To save their land from "forfeiture," they had to "clear and work" a portion of it or build a house and stock the land with cattle. Just as metropolitan colonizers submitted to an interview by the Board of Trade before the Privy Council issued an order for their townships, migrants seeking headright grants had to make a "personal application" to the governor and leave him "convinced of the probability of Cultivation" before they could be deemed "responsible Planters" and obtain warrants for their smaller grants.[33]

Metropolitan and vernacular expectations for the land converged along the St. Johns River to establish a distinctive plantation society in East Florida. After the 1765 search for inlets, De Brahm redirected the General Survey from Florida's Atlantic coast toward its interior riversides. Tasked to chart the course of the St. Johns, his surveyors fanned out along it in April 1766 as settlers began taking up tracts of land, some with warrants from the governor, some with orders-in-council from London, and others squatting without legal permission. At the same time, De Brahm oversaw the provincial surveyor's office in the capital, setting the machinery of East Florida's land system in motion by surveying tracts for prospective settlers. These two surveying activities under De Brahm's purview—charting the river, and measuring real property boundaries along its shores—proceeded side by side for more than two years.

In October 1768, he dispatched his completed "Geohydrographic Map" of the St. Johns and its lands, now lost, to London. In 1769, Board of Trade draftsmen John Lewis and Samuel Lewis compiled this and other De Brahm maps into "A Plan of Part of the Coast of East Florida," an elegantly rendered compilation made for presentation to the king. **[226]**

The "Plan" is a thirty-six-sheet manuscript map, five feet wide and nearly twenty feet in length. As it was unscrolled for the perusal of George III, Britain's monarch saw the river's many "creeks, branches, lakes & lagoons" delineated for the first time as a navigable waterway, a safe passage for shipping sounded through shoals and hazards. Colored by washes of brown, yellow, and green ink—now faded—it divided the riversides into pine barrens, hardwood highlands, salt marshes, savannas, and freshwater swamps, denoting each land type with its own iconography to make its part of the landscape legible for those seeking to cultivate it. De Brahm's maps organized the rush for land along the St. Johns, which "seem[ed] to be the Favourite Spot" for those bearing orders for townships. With General Survey maps in hand, Grant "pointed out" likely spots to the agents of English grantees when they arrived in St. Augustine, and De Brahm laid out the boundaries of these properties in great rectangular polygons in which he inscribed their names.[34]

One could reach the banks of the St. Johns by taking the muddy path from St. Augustine to Picolata, the blockhouse fort that guarded the ferry crossing and that had hosted a 1765 congress with the Creeks. Others made their way overland across the map's blank interior spaces from Georgia. These "People called Crackers" were already settling along the river without warrant, survey, or grant. Although the governor doubted that these "Strag[g]ling Woods Men" (who "hunt[ed] a great deal and Plant[ed] but little in the Indian Stile") could contribute much to his colony's improvement, he assured them that they would "be allowed to keep Possession and profit of their Labours," presumably by affirming their ownership of developed tracts with formal grants of land. Colonists with the means to do so set out for their St. Johns lands by sea from St. Augustine, sailing some thirty miles up the coast to the St. Johns (or St. Juans) Inlet and following its meandering course upriver. By 1772, De Brahm and his assistants in the provincial surveyor's office had "laid out in private Properties" about half of the land along the river's western side. They had marked off all the land south and east of the river, along the great arc it traced back toward the coast, to the point at which the General Survey had located the "Head of the St. John's River," just north of Cape Canaveral. Although the map ignores the presence of most of the smaller working plantations along this river, interspersed among the vast speculative holdings of London's elite sat the properties of migrating planters, estate overseers, and others who sought to claim their part of Florida's economic promise in increments of one hundred, five hundred, or a thousand acres.[35]

Two well-documented plantation failures reveal the consequences of the Board of Trade's unrealistic expectations for East Florida. In late 1764, Denys Rolle arrived in the colony, the first metropolitan colonist to come bearing a Privy Council order for twenty thousand acres. When he presented himself to the governor, however, a dispirited Grant reported that he was the "most Miserable Wretch [he] ever Saw" and was "convinced he never will be the Means of settling an Acre in this Country, but will be a Detriment to the Province by taking up Lands upon the River St. John's which might have been occupied by more usefu[l] Inhabitants." As his English indentured servants, recruited from among London's poorest, struggled to make a living from the land, Rolle "Wander[ed] the River St. John" during the winter of 1765, and "Whenever a place pleased his eye he built a Log house by way of taking possession." He dragged a deputy surveyor "over all the Swamps and Creeks round his Bluff" for two months, and yet after rejecting De Brahm's proposals for fixing the dimensions of his township, its boundaries remained undefined on the 1769 map. Grant concluded that the largesse of the Board had been misplaced and that a "Planter with a Dozen of Negroes would do more in Six Months" than Rolle would ever accomplish. Rolle amassed a paper empire of upwards of eighty thousand acres on which he ultimately settled a modest cattle-and-turpentine plantation, once he purchased twenty-two slaves and hired an overseer. Dr. Andrew Turnbull "deluded away" hundreds of Mediterranean peasants to come to his New Smyrna plantation as indentured servants and forced them to work "in the manner of negroes" in the Mosquitoes district, which the governor favored for settlement. When he failed to provide for his "thousand people living in a Wilderness," Grant relieved the New Smyrna migrants with timely shipments of cloth, tools, and food for the "distressed Greek settlement" and helped put down one abortive escape attempt. Although Turnbull's servants produced thousands of pounds of indigo, high mortality thinned their ranks, and in 1777, the survivors fled to St. Augustine, abandoning the settlement.[36]

Some metropolitan colonizers never invested a shilling to develop their East Florida properties, and others who did quickly abandoned them. Although Rolle's and Turnbull's failures revealed an especially dramatic distance between visions for reformed colonies and the demands of New World planting, they joined many bankrupts throughout British plantation America, whose aspirations were overturned by volatile weather, shaky finances, restive slaves, unpredictable markets, poor judgment, and bad luck. Their hapless attempts to populate large townships with poor European laborers stemmed less from their own misconceptions about what was possible in Florida and more from the script for reformed colonization handed them by the Board of Trade. The commissioners who evaluated their petitions assumed that African slaves would provide much of the labor needed to develop East Florida, but they also feared the reproduction of black-majority populations that had destabilized both the Carolina Lowcountry and the British Caribbean. They favored schemes that attempted to revive European indentured servitude and that made the settlement of white Protestants a condition—at one white settler for every one hundred acres—of retaining granted lands.

Confident in the province's generative environment, Florida's promoters encouraged the production of sugar, cotton, silk, and wine, but they looked askance at rice, the workhorse commodity of Carolina's economy, as something "common or unclean." At the same time, the Lowcountry's second most important export, indigo, seemed worthy of emulation as a high-value, tropical exotic. By buying slaves and making indigo, some metropolitan colonizers found a means of realizing the natural promise of these lands. John Perceval, Earl of Egmont and former First Lord of the Admiralty, obtained 65,500 acres. He set himself up at the head of a consortium of absentee "Adventurers" that established Mount Royal Plantation on the St. Johns. In addition, Egmont envisioned developing his ten-thousand-acre Amelia Island tract, complete with a scheme to build model slave villages aimed at keeping his human property "happy and contented." Mount Royal eventually foundered, but when Egmont died in 1770, he left behind a large indigo plantation on Amelia Island that persisted until its destruction during the War of Independence.[37]

Mercantile innovation likewise characterized Richard Oswald's Mount Oswald settlement, established in 1765 at the Mosquitoes. A prominent merchant at the head of a large Atlantic trading network, Oswald took James Grant's advice and set aside plans to recruit poor Germans to settle his twenty thousand acres. European servants "won't do here," the governor insisted. "Upon their landing they are immediately seized with pride[,] which every man is possessed of who wears a white face in America, and they say they won't be slaves and so they make their escape." When a hard frost killed off Oswald's sugarcane, he focused on indigo, purchased thirty-four Carolina slaves, and imported an additional seventy Africans directly from his firm's slaving factory at Bunce Island on the Upper Guinea Coast. From his London countinghouse, he leased land within his township, extending credit to his tenants, and imported many of the approximately one thousand slaves shipped to East Florida by 1771. Although Oswald struggled to make a profit out of his Florida enterprise through the transatlantic credit crisis of 1772, he had amassed a workforce of three hundred slaves at his Mosquito estate by 1781, when the disruptions of war encouraged him to disband the settlement and remove his slaves and tools to Savannah. In addition to the plantations established by Egmont and Oswald, large grants were developed with some success by Samuel Potts, Jacob Wilkerson, Edward Hawke, Richard Russell, Francis Levett Sr., Miller Hill Hunt, the Earl of Moira, and the Earl of Tyrone.[38]

Unlike resident planters whose prospects depended on improving their lands, many of the "adventurers in East Florida [grew] tired of their concerns there" and decided to "withdraw their Interest" when expected returns did not materialize. As surveyor Bernard Romans wandered across a landscape of abandoned plantations along the St. Johns River in the mid-1770s, he reflected on the meaning of these "sad monuments of the folly and extravagant ideas of the first European adventurers and schemers." When naturalist William Bartram returned in 1774, eight years after he had dissolved his own struggling rice plantation on the river, he described romantic scenes of an irrepressible nature reclaiming abandoned fields and buildings. Despite the evocative ruins they left behind, some of East Florida's metropolitan colonizers laid a foundation for development. With their trading backgrounds, they shared risk by pooling capital and used the practices of long-distance trade to manage remote properties from a distance. These absentees hired overseers to manage their enterprises, opening a path to plantership for a motivated group of men on the make.[39]

A group of London investors purchased seven hundred acres and established New Castle Plantation in 1769 under the care of an industrious Swiss manager, Francis Philip Fatio. After building up this property into a large indigo plantation, Fatio formed his own partnership with two London merchants to become a co-owner of New Switzerland Plantation upriver. The Earl of Egmont's manager, Martin Jollie, purchased a five-hundred-acre plantation, San Marco, to become a planter in his own right; he sold it to another manager, William Wilson, who amassed a labor force of twenty-five slaves as well as additional properties along the river. A number of other overseers invested their earnings in slaves and put them to work on plantations of a few hundred acres on the Pablo River, close to the inlet. As high land costs elsewhere in the Lowcountry raised a barrier to entry for overseers who sought to become planters, their counterparts in East Florida used their salaries from managing absentee estates and the headright system to subsidize their rise into the planting class.[40]

The planned colonization of the St. Johns watershed produced a distinctive new plantation culture focused on indigo production, which was made possible by the investment of capital from London as well as the emulation of established planting practices in South Carolina and Georgia. As some ambitious schemes collapsed, large- and small-scale landowners settled plantations along the river. De Brahm calculated that 2,588 people lived in East Florida in 1771, about 1,400 of whom were the Mediterranean servants "imported by Mr. Turnbull." The presence of more than nine hundred "Negroes," up from "about six hundred working Slaves" in 1767, revealed the intensification of conventional plantation agriculture. De Brahm cataloged 288 European inhabitants, including women and children, but only 73 could be counted as "planters." Those who succeeded in planting their lands did so largely by putting slaves to work making indigo. In 1771, East Florida exported 28,143 pounds weight of the dye, just a fifth of the quantity of South Carolina's exports for the year. Yet planters in this "infant colony" had, over the space of just a few years of production, managed to surpass the output of their counterparts in Georgia, who had been growing and processing indigo for a generation.[41]

Indigo had become South Carolina's second-most-important export in the second half of the eighteenth century, but its poor reputation dampened its price in London. Cold weather impaired the dye-producing compounds in the leaves of growing plants, and even the most scrupulous Carolina makers frequently failed to produce the high-quality dye cakes prized by British textile finishers. Florida's longer growing season and warmer temperatures rewarded planters who experimented with the crop, which grew to maturity so early in the season that the plants could be harvested once and then grown again to provide a second cutting. After steeping, fermenting, and agitating the leaves in large wooden vats, slaves produced samples of the dye that appeared to be "equal to the best Spanish Flor[a]," the highest grade of indigo "that ever was carried to the London Market." In 1769, South Carolina planter and merchant Henry Laurens told James Grant that he was glad to hear that "East Floridean Planters [were] successful" in their initial efforts. He condescended to add that he believed that the "fine quality" indigo his slaves made would "be entitled to the Medal" offered for the best produced in a British colony in recognition of his superior expertise. Neither he nor any other Carolina planter ever won this prize, however. The Royal Society awarded it only twice and, in each case, bestowed it on comparative novices in the new British territories of East Florida in 1774 and Tobago in 1778. Although still disparaged as "Indigo Crackers," hardscrabble planters made hundreds of pounds of the blue dye in the 1770s, joining their elite neighbors on the St. Johns in building plantations around the production of the colony's new staple commodity.[42]

Just as their predecessors had colonized Georgia in the 1750s, many "Gentlemen of worth and Substance, from Carolina and Georgia," came to East Florida to extend the lowcountry plantation zone farther south, intending to plant "Indigo, Rice and Cotton, all which, it is presumed[,] must answer well." They brought field hands "accustomed for Years to work in the Carolina Plantations," enslaved carpenters who knew how to put up houses and outbuildings, and enlisted black drivers and white overseers prepared to mete out the violence that subjugated enslaved field hands and kept them productive. When Grant himself took up land between St. Augustine and the site of Fort Mose, he aspired only to be the "best Farmer in the Province." "I cannot say Planter," he demurred, "for as my Ambition d[oes] not extend above Indian Corn, I do not deserve the name." But as he learned to put his "Negroes in the Field hard at Work," his identity as a colonist changed. At first he mustered them, like the soldiers he was used to commanding, but found that he could never "Assemble the Gang in the Field at a time" and that "They don't understand Roll calling." By hiring a Carolina overseer who assigned Grant's forty slaves individual tasks, as was customary to the north, his farm became a highly profitable indigo plantation and a showplace for teaching others the techniques of planting and processing the dye. "Like an old Planter," Grant could "even see them punished."[43]

Britain's administration of settlement along the St. Johns, made possible by geographic surveys and the transatlantic transmission of maps, accelerated the incremental process by which most colonies had previously developed. Rather than spend decades securing subsistence, searching for salable commodities, and amassing capital through farm building, East Florida planters entered a new colony primed by the state for social and economic innovation. Informed by the writings of William Stork (Florida's "Puff General") and the governor himself that Florida was an agricultural wonder in which anything planted in its prolific soils would grow to perfection, and encouraged by the Board of Trade to populate their townships with ineffective indentured servants, large ventures by several of its first planters collapsed. However, mapping the province based on rigorous surveys brought its factor endowments into focus for experienced planters from South Carolina and Georgia, who grasped its potential to support familiar forms of planting.[44]

Just before the War of Independence, East Florida was well on the path to becoming part of a common economic culture that spanned the Lowcountry from North Carolina's Lower Cape Fear to the Mosquito Inlet. As East Florida plantations intensified their production of indigo in 1772, the colony's slaves and servants produced enough food to provide for the "Subsistence of its Inhabitants," a rise to self-sufficiency that Lieutenant Governor John Moultrie believed was unprecedented. "East Florida has done more" in its first few years of existence, he declared to Colonial Secretary Hillsborough, "than any Continental Province ever did, since the first Institution of the British Empire in America." In less than a decade, he believed, its planters had "got over the greatest hardships and difficulties that occur in sett[l]ing new Countries; and they [were] begin[ning] to find themselves Comfortable and at home and also to have some hopes of getting rich."[45]

Mapping East Florida's Edges

In 1763, Britain did not have a single map of Florida that could enable its colonization. In 1773, William De Brahm submitted a portfolio of hand-drawn maps, plans, and charts of East Florida with his "Report of the General Survey in the Southern District of North America." Twelve of these described separate portions of the coast—inlets, islands, towns, and rivers—at one to two miles to the inch, a scale large enough to reveal the visible qualities of different kinds of soils, the shapes of sandbars, and the footprints of structures. The culminating accomplishment of the southern General Survey, however, was his "Map of the General Surveys of East Florida," a single image drawn across three separate sheets. [**227–229**] Once joined, these sheets formed a continuous terrestrial map and maritime chart that covered the colony's coastline from the St. Marys River, which marked the East Florida–Georgia boundary, to the Dry Tortugas, the last scraps of land at the end of the chain of islands known as the Keys.[46]

This map represented knowledge generated by astronomical observations, calculations of magnetic declination, and fixing the location of observed landmarks through triangulation—all techniques used to affirm the correspondence of forms on the map with real places on the surface of the earth. It summarized the work of five expeditions that established the contours of the coast, identified river inlets, and sounded those rivers and assessed the lands alongside them for agriculture. It derived from a more particular body of geographic information that pictured land and water in finer detail and higher resolution, a host of maps and charts that could be fitted together like pieces of a puzzle to form this general map. Once this interlocking corpus of images and description was shelved among the Board of Trade's papers in Whitehall and copied and circulated among other agencies of government, it promised to form a working archive that would permit imperial officials to see and understand this remote place with precision. De Brahm's map was one of several general maps of new territories that embodied Britain's quest for comprehensive knowledge of American places.

The "Map of the General Surveys of East Florida" also represented De Brahm's view of the colony as a place with limited potential for expansion. From the moment in 1765 when he had criticized the navigability of Mosquito Inlet, the place Governor James Grant singled out for new settlement beyond the St. Johns River, the two developed conflicting understandings of the nature of the colony. Every attempt Grant made to expand the territory for plantation settlement, De Brahm seemed to counter with information about its disabling deficiencies. Grant snubbed De Brahm by not inviting him to attend the first congress with the Creeks at Picolata in November 1765, although as provincial surveyor, he would be tasked with surveying the boundary line determined at this diplomatic meeting. Creek headmen proposed that the boundary should trace the curve of the St. Johns, making the part of East Florida open to settlers into a space that was "nearly an Island from [the river's] Source to its Entrance into the Sea." To avoid this reduction of his colony to the stature of a mere island and to ensure that both sides of the river fell well within colonial territory, Grant and John Stuart—superintendent of Indian affairs for the Southern District—pressed for an additional cession of land to the west. The Creeks had a clear geographic conception of a natural division of territory between Europeans and Indians. East Florida lands were to be located only along the "Salts & the White People" should "not settle on the fresh waters." At Picolata, they made an exception to this general principle that colonial land should extend only as far inland as the "tide flows," approving a "fine concession" that moved the colonial boundary west of the St. Johns "above 25 miles deep."[47]

On his general map of the province, De Brahm delineated the Picolata boundary line from a pine tree on the banks of the St. Marys, across from Sander's Indian Store, due south for ninety miles to its agreed end point at the head of Ocklawaha River. As the congress concluded, Creek leader Sempoiaffe remarked that they had "given the Governor some land to settle and hoped his heart was satisfied." It was not. Grant read the Picolata agreement to mean that all the land on the St. Johns south of this point was open to British settlement and assumed that the western boundary would be extended farther south in future negotiations. He did not hesitate to approve large grants of land in this diplomatically unresolved space and thought little of granting a "Twenty thousand Acre Tract" here and there, where its "back Line" might "pass the Boundary a Little." Since its planters would only be attracted to swamp and bottom lands "clos[e] to the River," he reasoned, their presence across the line "could give no Offence" because it would not "affect the Indians in their hunting Interest."[48]

De Brahm read the geographic terms of the treaty as a strict constructionist. In his view, all land that had not been formally granted by the Indians remained off-limits to European settlement, a reading affirmed by the Creeks, who harassed his surveyors in this contested territory to frustrate their acquisition of information required to grant it. He thought that the 1770 Creek raids on the Mosquito settlements were retaliations for "trespassing that very boundary, and taking violent possession of their reserved lands." By this interpretation, the 679,000 acres granted to forty-one metropolitan colonizers, including Andrew Turnbull, Richard Oswald, Charles Townshend, Lord Adam Gordon, and the Earl of Dartmouth, were illegal appropriations. De Brahm's conservative view of the location of the Indian boundary with East Florida limited the colony's legitimate domain to fewer than two million acres, a space he judged too big to be a "County, and too insignificant for a Province."[49]

In the heady early days of his governorship, when Florida's possibilities seemed limitless, Grant imagined that British surveyors would discover deep rivers behind Atlantic coastal inlets. As they followed them to their sources, their descriptions of fertile expanses of riverside land would draw planters from throughout the British Atlantic world to populate an extensive, valuable, and prosperous province. On the second sheet of the "Map of the General Surveys of East Florida," De Brahm summarized his view of the coast between Row's Hammock and the Cape of Florida. Presented with sobering evidence that it offered few unobstructed inlets, that the colony's rivers did not form a navigable interior network, and that "large Tracts of good Land [were] not easily found," Grant drew back from his unbounded optimism but still believed that the "Country will do very well."[50]

By bringing its natural limitations for commercial agriculture into sharp relief, De Brahm constrained Grant's plans for East Florida's settlement. At the height of their growing disagreement over surveying priorities, practices, and findings, the surveyor general produced his royal commission for the governor, who attempted to slap it out of his hand. Grant charged De Brahm with neglecting his surveys and collecting fees fraudulently, and in 1770 he dismissed him from his post as provincial surveyor. Ordered to "come to England, in order to answer to such Complaints," De Brahm sailed from St. Augustine in late 1771. The Treasury reinstated him to his position in 1775, and he made it as far as Charlestown aboard the newly outfitted armed survey ship *Cherokee* before the disruptions of the War of Independence made it impossible for him to return to East Florida to continue the General Survey.[51]

Back in 1771, however, before he departed for London to defend his conduct, the surveyor general completed his mission to comprehend the full length of East Florida's Atlantic coast. He led a final expedition around its southern islands, known as the Keys or the "Martiers," in recognition of the many mariners who lost their lives on that "heretofore unknown dangerous Coast." De Brahm sought, "by proper Observations on the Regular Courses and Dispositions of Winds, Currents, & countre-Currents," to clarify that "confused Labyrinth in Navigation." He presented his findings on the third sheet of his general map, the "Chart of the South-end of East Florida, and Martiers." Two kinds of appropriations threatened British control over this uncharted archipelago. First, sailors from the Bahamas and elsewhere descended on the Keys to exploit it as an Atlantic commons, "cutting mahogany and catching turtle" as they did in other undefended and underpopulated Caribbean places. De Brahm reported "no less than 25 Sails of Vessells cutting off mohagony" on his first expedition to southern Florida in 1764. Such depredations revealed Britain's inability to police this vast space, and by stripping the Keys of their valuable timber, illegal log cutting also lessened their prospects for future settlement. One of Grant's first acts as governor was commissioning the construction of a provincial schooner that could patrol the Keys.[52]

Second, Spain took advantage of the uncertainty about where continental Florida ended and the West Indies began to advance a claim on these islands. The "Spanish Governor of the Havanah," Grant reported in 1766, "looks upon the Keys of Florida to be the property of Spain." More alarming still, this was no passive claim. He gave "passports to Vessels to go to those Keys to Fish, not as formerly under the name of the Florida Keys, but to the Northern Keys." When Florida had been a Spanish province, "small Spanish Vessels from Augustine constantly passed that way by keeping between the Keys and main land 'till they turned the Point of the Peninsula" and entered the Gulf fishing grounds. Whichever power possessed the Keys commanded a passage into and out of the Gulf of Mexico through the Straits of Florida. Given the familiarity of Spanish sailors with these hazardous routes, there was reason to worry that fishing fleets from Havana might take de facto possession of southern Florida before British settlers had a chance to colonize it. In 1765, in fact, Governor Grant hired a Spanish fisherman to help the colony's pilot make the harrowing journey into the Gulf and back to St. Augustine to resupply the hungry soldiers of St. Mark's Fort at Apalache. By boarding these fishing vessels, scores of Creek Indians also gained passage to Havana, where they began forging an anti-British alliance to regain some of the diplomatic leverage they had lost in 1763.[53]

The Spanish made a case for this claim to the Keys with cartography. Manuel de Rueda's 1766 *Atlas Americano* included his "Plano y, Descripcion De Los Cayos del Norte." [**230**] The map's title affirmed this pro-Spanish nomenclature, referring to the islands as the "Northern Keys" in relation to Cuba. Rueda lists each of the Keys by its Spanish name, and pictures them abstracted from the Florida coast, as an independent island chain. Just below the title, he noted the disastrous wreck of the seventeen-ship flota in the hurricane of 1733, a loss of people and wealth that brought the Keys into a main narrative of eighteenth-century Spanish colonial history. In response to Grant's fear that the Keys might slip under Spanish control, Secretary of State Shelburne replied that the treaty was clear: "As to any pretentions formed by the Governor of the Havana that these Keys belong to Spain, it can require no other Answer than a Reference to the 20th Article" of the Treaty of Paris. "Florida <u>and in general every thing which depended on it,</u>" he emphasized, "[wa]s ceded." The legal question of who could lay claim to the Keys, then, hinged on their status as independent islands or continental dependencies.[54]

To prove that the Keys were in fact linked to North America, De Brahm imagined "ancient Tegeste," that is, the Florida peninsula as a prehistoric landmass. [**231**] Before the trade winds and the Gulf Stream carved channels between the Keys and the mainland, De Brahm discerned the "Antient Shape" of these scattered islands. He concluded that they had been "originally an Isthmus tor[n] into parcels." De Brahm's interest in the deep geologic history of the Keys and the flow of the Gulf Stream was not merely an intellectual exercise in natural history. Claiming that the Keys were once part of the mainland helped Britain establish title to these contested islands. The two paleo-peninsulas that once extended from the south Florida mainland in eons past helped make sense of the complex hydrography De Brahm charted with unprecedented precision in 1771. [**232**] Comparing the image of ancient Tegeste to that of early modern south Florida, he positioned the Keys in relation to a continuous reef that paralleled them to the south, which appeared to be the remnant of a once-unsubmerged isthmus. Naming the space between them Hawke Channel, he sounded it as a trans-peninsular passage in which vessels traveling south from St. Augustine along the Florida Stream could take shelter from the area's notorious storms. After naming seven "In & outlets" through the "Martier Riffs," he located three safe passages into the Gulf of Mexico, their routes indicated by a line of soundings that measured the depth of water in fathoms.[55]

De Brahm's "Map of the General Surveys of East Florida" revealed a territory whose protean geography was defined by natural extremes that made permanent occupation impossible across much of it. He envisioned Florida at the edge of creation, the "last in making of the whole North American Continent." As forests sprouted across the fractured Keys and convulsed inlets of the southern coast,

> the Land which is now covered with a sandy Soil will in time to come receive a Stratum of yearly dropping Leaves and Limbs, be shaded and consequently preserved against the Sun's Exhalation; whence a Putrefaction will ensue, an acid generated and the Sand corroded into a fine Marl; this will make its Impression yearly deeper by degrees, as the Stratum will increase to thicken itself with the deciduous Leaves &c: this is perhaps the Reason, that

> Providence has wisely prevented its Population to this day, in order to give the necessary time to this part, for making its Soil plantable.

De Brahm's appreciation of this environmental volatility, although it worked against Grant's attempt to settle as much of the province as he could as quickly as possible, contained the same prospecting impulse toward future improvement. His reflections on the process of geographic change attempted to comprehend divine scales of time and space by which land took form for human occupation. Prelapsarian Florida was inundated and ripped apart by the sea, but Britain took possession of it in the age of its reformation. When De Brahm returned to London to answer Grant's charges, he published this new knowledge of the East Florida coast in a slim volume with three original charts called *The Atlantic Pilot* (1772). Dedicated to Hillsborough, the book enabled the "safer navigation" of shipping around a "promontory" that, before the British arrived to map it, was "scarcely known to the world, save by its bare name and existence," or to those "who in their navigation near it became unfortunate by suffering shipwreck." After years spent examining this treacherous coast, De Brahm dispelled the myth that Florida was a mass of disconnected islands. As he left St. Augustine in disrepute, he made use of the voyage east to trace the violent force of the Caribbean's winds and currents and draft his "Hydrographical map of the Atlantic Ocean," which revealed the path of the Gulf Stream. [**233**] Displaying the full range of his erudition about navigation and hydrography, this dispassionate contribution to the world's stock of useful knowledge mentioned nothing of Britain's colonial project in the North American southeast.[56]

As they began to colonize East Florida, British officials struggled to understand it. The Board of Trade's vision for North American empire was a product of theories of political economy that had to be applied to the particular lands and waters that its surveyors described and that soldiers, settlers, and slaves encountered. Far from imposing a rigid abstraction, this plan was understood as a general framework that demanded particular maps be made to exacting standards to implement this vision. After a decade directing the colony's settlement, the War of Independence fatally disrupted Britain's experiment with improved planting guided by rigorous geographic surveys. Although the colony survived the war, and even gained a substantial new population of loyalist refugees, Britain retroceded East and West Florida to Spain in 1783.

To bridge the distance between this unknown land and the vision of a prosperous plantation society, Florida's new occupiers worked in text and image to establish the character of the British as a colonizing people and use the power of the state to accelerate and improve colonization as a process. England's first American colonists had faced just such unknown lands when they arrived on the shores of the western Atlantic, royal charters in hand, in the seventeenth century. They fortified outposts at the mouths of penetrating Atlantic rivers, planted the fertile bottomlands along their banks, and negotiated with and battled Indians to expand these early agricultural domains. Often acting in ignorance of the land and its inhabitants, they made fateful first choices by settling places they stumbled upon after a few exploratory journeys, locking their underpopulated and vulnerable societies into trajectories of underdevelopment that future generations had to work to overcome. This historic practice of taking possession of American territory while blind to its geography had been violent and wasteful, embroiling fragile societies in wars of occupation and fostering economic inefficiencies that squandered the blood that had been spilled and the capital that had been invested. The Lords of Trade wanted to avoid recapitulating this colonial history as they orchestrated the development of the new territories after 1763.

As a social experiment in improved planting, East Florida was a disastrous failure, but its short history demonstrated how an imperial state's command over geographic information could establish a class of improving planters on the land rapidly. The task of rigorously surveying the peninsula for the first time gave Britain the opportunity to present a cartographic image of the territory to the world—and, by representing it accurately, stake a strong claim to its possession. The Board sought to understand Florida comprehensively and use this knowledge to colonize it from a distance. De Brahm and his surveyors mapped the St. Johns from mouth to headwaters. Before most settlers arrived in the province, British negotiators agreed with the Creeks to place the entire length of this river under East Florida's jurisdiction, reducing the possibility of a frontier war. In both London and St. Augustine, officials vetted prospective landowners and granted hundreds of thousands of acres within a few years, unleashing colonists and merchant capital in a powerful surge. By 1768, just five years after troops had arrived to take command of St. Augustine, it seemed that "every foot of Land upon St. Johns River [wa]s taken up[,] and a great Deal [was] already settled."[57]

Regarding the area south of the St. Johns, about which De Brahm and Grant almost came to blows over how to describe the colony's natural endowment for growth, Britain did little to develop or defend this territory. In the General Survey's charts of inlets and keys, we can see how the Board of Trade attempted to master distant spaces in an effort to accelerate the process of social and economic development by which all previous colonies had changed from tenuous outposts to stable, self-reproducing societies over several generations. Its critique of the history of North American colonization extended to African slavery (which it worked to limit by encouraging the importation of European indentured servants) and the self-aggrandizing behaviors of entrenched provincial elites (whom it excluded from access to large grants of land, reserved for metropolitan insiders). The Board's efforts to reform the nature of American planting failed, and only when large grantees began buying slaves and making indigo did these efforts to colonize the St. Johns gain traction as viable enterprises. The General Survey's maps and charts conveyed the idea that because good land and navigable waterways could be mapped, they could be colonized immediately by putting wealthy investors in possession of fertile soils. Such an imagined process of economic growth discounted the ways in which previous plantation societies had taken command of space in incremental steps, which involved accumulating environmental experience, marshaling coerced labor, experimenting with new commodities, and adapting known agricultural techniques to new productions. Britain's cadastral perception of new American territory—that is, its use of maps to see the shape of private-property boundaries on the landscape—was a powerful visual representation of this perspective that denigrated, and sought to displace, American colonists as agents of empire.

Morrisania

Ship Channel

Brothers I.

Hunt Point

Hulst Island

Magnetic North

FLUSHING BAY

LONG ISLAND

{ SEVEN }

Atlases of Empire

Maps made American subordination within the Atlantic empire visible. As the War of Independence disrupted Britain's settlement schemes, London map publishers transformed the images surveyors sent home into published works of cartographic art and science. American colonists saw their societies bound and contained, and they voiced their concerns at the prospect of more centralized and intrusive imperial administration after the Peace of Paris in 1763. They criticized the limits of the proclamation as an unwarranted check on their aspirations, and chafed when they were overlooked in favor of metropolitan colonizers who gained privileged access to take possession of American land. The heated transatlantic pamphlet debates over the Stamp Act, explicitly focused on taxation powers and the rights of British subjects abroad, churned up long-standing disagreements about the political economy of colonization, the capacity of the state to gather reliable information about the material interests of distant provinces, and the ideal form of extensive territorial empire. These debates helped crystallize an emerging American corporate identity rooted in the history of North American occupation. Although the voices of colonial political protest touched only obliquely, at first, on the Board of Trade's scheme to develop the acquired territories through reformed programs of settlement, Britain's intentions for America took center stage with the passage of the Quebec Act and the new land system imposed by the king's "Additional Instructions" in 1774. In this fateful year, a critical mass of observers began to see in the newly surveyed maps of British America images of their own future subjugation. These less studied "intolerable" acts constituted a specific kind of provocation: the threat to bring the rigorous model of state control over land and development from the new imperial periphery to the center of settled

British America. By understanding "dependence" and "independence" as geographic as well as political ideas, we can understand more clearly the fears that drove American protest.

Geographies of the Stamp Act Crisis, 1765–1766

The "mood of national awakening" that linked Britons on both sides of the Atlantic during the Seven Years' War was only superficially unifying. After France's defeat, colonists faced the "demands of a powerfully self-confident imperial state" and experienced a newly forged British identity as an assertion of superiority that "radiated outward from the metropolitan center." In the final years of the war, Benjamin Franklin understood that leading British officials intended to exploit the peace to impose a program of intrusive administration. "The Prevailing Opinion, as far as I am able to collect it, among the Ministers and great Men here, is, that the Colonies have too many and too great Privileges; and that it is not only the Interest of the Crown but of the Nation to reduce them." John Carteret—Earl of Granville and president of the king's Privy Council—harangued Franklin at length in 1759 on the necessity for the "absolute Subjection" of colonial assemblies to the directives contained in orders to American governors. "'Your People in the Colonies refuse Obedience to the King's *Instructions,* and treat them with great Slight, as *not binding,* and *no Law,* in the Colonies,'" but these orders, Granville declared, were "'drawn up by grave and wise Men'"—particularly, he might have added, the Lords Commissioners of Trade and Plantations—whose reports and recommendations the Privy Council "'solemnly weigh'd and maturely consider'd, and after receiving such Amendments as are found proper and necessary, they are agreed upon and establish'd.'" Not only was the Privy Council "'*over all* the Colonies'" as a matter of law, he insisted, but the stature, knowledge, and wisdom of British authorities who shaped its decisions legitimated its right to issue "'his Majesty's Instructions'" as the "'the LAW OF THE LAND; *they are,*['] said his L—p, repeating it, [']the Law of the Land, and as such *ought to be* OBEYED.'" One did not have to listen to the anti-colonial bluster of the Earl of Halifax, who "preside[d] and govern[ed] all" on the Board of

Trade, Franklin noted, to understand this intention to take command of America. A review of the "fruitless Experiment" of "military Government" in Nova Scotia "sufficiently show[ed] what he thinks would be best for us." Those who raised their pens in protest against British policies joined Franklin in pained recognition that many Britons at home, especially those in Whitehall, regarded Britons abroad as a debased, creolized, and disloyal people in need of regulation.[1]

The Proclamation of 1763 began an incendiary process of tightening laws of trade, exercising taxation powers, and controlling how colonists took up, or were denied, American land. These initiatives exposed a dormant constitutional rift that widened across the controversies of the 1760s and 1770s. This bid to centralize imperial rule had as its objective more than simply the desire to demonstrate London's authority over Britain's overseas dominions. The central problem of the postwar moment was how to impose the principles of mercantilism on North American settler colonies that had grown into populous, materially self-sufficient societies geared to pursue their own material interests and demonstrating a high degree of economic, social, and cultural independence.

Board of Trade secretary John Pownall considered the project of forming and dividing the "new acquisitions" as an exercise in applied political economy. The Board sought to put knowledge of place to use by aligning each new territory's nature with the right system for granting land, regulating commerce, and organizing governments so that they would serve the "true interest and policy of this kingdom, in reference to its colonies." His efforts in framing the Board's plan for America mirrored those of his counterparts, working out of another office in the Treasury Building, to bring order to Atlantic trade. As joint secretaries to the Treasury, Thomas Whately and Charles Jenkinson undertook an analysis of customs data on which they based their proposal to reduce the duty on the importation of foreign sugars and molasses, commodities that colonists had imported illegally—and with the tacit approval of customs officials—for years. By setting this tax at a rate that they believed New England importers and rum distillers could bear—three pence per gallon—these reformers sought to lower the cost of colonial compliance. Parliament's Sugar Act (1764) and Stamp Act (1765) encouraged colonists to "sacrifice their own partial Advantage to the general good" and were prefaced with the intention of devoting new revenues to "defending, protecting, and securing the . . . colonies."[2]

Fear of American independence prompted these new regulations. The Seven Years' War disrupted a colonial system that had granted broad autonomy to provincial elites in negotiation with appointed officials. However, the calculative sensibility evident in the Earl of Shelburne's case for the Peace of Paris, in which he tallied the costs and benefits of enlarging the empire, put past precedents up for revision. Reformers understood the staggering outlays of underwriting the new empire's creation as an investment in the long-term prosperity of the North American mainland, and considered those expenses as debits that demanded repayment. Britons considered colonial dependence and independence with a rhetoric that drew its urgency from the enormity of numbers. On one side of the ledger of empire stood "three millions" of Americans (arrayed across a continent "of near three thousand miles in length"), and on the other, the national debt of "One Hundred and Fifty Millions" pounds sterling, amassed largely "on Account of [its] Colonies; that is, to save, defend, protect, and conquer for them; to enlarge their Territories, and put them out of all Fears for the future." After "viewing their numbers, and the extent of the country they possess," next to the "enormous Debt contracted by the last expensive War," advocates of reform demanded that the "interest of a part, ought to give way to the interest of the whole." For Americans to withhold financial "Assistance from our Colonies, to preserve to themselves their present safety . . . when this Country is almost undone by procuring it" seemed almost maddening in its selfishness.[3]

If Britain could not tax the colonies to answer the pressing financial needs of the nation, its colonies could not be considered part of Britain. Proponents and critics of the Stamp Act debated the rights and obligations of taxation and representation using the language of political economy and emphasized the geographic challenges of commanding British America's new extent. Echoing the points of those who had cautioned against the alluring grandeur of continental dominion before the Peace of Paris, critics anticipated the danger of imperial dissolution if growth were allowed to proceed without rigorous controls. With the acquisition of "so immense" a territory, Britain could no longer leave the task of managing American land to the colonists. "The most distant and desolate Part of the Dominions of any People," argued Cato in 1765, "has some Relation to them, either as a Benefit or Burthen." The ungovernable expanses of British America, stretching from the "Gulph of Florida to the North-Pole," Cato warned, would gradually but inevitably draw the nation into a fiscal and military catastrophe. Such an immensity of land open to British settlers would depopulate the home country, encourage the development of competing manufactures, and distract the commercial energies of its merchants. The work of "ten thousand Legislators constantly employed could not devise the Means" of reconciling the riot of interests in the "unwieldy Possessions" of this far-flung empire.[4]

After 1763, when Britons viewed "a map made on a large scale," they "became seized with fear and jealously of [their] fellow subjects in America," observing how they were "seated on an extensive continent," enriched by nature, to become, in time, an independent empire of "dangerous consequence." A review of the customhouse books revealed that "one third part of the commerce of *Britain* depends upon those Colonies," a trade that was "essential to the well-being and existence of her power." Resistance to the Stamp Act, conceived as a modest means to defray the costs of defending an enlarged frontier, seemed to reveal colonists' secret ambitions for independence.[5]

Soame Jenyns, one of the Lords of Trade from 1755 to 1780, advocated for the Board's mission to reform colonial administration and, like all other metropolitan contributors to the pamphlet wars of the 1760s, insisted on the importance of securing the dependence of American colonies to Great Britain. As he defended the Stamp Act against charges that it was "harsh and arbitrary," Jenyns took for granted that he viewed American affairs from a position of moral superiority. No matter how persistently colonists argued on behalf of their rights, he regarded their views as advocacy that could be discounted when compared with the broad view of the imperial common good that he and other metropolitan officials were exclusively qualified to take. "To comprehend the general trade of the British nation, much exceeds the capacity of any one man in America," agreed another Stamp Act proponent, for "trade is a vast, complicated system, and required such a depth of genius and extent of knowledge, to understand it, that little minds, attached to their own sordid interest, and long used to the greatest licentiousness in trade, are, and must be, very incompetent judges of it." It was true, some admitted, that metropolitan observers' general view of America might obscure the "*minutia* among them," but more important was the Americans' "inevitable ignorance of . . . the important concerns and interests of the State."[6]

Political rivals disagreed vehemently about the wisdom of new colonial taxes advanced by the ministries of George Grenville and Charles Townshend in the 1760s and about the best way to react to the rising intensity of colonial protests and armed conflict in the years that followed. Whether they viewed American colonists as fellow Britons or as wayward creoles, policy makers and pamphleteers tended to see the empire in the same way, looking down from a great height at British America's vast footprint across the Western Hemisphere. From this vantage, they searched for a method of attaching these territories across the Atlantic Ocean with fast bonds. There was perhaps no better friend to America in Parliament than Thomas Pownall, former Massachusetts governor, brother of John Pownall, benefactor of cartographer Lewis Evans, author of *The Administration of the Colonies* (1764), and steadfast advocate for reconciliation during the War of Independence. Few displayed more critical regard toward Americans and their "insolent licentiousness" than Henry Ellis, the primary draftsman of the Board of Trade's reforms and a hard-line proponent of strict regulation. Despite their contrasting sensibilities (the two behaved like "two great dictators" one evening over dinner when debating the question, Who lost America?), both men had entered government with Halifax's backing, and both shared their patron's view that rational state intervention could solve the problems of empire. Most of all, they believed that integrating American colonies as permanent dependencies of the British nation was essential to its security and prosperity. Just as an "increase of the quantity of matter in the planets" would, by the force of gravity, unbalance the solar system from its center in the sun, reasoned Pownall, unchecked American growth, as measured by the rising "magnitude of power and interest of the Colonies," threatened to destroy the equilibrium of empire.[7]

Colonists countered British assertions of superior judgment with their direct knowledge of American places. Not only was it their right to tax and govern themselves through elected assemblies, they claimed, but such a system made practical sense, since only those who lived in a colony could know the laws that were most "suitable to its various circumstances and occasions." Americans could not resist ridiculing the "geographical blunders" of prominent British statesmen who claimed to understand American affairs. MP Isaac Barré, a staunch friend of America in Parliament, noted with frustration that "very few" of his fellow members knew anything about the true "circumstances of North America." The Earl of Egremont, who as secretary of state for the Southern Department was Britain's most powerful colonial official, seemed not to know whether "*Jamaica* lay in the Mediterranean, the Baltic, or in the moon." Puzzled governors received letters from London addressed to the "*island* of New England." Against the idea that the king's instructions to his governors should be regarded as binding acts of law, Virginian Richard Bland observed that "ministers . . . from their distant situation from us cannot have so full and perfect [a] view of affairs in the colony." The very fact that those behind the Stamp Act conceived of it as a wise measure, designed to promote, as George Grenville wrote, the " 'ease, the quiet, and the good will of the colonies,' " revealed that true American interests and opinions must have been "strangely misrepresented in England." Benjamin Franklin patiently explained to the House of Commons that beyond the "sea coasts," the price of postage to send stamps into the interior would cost more than what the "stamps themselves" would produce in revenue.[8]

During the Stamp Act crisis, Americans suspected that this and other controversial measures were the brainchildren of a cohort of well-placed "Projectors, who consider[ed] the Honour of the Invention as a principal Object" of colonial policy, and who prized the abstractions of political economy over common-sense compromises that had sustained this growing transatlantic polity for the better part of two centuries. Without sound practical knowledge about America, members of Parliament had been encouraged to approve "wild projects" of taxation and regulation, enticed by the "fallacious reasonings of the projectors" behind their "loose impractical schemes." Far from being the staid and impartial judges of what constituted the common good of the empire, Britain's revolving set of competing factions pitted "one Set of Ministers" against "another Set of Claimants" to high office. Against charges of provincialism, Americans pointed out the battle for place and office in Britain and bemoaned these "Days of Venality."[9]

Their case against taxation without representation included the charge that it impeded the flow of "knowledge of America" to the central government. The Stamp Act's critics contended that because Americans were not represented in Parliament, it did not have the authority to impose direct taxes on them. Changing the British constitution to allow American constituencies to send members to the House of Commons would not only give Parliament this right, as some suggested, but would also serve as an expedient means of framing equitable laws, "giving those of both countries a thorough knowledge of each other's interests, as well as that of the whole, which are inseparable." As it stood, Britain gathered dubious hints and recommendations from self-interested advisers, limiting its "information in American affairs to every vagabond stroller that has run or rid post through America from his creditors, or to people of no kind of reputation from the colonies, some of whom, at the time of administering their sage advice, have been as ignorant of the state of this country as of the regions of Jupiter and Saturn." The patronage networks through which ambitious men on the make gained appointments to high office called into question whether metropolitan officials could best comprehend what was "for the evident good of the whole community."[10]

Without direct representation, there was no channel for Americans to explain the wants and conditions of their colonies before Parliament. Their governors—appointed military men for the most part—"seldom underst[oo]d the commercial interests of a trading people" well enough to do them justice. Americans opposed to direct taxation acknowledged that they were part of an integrated imperial system, but rejected the idea that those at its center knew best how it should operate at the periphery. The case they made against the stamp tax endorsed a negotiated form of governance that stood in contrast to the Board of Trade's focus on the centralization of authoritative knowledge about America. They argued that "whatever affects all, should be debated by all, so that knowledge and mutual interest will prevent mistakes and partiality." Some British Stamp Act supporters, however, ridiculed the idea of American-born members of Parliament. Their presence in the Commons chamber would see its "purity defiled, by the unnatural mixture of representatives from every part of the British dominions" and the crude "Powers of Speech, of which these *American* Gentlemen are possessed." Colonists could not but consider such reactions to be "an insult, the treating them as women, infants, and the dregs of the city of London" as if they were "without property or integrity, will or capacity." "Have we, say the Americans, no wills of our own? Are we not free agents?" Although some British commentators supported admitting Americans as members of Parliament, no major American figure who opposed the Stamp Act (with the exception of the eccentric James Otis Jr.) believed that colonial representation, and the subordination that gesture of inclusion implied, was desirable or appropriate given the legislative autonomy of the colonial assemblies.[11]

Not only did the British lack credible knowledge of American interests and conditions, but they also labored under the delusion that America was a collection of weak and vulnerable outposts that needed support. When they wrote of "what *we*, for such is their language, have done for them; what money *we* have spent; what blood *we* have lavished; and what trouble *we* have had in establishing and protecting" America, they again betrayed that they were "totally ignorant of the colonies," "presuming and uninformed" about their diverse histories, cultures, and accomplishments. The author of *The Constitutional Right of the Legislature of Great Britain to Tax the British Colonies in America* (1768) ridiculed the metropolitan narrative that Britain had endured the hardships of the Seven Years' War to procure the "security of the colonies." Like a wanton child who thoughtlessly betrays an indulgent parent, the story went, America had the gall, "after having been reared into maturity at the boundless expense of her best blood and treasure, to spurn and reject her in her state of desolation; and springing fresh, young, and vigorous, into a reign of absolute independent and national government; start for the prize of preeminence, with all other powers of the earth." This writer rejected the sanctimonious refrain of "blood and treasure." Every nation has lavished "oceans of blood and treasure in every age; and the blood and treasure have upon the whole been well spent. British America hath been bleeding in this cause from its settlement," and during the war, the colonies "have spent all we could raise, and more."[12]

"One single act of Parliament," the Stamp Act, "set the people a-thinking in six months more than they had done in their whole lives before" about the nature of colonies. Stamp Act advocates viewed American colonization as a "modern" innovation—one quite distinct from ancient Greek or Roman expansion—by which European states authorized their subjects to improve the lands of "uncultivated and uncivilized countries" with the aim of extending their commerce. Expectations for territorial acquisition were written into the charters of the first colonies and became the means by which tenuous coastal outposts grew and became prosperous. From this perspective, American colonies did not exist as their own corporate entities but rather appeared as a "field of adventure belonging to Great Britain, fitted to the geni[u]s, industry, and enterprize of her people."[13]

A decade later, the author of *Colonising, or a Plain Investigation of that Subject; with a Legislative, Political and Commercial View of Our Colonies* (1774) codified this understanding of the American colonies as indissoluble extensions of Great Britain. A rising empire was like a growing family: territorial expansion was the natural act of a benevolent, "parental" people by which a nation might reproduce its power into the future and across space. Colonies "owe[d] their establishment to the conception, maintenance, or protection of a parent government." Like the boughs of a tree, they spread "themselves to the sun" and "wrestl[ed] with the Wind," but the fruit they bore drew life from the "Stock" of the mother country and "both grow together." One should marvel at the well-balanced political ecology that had flourished from early and tenuous plantings in North America. The British empire was a "self-moving System, effecting constant order among so many new, various, wide-settled, differently occupied, and differently circumstanced Communities as our Colonies consist of." Relax the natural relation of superiority and inferiority, however, and colonies would become parasites rather than fruit-bearing branches. As they grew, colonies must be tended and pruned; they were a "work of culture" to which the state must devote "science and attention" or face "want and weedy mischief" from "ignorance and neglect."[14]

Against this idea of natural dependency, Americans asserted their independent capacities, not yet as a political aspiration but rather as a commonsense description of the magnitude of their territories, the volume of their commerce, and the competence of their governments that should be acknowledged in the form and tenor of the political bonds that linked them to Britain. James Otis Jr. defined a "plantation or a colony" as a "settlement of subjects in a territory *disjoined* or *remote* from the mother country." Those "*modern* colonists," whose intrepid migrations created "*wealth* and *plenty*" from these lands, should be regarded as the "noble discoverers and settlers of a new world." The charters they received "were given to their ancestors in consideration of their sufferings and merit in discovering and settling America." Americans looked back on their history as the story of how several generations of colonists, from the first founders to those who patented western lands, secured the "independence which a bountiful nature" provided in the form of an extensive continent, open to improvement. This legacy of imperial agency seemed to them incompatible with the "slavery" expected of them within a strictly defined political economy of dependency. Having developed the "strongest sense of liberty" through colonizing, colonists understood themselves to be a people who could not legitimately be subjugated.[15]

The idea that American subjects of the king might be considered in legal terms as a conquered people provoked furious historical rebuttals. That Virginians, whose ancestors "at the expense of their own blood and treasure undertook to settle this new region," should be classed alongside the "savage ABORIGINES of this part of America" was anathema to Richard Bland, who imagined the Old Dominion's first settlers subduing the Indians on behalf of the "parent Kingdom." From 1763, as Britain took possession of newly conquered territories—in which the Crown "retained pre-eminent rights to the land"—and attempted to mold them into perfect colonial dependencies, Americans on the mainland positioned themselves in their own colonial histories as the heirs of conquerors. It seemed profoundly unjust that they should "by making conquests . . . become slaves." Never conquered, American colonists "came from England to *colonize*" as "the first adventurers into these uncultivated desarts." These "Men who came over voluntarily, at their own Expense, and under Charters from the Crown" to lands inhabited by a "savage People, scattered through the Country," believed themselves to be partners in the imperial enterprise.[16]

As British writers demanded affirmation that American colonies were subordinate dependencies, colonists were galled by so many "imperious expressions" of possession articulated in "this lordly stile, *our colonies—our western dominions—our plantations—our islands—our subjects in America.*" Britain's "great men and writers" seemed to think of American colonies more as a "parcel of *little insignificant conquered islands* than as a very extensive settlement on the continent." The idea of "America" emerged during the Seven Years' War as an imagined community whose vast extent suited the colonists' new self-confidence in their economic and social stature. If Britons viewed the "Colonies as Children, easily held in Leading Strings whilst in their Infancy," they should also acknowledge that they had now "grown to Maturity" and could only be bound by the "stronger Bands" of "mutual Friendship" and "reciprocal Advantage" through which grown children honored their parents.[17]

Even when metropolitan writers tried to diffuse the Stamp Act crisis by devising new administrative systems to placate colonial concerns with representation, they still imagined that Britain possessed the right and the power to reconfigure the political map of America. "The colonies may be divided into circles, or provinces; so that three colonies may make one circle, or province," wrote the creator of a proposed system to "new model" the political structure of "North America upon a plan of liberty suitable to the nature of colonies, and the dominion of the mother country." Even those most sympathetic to American perspectives could not resist the impulse to see American polities as arbitrary forms that could be erased, overwritten, and redrawn. Rejecting the emergence of eighteen separate colonies with their "eighteen different Parliaments" as an "accident" of history, a scheme for another "new-modelled" political geography featured a single colonial legislature, over which a single "Lord Lieutenant" from somewhere in the "centre province" would preside. As the imperial crisis deepened in the late 1760s, William Knox mocked Pennsylvanians when they insisted on calling their colony a "country." "I must not henceforth call it *province,*" he remarked snidely, "for that term implies dependence." An architect of the Board of Trade's vision for American empire in 1763 and joint undersecretary of state for the colonies from 1770 to 1782, Knox believed that thriving empires subordinated colonies and could not long survive if these were allowed to become independent countries, each with its own interests and developmental agenda.[18]

For protesting colonists, these extended meditations on the "great general principles of colonization" created an imagined union of sufferers forced to reclaim the credibility of their origins as a people. The central question of the Stamp Act crisis was whether colonists abroad possessed the rights of English subjects, and in answering it with a narrative about how they colonized America, Americans laid a foundation for a continental consciousness. "America" and the "continent" came to stand for this perceived historical accomplishment of occupying and improving the land for which they took collective credit. This geographic dimension of political identity emphasized that Americans were a "distinct People from the Inhabitants of Britain." The more colonists reached for natural rights arguments to establish their claims as subjects, the more they bound these rights to their capacities—in the past, present, and future—to occupy American territory. These "first adventurers," by leaving England to colonize America, "recover[ed] their natural Freedom and Independence" and reconnected themselves to England through the imprimaturs of their charters.[19]

In resisting the strictures that might keep them in a stable state of dependence, colonists revealed a "high and imperious ambition" of becoming a "nation of independent states; the accomplishment of which [would be] fatal to the prosperity of Great Britain." With its assertions of absolute power over the colonies, and intimations that it would consider "Enforcing Obedience by Military Measures," Britain now seemed to "despise the colonies" and consider colonists as a "set of vagabonds and transports" rather than "industrious, honest, and free people." Colonial responses to the Stamp Act confirmed British suspicions that, driven by "their passions and prejudices," they refused to see how the tax might be "beneficial to the general good," even as it might be "less commodious to their particular views." "America is like a young man growing up into his strength," one wrote, returning yet again to the well-worn metaphor of well-ordered families. "By good discipline and careful management; he becomes an honour, an ornament, and an addition of strength and security to his family. By being neglected and permitted to follow every humour, and indulge every passion, without controul; he l[o]ses his natural, social and filial affections; considers himself as the sole and only object of his concern; gives vent to every forward passion, and promotes his own single and partial interest, in opposition to every generous, laudable, and public consideration; and becomes, at length, the shame, the scourge, perhaps, the ruin of his family."[20]

If colonists could not be reached by reason, Parliament must "awe them into acquiescence." Those who had drawn up Britain's plan for America in 1763 did not advocate stationing troops in America primarily as an army of occupation, but they recognized that a show of military force might inspire obedience, particularly in conjunction with the "Savage" threat. "Under pretence of regulating the Indian trade," advised Maurice Morgann in a Board of Trade planning document in 1763, a "very straight line [should] be suddenly drawn on the back of the colonies and the country beyond that line thrown, for the present, under the dominion of the Indians." This coercive rationale for the proclamation's prohibition on interior land grants imagined American colonies "surrounded by an army, a navy, and by hostile tribes of Indians," and although this measure was not intended to "oppress or injure them in any shape," in this vulnerable position Britain could "exact a due deference to the just and equitable regulations of a British [P]arliament." The authors of advisory "hints" imagined laying siege to American cities and seizing colonists' property in retaliation for imagined future acts of disobedience. Their confinement to the "Sea Coast or on the Banks of Navigable Rivers" would expose their settlements to the guns of British vessels and forts that might "subject their Houses to be pillaged and Destroyed." Thus "moulded & limited" by new boundaries, the "Arm of Government" would "be able to stretch itself over" mainland colonies to ensure that the scattered territories linked together across an ocean into a formidable nation would not fall to pieces.[21]

In 1765, some pretended to "ardently wish" for what they most feared—the takeover of the colonies by a rival power—so that these "spurious, unworthy sons of Britain could feel the iron rod of a Spanish inquisitor, or a French farmer of the revenue; it would indeed be a punishment suited to their ingratitude." The first Lord of the Admiralty "spoke freely of driving them with a few hundred soldiers all through their country." When the ten thousand troops the stamp revenue was earmarked to support arrived in Boston, some urged that the "ringleaders of the noisy rabble" should be "brought to England, loaded with irons," to be "publicly executed, and their bodies mangled and burnt as traitors." Soon after the celebrations of 1763, British planners took America through a cycle of growth and rebellion in their imaginations. Beneath the reasoned discourse of Board of Trade and Treasury officials, with their philosophical talk of reconciled interests and the burdens of government made palatable, lurked these darker fantasies of military subjugation to keep American colonies bound to Britain.[22]

The Right to Colonize and the King's "Additional Instructions" of 1774

As the controversy over the Stamp Act intensified, the Board of Trade began "peopling, & settling, the new Governments with useful and industrious Inhabitants." As colonists on the mainland challenged the constitutional basis of new taxes, they voiced no direct objections to settlement schemes in Quebec, the maritime northeast, the Ceded Islands, or the Florida colonies on legal or philosophical grounds, but they knew that, far from their settled districts, Britain was working to reinvent the colony as a unit of empire. When Martin Howard Jr. penned his pro–Stamp Act tract, "A Letter from a Gentleman at Halifax," he was in fact a loyal subject of the king living in Newport, Rhode Island. His persona hailed from a colony he admired, governed from London in a perfect state of dependence, and so he took aim at the infamy of American disobedience from the fictive shores of well-ordered Nova Scotia. Established South Carolina planters voiced skepticism over East Florida's experimental version of subtropical plantation society, and their counterparts in the West Indies warned those who expected immediate sugar fortunes in the Ceded Islands of the astounding mortality that came with the first clearing of land for cane. Richard Bland observed that "planting Colonies from Britain is but of a recent Date." Before 1763, independent colonists had settled all the colonies "except those of Georgia and Nova Scotia," the only two enduring experiments in state-directed colonization prior to that year. He had no reason to disagree with the "Regulations lately made to encourage

Population in the new Acquisitions," but he wondered what the colonists there would do if their new governments did not honor the political customs of the old colonies. These places would probably be avoided by "Englishmen, or even by Foreigners, who do not live under the most despotic Government," he suspected, because under these new "Principles of Colony Government," "such Constitutions will not be worth their Acceptance." The "Colonies in North America, except those planted within the present Century, were founded by Englishmen," Bland insisted, who had "established themselves, without any Expense to the Nation, in this uncultivated and almost uninhabited Country."[23]

Although the colonists believed that they could not be regarded as a conquered people, some of the postwar colonies had in fact been conquered by force of arms, and all were ceded by treaty; thus, the king had the authority to settle and rule them as he wished. Colonists questioned Britain's wisdom but not its right to regulate the transatlantic economy through the Sugar Act and the Currency Act, to deploy troops to western forts and garrisons, and to impose new rules for trade with the Indians. Although they had strong opinions on the matter of western settlement, they would "not presume to say to whom our Gracious Sovereign shall grant his vacant lands." However, because each one of these new initiatives not only sacrificed colonial interests in pursuit of the idea of a common imperial good but also ignored on-the-ground realities that threatened to undermine its effectiveness, Britain's new map of empire produced a current of disquiet among those colonists who took note of it.[24]

A decade after the Peace of Paris, what had begun as a grand experiment in colonization at the periphery formed the ideas and practices that the imperial state attempted to impose as part of a "general Plan for the future Regulation" of the whole American empire. This distance between the new territories and the old settled colonies collapsed in 1774, the year Parliament passed the Coercive Acts. One of these, the Quebec Act, dramatically enlarged the jurisdiction of British Quebec so that it encompassed all the territory north of the Ohio River and set it permanently outside of future jurisdictions of other colonial governments. At the same time, the king-in-council imposed a new system for granting land that sharply regulated its distribution, increased its cost, and enforced new administrative scrutiny over landownership. Together, these metropolitan actions brought the hitherto distant threat of reformed colonization dangerously close to home.[25]

After Parliament repealed the Stamp Act in March 1766, Secretary of State Shelburne instructed North America's military commander in chief, General Thomas Gage, to prepare to enforce another means of producing American revenue. "Nothing can be more reasonable," he thought, "than that the Proprietors of large Tracts of Land . . . should either Pay their Quit Rents punctually for the time to come or relinquish their Grants in favor of those who will." Every acre of land granted under the terms of free and common socage tenure came with the obligation to pay an annual quitrent, typically two to four shillings per hundred acres, in acknowledgment that the land had once belonged to the Crown. The following year, he proposed imposing the land system that was being established in the "New & Conquered Provinces" on the long-settled mainland colonies. This initiative would begin with comprehensive surveys to divide unpatented land within existing counties and parishes as well as ceded Indian land that had not yet been incorporated into colonial jurisdictions. By gathering precise information about where these new tracts were and to whom they were granted, the Treasury could collect quitrents due to the Crown directly and far more effectively than provincial governments, whose past quitrent payments were notoriously in arrears. Such a revenue system, Shelburne believed, would "promote the good of the Colonies, & lighten the burthen which lies upon the Mother Country" without provoking another constitutional crisis, since these lands, all agreed, belonged to the king.[26]

Gage prepared to implement this directive by notifying the surveyor general of the Northern District, Samuel Holland, that "when you [s]hall have finished the Coast, you will be employed in Surveying the Interior Parts of this Country." A general survey of the continental colonies would be the first step toward implementing Shelburne's quitrent plan. It also offered the general an opportunity to clarify a geography that was, from a strategic standpoint, opaque. "There is no Map of the Inhabited Provinces of any use," Gage declared, "for there is none correct, even the Roads are not Marked. A good Military Sketch of these Provinces, was it not more, might be of great use." After Shelburne left the ministry in 1768, little came of this plan until April 7, 1773, when the Privy Council ordered North American governors to halt all grants in anticipation of a new policy for surveying, selling, and regulating North American land. Once it was proposed by the Board of Trade on June 3, 1773, and approved by the Privy Council on February 4, 1774 colonial secretary Dartmouth dispatched the king's "Additional Instructions . . . for the Disposal of His Majesty's Land" to the royal governors of Nova Scotia, New Hampshire, New York, Virginia, North Carolina, South Carolina, Georgia, East Florida, and West Florida to put them into execution.[27]

A look back at the checkered histories of colonial encroachment, squatting, and speculation proved that America's land system was "inadequate[,] improper[,] and inconvenient." Henceforth, surveyors would "cause actual Surveys to be made" before the first colonist was authorized to take possession of an acre of new land. Once they had divided a new area of settlement into tracts of one hundred to one thousand acres, prospective purchasers could inspect a "Map of the district so Surveyed, with the several Lots marked and Number'd," which was to be "hung up" for public view in every provincial secretary's office. Armed with this information, they could then bid on a tract and expect to pay at least the minimum price that had been fixed to reflect the tract's value at a well-advertised public auction. Once the surveyors drafted these district maps, they were to send copies along with a written report that described the "Nature and Advantages, not only of the whole district in general, but also of each particular Lot," to the secretary of state, the Treasury, and the king. The power to represent land at this salient scale for empire promised to open even the most remote North American frontier areas to view. By setting the minimum purchase price at six pence sterling per acre and the annual quitrent at a half penny sterling per acre, the costs of land under the "Additional Instructions" were no higher than the official rate on previously purchased grants. What was different about this measure was that Britain now declared its intention to eliminate headright grants without fee as a standard expectation for immigrants and slave purchasers, prohibit speculators from amassing huge tracts at low cost, and collect quitrents directly and systematically by holding owners of each tract to account with the threat of forfeiture if they did not comply.[28]

Faced with carrying out these new rules, Georgia's surveyor, Henry Yonge, reported that his government's collection of plats—the documents that recorded the boundaries of tracts granted to date—gave no indication of their precise geographic locations and so offered no guide to where his colony's vacant lands might be. To implement the "Additional Instructions," he would need to initiate a general survey of the province to map the whole—a three-year undertaking, he estimated—which would cost at least £60,000. When he attempted to follow the king's instructions as he prepared to grant tracts in the "New Purchase" acquired from the Creeks, the settlers who sought to claim lands between the Savannah, Oconee, and Little Rivers objected when he told them they would have to bid on the tracts that he had laid out in a comprehensive survey. The "people," he reported, "were so fond of choosing (as they call it) for themselves." South Carolina settlers to whom the provincial council had already issued warrants refused to return their plats to Charlestown in order to avoid paying new fees. In the far reaches of Virginia, settlers who had taken up lands that the proclamation made it impossible to grant continued to occupy them without sanction.[29]

Despite these instances of noncompliance, news of the regulations stalled the initiatives of speculative land companies, because backcountry settlers "would rather purchase even at a higher price from the crown and be assured of a good title than run any risque" that land they were busy improving might be taken from them by law. In reaction to the prospect that others might follow their lead and legitimate this metropolitan takeover of the colonial land system by making use of it, revolutionary committees ordered citizens to refuse to take part in its implementation. Georgia's Darien Committee cited the "shutting up [of] the land offices, with the intention of raising our quit-rents, and setting up our lands at publick sale" as a "principal part of the unjust system of politicks adopted by the present Ministry, to subject and enslave us, and evidently proceeds from an ungenerous jealousy of the colonies, to prevent as much as possible the population of America" from extending itself across the continent. Among the questions taken up by Virginia's revolutionary convention in 1775 was "to inquire whether his Majesty may, of right, advance the terms of granting lands in this colony." It passed a resolution that forbade Virginians from purchasing grants under the new system and enjoined surveyors not to lay out new lands dictated by its terms.[30]

Although historians have generally ignored the "Additional Instructions" of 1774 as an important metropolitan provocation, its implications did not go unnoticed by American colonists. Thomas Jefferson focused on the power it assumed to regulate American expansion as one of Britain's "many unwarrantable encroachments and usurpations." He singled out the new land regulations for censure in his *Summary View of the Rights of British America* (1774) and predicted that the new system would make land twice as expensive as it had been when colonists claimed tracts where they wished by headright, at the nominal price of surveying and registering them. The inevitable result of the "Additional Instructions"—and perhaps, he thought, the secret goal of making the "acquisition of lands" more "difficult"—was the reduction of the "population of our country," an effort to diminish America's stature so that its independent societies could be more easily made to play the role Britain envisioned for them as dependencies.[31]

Americans had spilled their "own blood . . . in acquiring lands for their settlement," and they had expended "their own fortunes . . . in making that settlement effectual; for themselves they fought, for themselves they conquered, and for themselves alone they have right to hold," argued Jefferson. It was time, therefore, to discard the "fictitious principle that all lands belong originally to the king" and embrace the idea, borne out by historical experience, that colonists as "individual adventurers" had created their own right to the "wilds of America" by risking their "lives, [their] labours, and [their] fortunes" as they transformed it into a civil and productive space. This theory of possession through improvement, supported by the natural rights philosophy of John Locke, assigned the ownership of New World land to those who had cleared and cultivated it. As each American colony developed from an isolated outpost into a full-fledged society with its own legislature, its representative assembly gained the exclusive right to dispose of "all the lands within the limits which any particular society has circumscribed around itself." The "Additional Instructions" ran roughshod over this right to colonize.[32]

Completing the Map of America

Quebec's changing territorial limits illustrate a problem of empire before the American Revolution. How could Great Britain establish its sovereignty over eastern North America in relation to the opposing claims of Native Americans and existing colonies? The Board of Trade drew Quebec's boundaries in 1763 to put in place its vision for dividing eastern North America into a limited settler empire along the Atlantic coast and an autonomous domain of Indian nations in its interior. As this plan fell apart under the weight of boundary erosion, uncontrolled migrations, and frontier conflicts, Britain enlarged Quebec's boundaries. This change was its most explicit revision of the Board's plan of 1763, and it was justified on the grounds that every acre within the empire should be located within a clear jurisdiction. Armed with new knowledge about America as a place, a new generation of imperial officials sought to complete the map of America left open in 1763. The geographic provisions of the Quebec Act made visible Britain's intention to take direct command of territory beyond the pale of British settlement and thus amounted to a "fundamental assault on the basic tenets of settler life."[33]

In 1763, the Board of Trade shaped Quebec to fit the proper geographic form it dictated for mainland colonies in general: they should all be Atlantic places whose boundaries were to extend only as far into the interior as the last river that emptied into the Atlantic Ocean. Following this logic, the Board urged that the "new Government of Canada . . . be restricted" so that "all the Lands lying about the great Lakes and beyond the Sources of the Rivers which fall into the River St. Lawrence . . . be thrown into Indian Country." As "possessed and claimed by the French," the Board's commissioners explained, Canada "consisted of an immense Tract of Country" that had few European inhabitants. By restricting it to lands within the St. Lawrence watershed, the commissioners identified the space in which almost all of the king's seventy thousand new subjects lived, as well as the space that was suitable for "Planting, perpetual Settlement and Cultivation."[34]

These "proper and natural Boundaries" connected every tract of land to rivers that flowed toward the Atlantic Ocean. Quebec's compact form also sought to prevent French Canadians from "removing & settling in remote Places" and keep them fixed in places where they would be "subservient to the Interest of the Trade & Commerce of this Kingdom by an easy Communication with & Vicinity to the great River St. Lawrence." Enclosing Quebec meant leaving the lands that fell beyond its colonial borders outside of any formal civil order. The Privy Council's only substantial objection to the Board's "Report on Acquisitions in America" of June 8, 1763, was the creation of an Indian country that Britain claimed but did not occupy. The king and his councilors feared that "so large a Tract of Land being left, without being subject to the Civil Jurisdiction of some Governor," might become truly ungovernable. Such a space might become a refuge for criminals, and its undefined and unguarded borders invited future land disputes among rival colonies contending to stake their claim to it. Neglecting to establish any kind of European civil authority over such an immense territory made it appear "abandoned or d[e]rel[i]ct" in the eyes of the law of nations, laying a legal foundation by which France might return to reclaim its lost North American empire.[35]

To demonstrate the Crown's sovereignty over all the lands ceded by France, the king and his council countered the Board's image of Quebec with a vision of an immense American province, which occupied all of British North America that fell outside the jurisdictions of the other colonies. Secretary of State Egremont urged the Board to reconsider Quebec's modest dimensions and "assig[n] to the Government of Canada" the vast swath of unincorporated interior territory from Hudson Bay to the Mississippi River, including the Great Lakes, and the lands around them. The Board responded with a powerful justification for a delimited Quebec: "nothing is more necessary," the commissioners wrote, "than that just Impressions of this Subject should be carefully preserved in the Minds of the Indians, whose Ideas might be blended and confounded, if they should be brought to consider themselves as under the Government of Canada." Any attempt to include the land of powerful indigenous nations within colonial boundaries would confirm their fears of imminent dispossession. In response to legal concerns created by undefined jurisdictions, the Board worked to redefine this Indian country in positive terms as an imperial commons "free for the hunting Grounds of those Indian Nations[,] Subjects of Your Majesty, and for the free trade of all your Subjects." The king assented to this novel experiment in conditional sovereignty and agreed to the boundaries recommended by the Board as drawn on Bowen's map.[36]

In the decade before the passage of the Quebec Act, Indian diplomacy transformed the Board's map of North America. Negotiations between Native headmen and the Indian superintendents drew new boundaries for Indian states across the undifferentiated territory between the Appalachian Mountains and the Mississippi River. This diplomatic enterprise erased the abstract idea of a proclamation line and replaced it with a surveyed boundary. At the conclusion of these efforts, however, large expanses of territory still remained outside any legally constituted jurisdiction. After the violence of Pontiac's Rebellion, the British demanded that Great Lakes tribes accept treaty terms at the point of a sword, but these came with no guarantees of extensive territorial rights. In 1768, Britain affirmed the Ohio Indians' status as dependencies of the Six Nations and not a nation capable of governing an extensive territory. As a result of this history of violence, there would be no recognized Indian states in the Canadian Upper Country, leaving an unchartered void in which Indians, traders, officials, speculators, and squatters competed for land and influence.

With the Quebec Act, the ministry headed by Frederick North, Earl of Guilford, demonstrated the pragmatism and expediency that characterized his approach to America in contrast with the Grenville ministry's far-reaching schemes for imperial expansion, regulation, and integration. From 1767, the working assumptions that the Board of Trade had established about the future of the North American interior began unraveling. In that year, Shelburne ordered the withdrawal of most of the military garrisons, downgraded the diplomatic authority of the Indian superintendents in favor of the governors, and began entertaining proposals for new interior colonies. The vast Indian country created in 1763, although imagined as a place through which British people would pass to engage in trade, became a place where, as North put it, "settlers are going to the inward parts from time to time." As he promoted the act, he noted that communities had formed around the "scattered posts" maintained by the military. It was intolerable to leave such people outside any system of civil order, especially when most were French Catholics who might invite French rule should Britain fail to establish its own. Once the British had made their treaties and set their boundaries with the larger Indian states, there remained only two viable options for the rest of the undefined interior: annex it to an existing government or "make separate governments" by founding new colonies. Lord North and the other proponents of the Quebec Act believed they were completing the map of North America in a way that followed from the peace of 1763 and recognized what the British had learned about America in the years since. Following a brief preamble, the text of the Quebec Act begins with a detailed geographic description of the colony's new boundaries: a litany of latitudes, riverbanks, and abstract lines that established its new shape in law. Enlarging the boundaries of Quebec filled in the jurisdictional void left by the Proclamation of 1763. Although these boundaries did not trespass on lands formally granted to a few large Indian nations, they threatened the corporate autonomy of the Ohio and Great Lakes Indians, who now fell within the colony's jurisdiction.[37]

As drafts of the Quebec bill circulated among senior officials, the debate between former Board of Trade president and colonial secretary Wills Hill, Earl of Hillsborough, and current Board of Trade president and colonial secretary William Legge, Earl of Dartmouth, shows how much had changed between 1763 and 1774. Dartmouth pronounced the Board of Trade's vision a failure and defended the plan for Quebec as the solution to the problem of establishing British sovereignty in North America. He declared before the House of Lords that there was "no longer any Hope of perfecting that plan of Policy in respect to the interior Country, which was in Contemplation when the Proclamation of 1763 was issued." Privately, Dartmouth explained that Britain had accumulated more information about the "Inhabitancy of parts of that Country" that was "then unknown." The presence of settlers in the space that Quebec's new boundaries encompassed convinced him that "restraining the Colony to the narrow Limits prescribed in that Proclamation" had always been impossible. It was vital to place the "numerous settlements of [F]rench subjects"—particularly those in the Illinois Country—under "civil government." Moreover, this establishment of formal civil authority was the only hope left for Britain to regulate illegal settlement in the region, which "in the present state of that Country . . . it is impossible to prevent." Dartmouth abhorred the vacuum of authority in the unregulated west and sought to impose clear administrative control over British land.[38]

Hillsborough, however, remained a champion of the Board of Trade's vision of American empire, grounded on the idea of limited Atlantic colonies and Indian autonomy. When he began a four-year tenure as Britain's first secretary of state for the colonies in 1768, he rejected Thomas Gage's suggestion that a government be established in Illinois to impose order at Vincennes. Instead, Hillsborough remained true to the spirit of the proclamation and ordered—ineffectually—that its residents be forcibly evacuated and resettled within the 1763 boundaries of Quebec. After reading a draft of the Quebec bill, he noted "insuperable objections to the extension to the Mississippi and Ohio." It was inevitable, he thought, that Canadians would migrate in greater numbers to Ohio and Illinois to take possession of new lands within Quebec's jurisdiction, encroaching on Indian lands and provoking violent resistance. For ignoring these realities, Hillsborough criticized the Quebec Act as the "work of a child in politics." Dartmouth dismissed these reservations as the "objection of an old dotard," part of a generation of reformers that had failed to impose order with their schemes for conciliating Indians and whose time had passed.[39]

The Quebec Act's proponents presented this revision of the plan of 1763 as a rational response to American realities that a decade's experience had revealed. Few now believed the Board's presumption that large numbers of settlers would soon transform Quebec demographically into a predominantly Protestant settler colony. Although the proclamation's halt to new western land grants combined with active Indian diplomacy had secured parts of the frontier, encroachments across the Ohio River had intensified. The act was meant to strengthen the expanded Indian boundary as a barrier to settlement. In the words of Solicitor General Wedderburn, the establishment of a colonial jurisdiction north of the Ohio River said to squatters and speculators: "this is the border, beyond which, for the advantage of the whole empire, you shall not extend yourselves." The promise of making the Indian country a free-trade zone had proved difficult without institutions of law and order. The "Extension of the Province" of Quebec, however, made the entire Upper Country subject to British law, a new status that Dartmouth hoped would entice the "British Merchant" to risk his stocks in "so many new Channels of important Commerce" that were now at least formally protected.[40]

The Quebec Act, MP John Dunning observed, "carries in its breast something that squints and looks dangerous" to American colonists. With its government ruled by an appointed governor and council without an elected assembly, Americans saw in Quebec's new form a preview of what Britain intended for the rest of colonial America. "The King has signed the Quebec Act," recorded New England cleric Ezra Stiles in his diary, "extend[ing] that Province to the Ohio & Mississippi and comprehending nearly Two Thirds of the Territory of English America, and establish[ing] the Romish Church & IDOLATRY over all that Space." For decades, Britons observed the growth of New France and Louisiana into a malevolent zone for French intrigue across published maps of North America. Stiles drew his own map of the same space—surrounding, outflanking, and threatening Britain's seaboard colonies—colored it red, and titled it "The Bloody Church." **[234]** Tyranny and popery, the twin pillars of British Francophobia during the Seven Years' War, now appeared as weapons aimed at Protestant America by a hostile British state.[41]

The First Continental Congress's Articles of Association condemned the act because it deployed a threatening population of French Catholics at the very "frontiers of these colonies," poised to invade at the bidding of a "wicked ministry." On the last day of its final session on October 26, 1774, the Congress sent an "Address to the Inhabitants of the Province of Quebec." Written by John Dickinson, it appealed to them on the grounds that they were being deprived of the rights to which they were entitled: representative government, habeas corpus, trial by jury, freedom of the press, and free land tenure without obligatory duties to the government. The act gave colonists "great reason to fear the loss of [their] liberties," wrote Philip Livingston, "when surrounded by a multitude of slaves." Colonists objected to efforts to equalize America's inhabitants. They saw the recognition of Indian polities and the observance of civil and political rights for French Catholics as a diminishment of their standing as British subjects. Against British attempts to reconcile contested spaces by "accommodating multiple claimants," colonists resisted anything but their own "unitary control over the spaces in which they lived" in the place they increasingly called "America"—a continent they imagined to be theirs by right. The Quebec Act concluded a process of dividing the interior into Indian states, making space within it for possible new colonies, and, finally, establishing the largest colony in North America without any of the basic guarantees of English law. One glance at the new map of Quebec as represented by maps such as William Faden's *The British Colonies in North America* (1777) made it clear that American colonists were to be pushed aside as meaningful agents of colonization. [**235**] The Quebec Act codified an abstract fear of future tyranny and made it visible in the present.[42]

The Second Continental Congress's justifications for breaking the "political bands" that had connected the colonies to Great Britain included the charge that the king had "abolish[ed] the free System of English Laws in a neighbouring Province, establishing therein an Arbitrary government, and enlarging its Boundaries so as to render it at once an example and fit instrument for introducing the same absolute rule into these Colonies." Several other clauses of the Declaration of Independence took aim at the colonial order Britain attempted to impose after 1763. The king "kept among us, in times of peace, Standing Armies without the Consent of our legislatures" and made the "Military independent of and superior to the Civil power." Not only did Britain station troops in Boston to enforce two other "intolerable" Acts—the Boston Port Act and the Massachusetts Government Act—but it had also deployed forces in newly acquired territories and concentrated ships and men in new military centers, such as St. Augustine and Halifax. When civil and military officers came into destabilizing conflict, particularly in West Florida, Britain gave ultimate authority to military commanders. Lacking assemblies to check their governors and commanders, noted Ezra Stiles, the "military Provinces of Quebec and the two Floridas" predictably failed to send delegates to the Continental Congress in 1774.[43]

By mobilizing Cherokee and Iroquois fighters, George III had "endeavoured to bring on the inhabitants of our frontiers, the merciless Indian Savages." Imperial planners viewed diplomatic agreements with Native nations as having strategic value in keeping the threat of Indian attack alive in the minds of fearful colonists. After negotiating treaties that promised civil autonomy, British officials gave Indians surprising latitude to enforce "justice" within their territories. Fomenting new frontier attacks during the War of Independence was the latest and most serious instance in which Britain aligned itself with Native interests at the expense of the colonists since the Peace of Paris. By "raising the Conditions of new Appropriations of Lands," the king's "Additional Instructions" of 1774 "endeavoured to prevent the Population of these States." Hindering their ability to grow and expand sought to force them into unnatural states of dependency to better conform to the British ideal of proper subordination of colonies. In a phrase excised from the Declaration of Independence's final draft, Jefferson upheld the image of the intrepid first colonists who had spent their own "blood and treasure" to colonize America and, by doing so, generated their own rights to rule the land by occupying and improving it.[44]

Before Jefferson summarized the "long train of abuses and usurpations," which "pursuing invariably the same Object evinces a design to reduce them under absolute Despotism," he wrote of a "series of oppressions, begun at a distinguished period, and pursued unalterably thro' every change of ministers, to plainly prove a deliberate, systematical plan of reducing us to slavery." Although most of the Declaration's grievances were of recent vintage—acts of war committed by British forces on America soil after 1773—Jefferson dated this conspiracy to deprive colonists of their liberty much earlier, some "twelve years" before 1776, to a moment when the king, Board of Trade, secretary of state, Privy Council, army, Treasury, and Admiralty worked together to begin reshaping the geography and commerce of American empire to better suit the interests of metropolitan Britain.[45]

Atlases of Empire

London's mapmakers produced complex images of American places, and none were more impressive than those collected in the great atlases that celebrated British knowledge and command of a world that was being torn apart by war. In the second half of the eighteenth century, surveyors sent home hundreds of hand-drawn maps, plans, and charts that represented British America in stunning detail. As map publishers gained access to this growing archive of manuscript images, they engraved new maps that revealed striking geographic forms. From 1768 to 1781, they republished maps from deep catalogs and engraved new images that integrated the latest surveys to produce six important published atlases. Rigorously surveyed representations of the northeastern, southeastern, interior, and Caribbean frontiers distinguished these collections. The original purpose and plan for these maps was to clarify the geography of poorly understood places, but during the war, Britain used them to open American places to attack and to illustrate for a broader British public, perpetually interested in visualizing distant sites of battle, the progress of the campaigns against the colonists. Although many of the same images persisted in these American atlases, their meanings changed as the colonial system they were created to enable came undone.

Naval officers began taking episodic surveys of the maritime northeast at the close of the Seven Years' War, and surveyors James Cook and Michael Lane continued them more intensively under special Admiralty commissions throughout the 1760s. Based on the manuscripts they produced, Thomas Jefferys engraved sixty charts across thirty-six copper plates. The partnership of Robert Sayer and John Bennett acquired these after Jefferys's death in 1771 and used them to publish *The North-American pilot for Newfoundland, Labradore, the Gulf and River St. Laurence* in 1775. The Admiralty's decade-long survey examined thousands of nautical miles of coastline in the space that had once constituted a great breach in the contiguity of British America. **[236]** Mariners could purchase this atlas, which charted safe passage from the fishing harbors of Newfoundland to the great bays and islands of the western gulf, along with separately printed sailing directions, to enable their safe passage into a largely untapped fishery. Samuel Holland's map of the island of St. John, divided into townships, appeared as the lone terrestrial map within this maritime world. *The North-American pilot* delineated one of the century's most contested regions in clear lines and a common graphic style that presented the Gulf of St. Lawrence as a known place, from the developed fortified naval station at Halifax, Nova Scotia, to previously remote and uncharted Chaleur Bay and the Magdalen Islands. These published pilots maintained the Board's vision of opening remote and obscure places to settlement and development through the clarity of well-organized geographic knowledge. Although nowhere reflected on these charts, however, the war had begun unraveling the idealized spatial order they depicted. In 1778, Congress considered sending "Expeditions for the Relief" of isolated St. Pierre and Miquelon, two islands granted to France by the Treaty of Paris, which Cook had declared incapable of sustaining a resident population. A stream of migrants from the mainland arrived to populate St. John, as the Board had hoped, but as loyalist refugees rather than as aspiring fishermen and farmers responding to new incentives to lease the island's land from the proprietors of its townships.[46]

Sayer and Bennett published Thomas Jefferys's *American Atlas* posthumously in 1776. The twenty-nine images in this "factice," or false atlas (so called because it was made up of previously published sheets), included new regional maps that showcased the transformation of British North America since the Peace of Paris. Samuel Holland's map of the Middle Colonies frames New York, Pennsylvania, and Delaware as a new space, defined by the watersheds of its four great Atlantic rivers. [**237**] To the west, another newly drafted regional map—provincial surveyor William Scull's map of Pennsylvania and its "Extensive Frontiers"—articulates the colony's western mountains with new clarity and complexity. It shows the vast extension of Berks County into the Susquehanna River valley, where Scull had taken up residence to lay out new tracts for settlers seeking land ceded by the Iroquois in 1768. [**238**] To the south, it featured Henry Mouzon's "An Accurate Map of North and South Carolina with Their Indian Frontiers," which pulls back from the coastal plain to bring the emerging Carolina backcountry into focus. [**239**] Mouzon's map was one of the only published maps of America that printed the course of the negotiated Indian boundary line that John Stuart, William Johnson, and Indian headmen had negotiated and surveyed since the first congress of Augusta, Georgia, in 1763. Maps and charts by British surveyors derived from surveys taken since the Seven Years' War extended the atlas's coverage of British America to include the Gulf of St. Lawrence, St. John Island, the fishing banks south of Nova Scotia, the coasts of Florida, and the Mississippi River. These regional maps in the *American Atlas* reflected new conceptions of space as well as the changing geography of settlement since 1763. The atlas's new edition of Emanuel Bowen's "An Accurate Map of North America" etched the boundaries the Board had drawn by hand on a previous version of the map so that they could be inscribed on its surface, capturing North America's provisional political geography at the moment of the new empire's inception in 1763. [**240**] Betraying no sign of the sharp debates over American dependency or the military struggle underway on the continent and throughout the islands, the *American Atlas* pictured the spread of detailed knowledge of American geography for an empire still poised for expansion.[47]

William Faden published *The North American Atlas* in 1777. His own map of *The British Colonies in America* was one of a few to show the enlarged shape of Quebec dominating the center of the continent after 1774. New maps and plans by army engineers John Montresor, Bernard Ratzer, and Claude Joseph Southier annotated North America's landscapes to record places of military conflict that shifted the scale of the atlas from broad regions to strategic locales. Charles Blaskowitz, a long-serving assistant general surveyor under Holland, charted the coast of Rhode Island in 1764, employing the general survey's rigorous standards for the representation of coastlines. He returned with a surveying team in 1774 to perfect the chart's plan of Newport and map its countryside at the invitation of Rhode Island's "Principal Farmers," whom he listed by name. [**241**] Below this acknowledgment of the local gentry, Blaskowitz drafted a narrative that praised the colony's religious toleration; its "delightfull" summers; and its "many advantages" for commerce, husbandry, and agriculture. On December 6, 1776, following years of violent resistance to the British naval and customs officials in Rhode Island, British major general Sir Henry Clinton's expeditionary force seized Newport and used a copy of Blaskowitz's chart to organize its occupation by seven thousand soldiers. Faden inscribed his published version of it with a dedication to Hugh, Earl Percy, the commander Clinton left behind to defend the town, and annotations that located the "Works & Batteries raised by the Americans" in their failed stand against the assault. These and other images of revolutionary engagements in *The North American Atlas* showed how quickly the Board's new maps, commissioned to settle and develop the empire, could be repurposed to aid in the attempt to conquer it.[48]

As Britain invaded America, the army's engineers remapped it. They visualized engagements at Lexington and Concord, Bunker Hill, Halifax,

Charlestown, New York, Quebec, Lake Champlain, Fort Ticonderoga, Trenton, Brandywine, Saratoga, Newport, Savannah, Camden, and Yorktown, picturing these contested places at the salient scale for colonization, from a half inch to two inches to the mile. The same military eye that shaped the look of maps produced by George Gauld, Samuel Holland, John Stuart, James Cook, John Byers, and William De Brahm persisted during the War of Independence. These new maps were the product of the same methods used to render the shape of the land in true proportion across distances combined with impressionistic sketches that made the two-dimensional surface of a paper map simulate the look of a real landscape. Like the new maps of empire, composed between 1763 and 1775 with the optimism that they would be the means of planning ideal colonies from the ground up, the military maps produced from 1775 to 1781 pictured the same signs of human occupation on the land, from the rectangular geometry of cultivated fields to the squares and lines that stood for houses, bridges, and roadways. The additions that made these wartime maps distinctive were the jarring iconographies of battle. From the thin colored rectangles drawn to locate the positions of massed troops to the jagged siege lines erected across American cities targeted for attack, army mapmakers updated old maps with new surveys to document the war's progress. They populated North American harbors with meticulous drawings of men-of-war, complete with their names, sails, rigging, and flags, and drew lines to reveal the vectors traced by shells blasted from their cannons.[49]

Thomas Jefferys's final masterwork went beyond producing engraved copies of original manuscripts and gathering together previously printed sheets. Published by Sayer and Bennett in 1775, *The West-India atlas* integrated new and old geographic knowledge to create a coherent, intentional, and complete work of cartography. It presents the Caribbean as a multinational space that Great Britain controlled through the geographic reach and economic dynamism of its maritime commerce, a vision of integration and rising prosperity that was still possible at the time of his death in 1771. A few maps in *The West-India atlas* follow the convention of the English island map, showing each colony floating in its own portion of a seemingly trackless sea. At the heart of the atlas, however, sixteen separate sheets joined to form a single image of an interconnected circum-Caribbean region. An index map showed how these discrete parts fit together to cover not only the islands but also portions of the North and South American mainland. [**242**] These interlocking images are hybrids of terrestrial maps and sea charts, both an "Atlas and Pilot for the West-Indies, shewing both the Geographic and Hydrographic parts." They depicted British and non-British spaces at equal density and resolution, covering all of the islands as well as the encircling coastal plain of Latin America, including the "habitations of the wild nations" as well as the networks of roads that linked towns across the Spanish Main.[50]

To show how the sea bound this West Indian world together, Jefferys inscribed a single route across these maps: the course taken by the Spanish galleons, whose "constant track may be considered as an interesting object in the history of commerce." Created to protect the treasure fleets from English, Dutch, and French piracy and privateering in the sixteenth century, this convoy system known as the *Flota de Indias* persisted to the end of the eighteenth century. But for Jefferys, the tiny figures of ships he etched into these sixteen sectional maps were more than a curiosity. No longer was this space primarily interesting to Britons for its plunder. Instead, the maps testified to a "complete knowledge of the country . . . and shews its communications and connections, whether natural or acquired, without which there could be no commerce." Joined together, the maps laid bare a geographic sphere of influence commanded by the power of commerce. British dominance in American trade, he declared, was driving the French empire to ruin. Note the lost "splendour" of Martinique, once the center of a "lucrative contraband trade with the Spaniards." Its planters were now deep in debt, surrounded by British islands and seafarers on every side, and this once envied island was "reduc[ed] . . . to itself" and, in its isolation, became "diminished" in its "trade and riches."[51]

The *West-India atlas* presented the most contested place in the New World as a British sea. Compiled from seized charts, firsthand accounts of travelers in the "Spanish and English Settlements," and the "abundance of new materials" that arrived in Jefferys's London workshop "when the victorious fleets returned to England," this new collection strove to displace the "ancient" works of John Seller, featured in the Blathwayt Atlas, as well as the "undigested compilation of Charts" derived from Dutch originals, too "much in the hands of British sailors," that pictured the Caribbean at a jumble of "different scales" and were filled with "errors, or absurdities." Jefferys's new charts were grounded on "thirty astronomical observations" and the records of voyages kept in ships' logbooks to establish true distances across space. They integrated new information recorded on "English Maps which have appeared since the Peace," and displayed place names, so frequently "mangled and mis-spelt," in English, French, and Spanish. Against the convention of splitting this maritime zone into two parts—the Gulf of Mexico and the Caribbean Sea—Jefferys urged his readers to see this maritime world bound together, comprehending "all the coast of the main land, which lies adjacent to it, as well as all the islands, the chains of which seem to keep back the sea, which beats with violence against this part of America."[52]

Robert Sayer dedicated *The West-India atlas* to William Young, head of the Commission for the Sale of Lands in the Ceded Islands, but the atlas's vision of British supremacy across the Caribbean was published during a year in which American privateers menaced British islands and shipping. From 1775, New England merchant vessels that had brought lumber and provisions to island port towns descended on British islands to burn plantations, seize slaves, and capture vessels as prizes of war. As the Revolutionary War spread to the British Caribbean, Attorney General Charles Winstone wrote desperate accounts from Dominica, describing American privateers "swarming about Us." Tobago's position as Britain's southernmost colony in the hemisphere left it in a "defenceless State," so "exposed to be[ing] unsettled and plundered" that it practically "Invited their Enemys to these Depredations." As "Rebel Privateers" descended on its northern bays to intercept supplies from Britain, Spanish pirates raided its southern coast, taking slaves and selling them in Trinidad. Since Trinidadian traders frequently crossed the Galleons Passage with livestock to sell to British planters, they knew "every Creek and Landing place in the Island," and being an "uncivilized, lawless, and barbarous set of People," they raided the island with impunity in small launches that could "approach the Coast" of Tobago "unperceived" in just two or three hours.[53]

After cruising against privateers in the southeastern Caribbean in the sloop-of-war *Favorite,* Commander William Fooks sketched a "Map of the Bearings of Grenada & Tobago to the Spanish main & Trinidada." [**243**] It made clear why the British frigate stationed on the north side of Grenada to defend the region against attack offered little protection. Sailing south against the "Currents and Winds" took "at least four or five days" and as long as a month to reach Tobago's provisional capital at Scarborough along the southern shore. Privateering raids depopulated Tobago, leaving only a few working sugar estates behind before its 1781 capture by the French, who retained it at the Treaty of Paris in 1783. Britain regained Tobago at the Treaty of Paris (1814), but by 1828 there were only three resident proprietors left in the colony. Britain had surveyed Tobago to establish a structure in which a perfected plantation society could take root and flourish. The war violently dismantled this vision of spatial order, which lived on only as images on maps of the island.[54]

After an "Honorable Capitulation" to French forces on September 7, 1778, privateering raids against Dominica stopped, along with much of the island's commerce, plunging the colony into the "utmost Distress." Dominica's slaves, "much in want of Cloathing," now spent their time growing cassava, manioc, and coffee, reversing the intensive specialization in sugar that had distinguished British plantations before the war. While "every Person who ca[n] leave the Island is doing it," those who remained behind found "Provisions and all other Necessaries scarce, at treble the former prices." Many planters managed to pay the costs of their expenses during France's occupation, but the war's disruptions had halved the value of their land and slaves. Wartime privations encouraged slaves to run away to the protection of maroon communities in Dominica's mountains, from which they descended on the plantations in daring raids. The managers at the Rosalie settlement struggled to keep the Sea-Side Estate functioning after maroons staged a night-time attack, setting fire to the boiling house and cane fields and killing the manager, the carpenter, two overseers, and the "chief negro driver." When Dominica was restored to Britain in 1784, slaves sold briskly at "amazingly high" prices, a sign of economic confidence in its postwar prospects, and Rosalie was one of about fifty working sugar plantations that remained on the island in 1791 as sugar production resumed on a large scale. The crisis of the war provided an opportunity for the wealthiest planters to consolidate their command of plantation wealth. Although the Board of

Trade had hoped to spread productive land among a large number of resident planters with its land regulations, Dominica followed a more typical West Indian pattern of development and social formation into the nineteenth century, featuring an elite planting class with many absentees. The battle waged to sustain the Ceded Islands during the war exposed the volatile realities of navigating Caribbean plantation slavery, which were nowhere visible on the engraved sheets of the *West-India atlas*.[55]

Britain had devoted men, money, and matériel on an unprecedented scale to survey America in the generation before the American Revolution in support of new colonization, and then repackaged that knowledge to make war on America. Urged by senior officials to publish a compact geographic reference work, Sayer and Bennett condensed American geography onto six maps, which folded out of a six-by-nine-inch octavo volume, bound in calfskin and protected by a cloth slipcase. *The American Military Pocket Atlas* (1776), also known as the "Holster Atlas" because some wore it attached to leather cases on their belts, provided officers with a common geographic work of reference. At the Board of Trade's Whitehall offices, clerks examined unwieldy, multisheet manuscript maps, which surveyors sent from America. But far from London, where opposing forces maneuvered across vast theaters of war, commanders in the field turned to this compact edition, which compressed geographic space for the purpose of battlefield clarity. In addition to "A Compleat Map of the West Indies," assistant surveyor general Bernard Romans divided North America into three sections, derived from the "Modern Surveys" of the 1760s and 1770s. [**244–247**] These maps gave commanders a common language to describe the hundreds of places spread across the continent and the islands as land and sea forces coordinated their attacks across spaces "which now are, or probably may be[,] THE THEATRE OF WAR."

"On the map of America, the ministry may form a plan of operations" to bring the colonies to heel, cautioned the author of *The Critical Moment*, but once dispatched to the continent, soldiers could not "travel so easily in America, as our wise politicians on a sheet of paper." A new appreciation for the vastness of the American mainland gave commentators pause before imagining a war to suppress independence. If "twenty or thirty thousand soldiers were to land in Canada, and the same number in the southern provinces, either in Georgia or South Carolina," was it "probable that these armies could ever meet?" Once British troops ventured beyond the well-settled coastal plain, they would find themselves bogged down in a wilderness of "thick woods."[56]

When news of fighting in Massachusetts came to the attention of Prime Minister Charles Townshend, he asked his colonial secretary, the Earl of Dartmouth, to show him Samuel Holland's map of Boston Harbor to "gratify my curiosity as to this cursed Height above Charles Town, which I was anxious about for [the] moment I heard the Rebells seized much artillery." British officials reached for their maps to understand American geography, but frequently found them wanting. No chart could capture the shifting sandbars of Charlestown harbor, on which an attempted 1776 invasion of South Carolina foundered. One British officer, when he learned the news of General John Burgoyne's failure to take Albany, New York, criticized the hubris of metropolitan battle planners. These deluded leaders, "sitting in their closets, with a map before them," he wrote, "ridiculously expect[ed] the movement of an army to keep pace with their rapid ideas," as if they could direct troops "at a distance of three thousand miles" across the real terrains of North America from London.[57]

As Sayer and Bennett and William Faden compiled new atlases of a disintegrating empire, American revolutionaries attempted to imagine their incipient nation as something more than a collection of former colonies. As head of the Second Continental Congress's Board of War and Ordnance, John Adams saw the same need that British commanders did. In the hope of making Continental "Officers . . . perfect Masters of American Geography," he began "making a Collection of all the Maps, extant, whether of all America or any Part of it, to be hung up in the office, So that Gentlemen may know of one Place in America where they may Satisfy their Curiosity, or resolve any doubt." He asked the army's judge advocate general, William Tudor, to contribute to this effort by "inquir[ing] at every Print sellers shop in New York, and of every Gentleman, curious in this Way concerning American Maps." In Philadelphia, Adams had seven images "framed and hung" on the walls of the War Office. "America is our Country," he wrote to his wife, Abigail Adams, "and therefore a minute Knowledge of its Geography, is most important to Us and our Children." As he examined this nucleus for a growing national map collection, he imagined a rising generation scrutinizing these maps until there was not "a State, a City, a Promontory, a River, an Harbour[,] an Inlett, or a Mountain in all America, but what should be intimately known to every Youth." What were these maps that every child born under a free government should master as part of an essentially American "liberal Education"? So far, they amounted to a handful of aging artifacts from the era of the French and Indian War and a few images of the Pennsylvania frontier, already made out of date by recent settlement.[58]

To "turn the attention" of his own "Family to the Subject of American Geography," John transcribed for Abigail a full list of these "dry Titles, and Dedications of Maps": Green's six-part chart of the hemisphere as well as his map of New England, Popple's and Mitchell's landmark maps of North America, *A New and accurate Map of North America* drawn from a French original, and three maps of Pennsylvania. The War Board's first seven maps, available to any buyer in the print shops of New York, Philadelphia, or London might at first glance seem like poor material out of which to form an independent image of the new nation. Yet these old maps, by being singled out and included within the first official map collection of the United States, illustrated a new continental destiny for the former colonies.[59]

Former colonists could visualize Thomas Paine's common-sense notion that it was "absurd for an Island to rule a Continent" by seeing how Green's chart of the Atlantic revealed the British Isles to be small and marginal compared with a North American landmass so vast that it could not, in the natural order of things, remain dependent forever. Although it was composed to proclaim the range and superiority of British geographic knowledge, to revolutionary eyes, this chart positioned the new American nation at the center of the hemisphere, from which it might orient its trade to the west as well as to the east, stretching out to the Pacific as well as the well-traveled routes of the Atlantic Ocean. The Declaration of Independence charged the king with the crime of cutting off American "trade with all parts of the world." By closing the port of Boston in 1774 and blockading American shipping in 1775, Britain attempted to isolate the colonies. On this chart, revolutionaries could imagine the global sweep of a new independent commerce, unburdened by navigation acts, onerous customs duties, or punitive port closures. In the language of the Declaration, the "Free and Independent States" that this chart placed in the center of the western world would exercise their right to "contract Alliances, establish Commerce, and do all other Acts and Things which Independent States may of right do."[60]

Adams's next War Office map, John Mitchell's *Map of the British and French Dominions in North America,* was still regarded as a state-of-the-art image of the continent. Indeed, when the war ended and British and American negotiators hammered out a peace plan in 1783, they drew the boundaries between Canada and the United States on a copy of this map, first published in 1755. This map's original purpose was to make the strongest possible sovereignty claims for Britain against a threatening France. But twenty years later, as Americans fought against British policies, Mitchell's image of expansive American provinces resonated in a new way. In 1763, British authorities worked to constrain expansion. The king's proclamation prohibited new land grants in any place that might be regarded as Native territory. British officials negotiated new boundaries with the Creeks, Cherokees, and Iroquois that terminated long-standing charter claims for western lands. In 1774, Britain cordoned off a vast part of this territory by putting it under the formal jurisdiction of Quebec, alienating colonists who thought they had fought the French in North America for the right to colonize these very lands. When Adams looked at Mitchell's map, he could imagine the future expansion of an independent nation, unconstrained by British limits.

Another map in the collection, Peter Bell's *New and accurate map of North America,* could be seen in the same way—as a map of a continent prepared by its ancient charters for ongoing colonization. **[248]** Former colonists believed that their title to American land was not a favor granted by the king but rather a right they had earned by risking their own lives and fortunes in taking possession of the New World. Disputes with Britain brought this assumption to the forefront of political discussion. John Green's *Map of the most Inhabited part of New England* could be reimagined through this lens of settler expectations. **[249]** From Boston to Fort Frederick on the Pemaquid River and from New Haven to the wilds of the upper Connecticut River valley, enterprising townsmen, empowered by their legislatures to found townships across an ever-expanding frontier, had secured this territory for themselves and for Great Britain.

Perhaps because Adams assembled this map collection in Philadelphia, three of its maps describe a sequence of Pennsylvania's expansion. When Lewis Evans visited the fork of the Susquehanna River in 1742, it was the site of Shamokin, a Delaware Indian town. Nearly three decades later, on William Scull's 1770 *Map of the Province of Pennsylvania,* Shamokin had vanished from the map. [**250**] In its place, along Shamokin Creek, was the Euro-American town of Sunbury. This frontier town anchored a rapidly settling farming district just within the boundary negotiated with the Iroquois. British critics had rebuked provincials for regarding their colonies as "countries" within a composite British nation, but across the surface of these maps, citizens of a new nation could see their own stature as a colonizing people validated. Adams's maps revealed the power of settler colonialism to make every North American state its own empire.[61]

Such maps, all anthologized in previous published atlases, revealed an ever-expanding national domain rather than a distant colonial periphery. In New England, Pennsylvania, and Virginia, the same story of expansion and settlement gave Americans a common cause, a common sense of history, and a common grievance against British limits and interventions. Such maps made visible the nation that Americans were fighting for. Just days after Congress issued its Declaration of Independence, Philadelphia physician Benjamin Rush urged his fellow Americans to give "up colony distinctions" and instead think of themselves as "now One people—a new nation." Adams himself had started referring to Americans collectively as "the Continent." Even as they put armies in the field, many feared that if the North American settlers managed to achieve political independence, the new nation would fragment into a collection of ungovernable provinces or two or three regional confederacies. These maps held out a vision for integration through expansion that might bind the former colonies together as united states. Unlike the visual uniformity of the most ambitious of Britain's atlases of empire, however, Adams's eclectic map collection combined a riot of colors, forms, and competing styles, a visual image of an unstable confederation that might soon break apart.

The Atlantic Neptune was the only great British atlas of the eighteenth century that came close to realizing the vision of perfected geographic knowledge to which all of them aspired. By the war's end, it was clear that Britain had failed to gather a comprehensive body of knowledge about America that could be deposited in a dynamic imperial archive through which the imperial state could direct colonization or wage a war to retain its colonies. Nevertheless, the publication of this great atlas proclaimed to the world the acumen of its engineers, surveyors, diplomats, and cartographers. J. F. W. Des Barres populated *The Atlantic Neptune* with charts from his own naval surveys of the Nova Scotia coast, to which he added scores of charts generated by other Admiralty, army, and Board of Trade surveys. Focused intensively on the maritime northeast and New England, the atlas made a bid for continental coverage by including maps, plans, and charts of Gulf Coast ports, southeastern cities and rivers, two maps of Jamaica, and a chart of Havana harbor. Released in four volumes from 1777 to 1781, the complete atlas featured 176 separate plates. When the Board of Trade launched the General Survey of North America in 1764, its work was meant to continue until all of British America was mapped to the salient scale for empire. But after just a dozen years, the war disrupted surveys that were never resumed. American forces captured William De Brahm, the southern surveyor, in 1776 and deported him to France as a prisoner; Samuel Holland departed Nova Scotia for Britain in the same year, leaving behind two disabled soldiers to look after surveying equipment left behind in Quebec. Although Britain's American surveys were abandoned before they could be completed, Des Barres gathered together the manuscripts they had produced and etched them on new copperplates for publication. The result was *The Atlantic Neptune,* an epic of American cartography, published with money Parliament had earmarked for the now defunct General Survey.[62]

Des Barres engraved *The Atlantic Neptune* in a common style. Leafing through its images is an immersive experience that underscores the immensity of American landscapes. Perhaps this effect is more pronounced along the edges of the complicated coastlines of the northeast, inscribed to make every shoal, rock, and promontory stand out in relief, than it would be anywhere else. Beyond the shore, we can see buildings dot the coast, the edges of fields, and the rolling topography of the highlands beyond. These intricate coastal landscapes stretch on for mile after mile, sheet after sheet, in remarkable detail. Des Barres presents this high-resolution world, for the most part, as one of peace and silence, curiously empty of human beings. This immensity and vacancy imparts an elegiac tone of loss to the images. Similar to the maps of Florida's coast by William De Brahm, Des Barres enhances this wistful view of American geography with trompe l'oeil graphic effects that make it seem as if the printed page were in fact a jumble of found images and documents scattered across a table, or that the edge of the sea flows across the margins of the printed page. [**251**] In an even, flowery script, handwritten note cards seem to rest on top of charts and seascapes. The effect is playful, artful, and dreamlike, as if the viewer were sorting through a collection of mementos of past voyages. [**252**]

Sea charts often featured profiles of the coastline, to alert ship captains to the shape of the headlands they should see if they had correctly navigated to their destinations. In 1759, Admiral Edward Hawk complained to the Admiralty that he had lost two ships for want of a copy of the famed *Neptune François* and suggested that this standard work be "printed on a large scale, and translated, at the King's expense, and given to all the ships that are employed, as well as the best draught that can be had of the different coasts in any part of the world where our ships may be sent." *The Atlantic Neptune*'s 145 views and coastal profiles serve this navigational need, but they also add to the general tone of calm that pervades the atlas. [**253, 254**] There were no people to be seen in these coastal views, only distant vessels with their sails puffed by generous winds sailing across untroubled waters. The cumulative effect of these images is to see British America as an expansive space linked everywhere by the sea. These peaceful shores, their tidy port towns and farmsteads, the busy maritime traffic that bound even the most remote places together—it all seemed so benevolent and self-sufficient that, after looking at the maps for a while, it is easy to forget that a war was being contested in these very spaces when Des Barres was creating *The Atlantic Neptune* in London.[63]

After a final Admiralty expedition to obtain soundings along the Nova Scotia coast concluded in mid-summer 1775, the naval vessel assigned to the task was dispatched to Boston to support the British blockade of the harbor. A chart of Falmouth Harbor, published in 1781, bears no trace of the violence that took place there six years earlier. [255] On October 18, 1775, British Captain Henry Mowat, commander of the *Canceaux*—the hydrographic vessel that worked for several years supporting the General Survey—led a naval fleet against the town of Falmouth, Massachusetts, site of the modern-day city of Portland, Maine. After Americans seized ships and stores intended for British forces at Boston, Mowat fulfilled his orders to take retribution. He notified Falmouth's residents that they had two hours to evacuate, after which he intended to "execute a just punishment" for their behavior. After firebombing the town, Mowat sent in troops and set fire to what artillery shells had not yet burned. Mowat himself had overseen some of the coastal surveys that appeared in *The Atlantic Neptune*'s sheets, and before putting Falmouth to the torch, he surveyed it, generating the data that produced this chart.[64]

Britain's atlases attempted to reconcile American violence and American vastness. A number of atlas compilers formed their volumes from the maps derived from military surveys of the Seven Years' War. Every atlas that drew on this well of cartographic material inevitably told a war story. In one sense, the maps figure as material tokens or patriotic artifacts of the conflict, allowing Britons to remember the war; meditate on early, devastating defeats in the American woods; and celebrate a series of surprising, nationally uplifting victories: the conquests of Louisbourg in 1758 and Quebec in 1759, and a string of naval victories in the Caribbean that took Havana, Martinique, Guadeloupe, and Grenada in 1762. The atlases recounted the story of the war, tallied its costs, and acknowledged the great sacrifices Britain had made to triumph over France in America. Atlases produced after 1775 included not only maps derived from the Seven Years' War but also new maps made to take control of rebellious colonists. They told the story not of a single war but of two contests to claim and reclaim America. These tempered patriotic celebration with more complex nationalist emotions about the American uprising as a betrayal; their images could sound notes of cynicism about whether it was worth the effort to forge these fractious colonies into a stable empire. They became unintentionally ironic when readers viewed their hegemonic visions against the chaotic news of a war that Britain proceeded to lose.

In addition to the war maps, these atlases of empire featured maps commissioned following the Peace of Paris to represent places deemed vital to empire. Britain took possession of this immense territory by commanding exquisite and detailed knowledge of it; it invested resources, effort, and expertise in the task of visualizing distant lands with precision. The very existence of these maps and the countless hours of surveying that produced them testified to Britain's authoritative knowledge of American spaces. Especially in the graphic images engraved to illustrate coastal landscapes and bustling ports, we can see how this huge space could be held together for the benefit of empire. Maritime commerce—the foundation of Britain's economy and its singular advantage over Spain and France—these images seemed to say, could bind Britain to the resources of a vast American continent and the lucrative West Indies and, in the process, enlarge an island nation's population, trade, and wealth far beyond its natural capacities.

Patriotic Britons could look at these maps and see America as an extension of Britain, and patriotic Americans could look at the same maps and see an independent continental nation taking form. For two decades following the Peace of Paris, mapmakers engraved surveyors' manuscript images onto copper plates and compiled the sheets that came off their presses into published atlases; collectors on both sides of the Atlantic stocked their libraries with volumes and wall maps of American lands and waters; and the king's subjects in Britain and America who saw these new maps, whether or not they could afford to possess them, filled their understandings with new geographic forms that described a shared Atlantic empire. Although individual maps communicated powerful messages to early modern viewers—declaiming imperial sovereignty, recasting histories of discovery, pinning down indigenous societies beyond the edges of expanding frontiers, and celebrating the transformation of wilderness into colonized space—the meanings of an individual map were limited by the bounds of its particular frame as one view of one place at one particular moment. Multiple maps viewed in sequence spoke in complex and open-ended new narratives. The sheer volume of images contained in these new collections offered the illusion that they depicted the world as it was, in all of its complexity, at multiple scales of representation. Although Britain's imperial state commissioned these very images to create rigorous representations of America, it could not manage this surging influx of geographic information for the purpose of directing colonization from across the Atlantic Ocean. Although the Board of Trade's wished-for metropolitan archive of American places failed to function as a dynamic instrument of empire, the great atlases produced by London's prolific map publishers simulated the comprehensive spatial knowledge that this program had sought to create.

The tension between the Board of Trade's vision of empire and the "fragmented and loose" collection of colonies arrayed across the northern Atlantic was one of an idea of coherence and integration versus a reality of autonomy and social divergence. This "disparity between structure and theory" did not sit well with those charged with the task of assessing the empire's vulnerabilities and advocating for uniform systems of governance. From 1713, when the Treaty of Utrecht entrenched the idea of a contiguous "British Empire" along the Atlantic seaboard, metropolitan agents put forward policies of "redefinition and reconstruction of empire," seeking its fundamental transformation. This decades-long struggle to reform the colonies "registered a deep and abiding unease about the ultimate consequences of the failure of metropolitan efforts at centralization and betrayed a mounting conviction that the future viability of empire depended on tighter metropolitan controls."[65]

Colonists understood the vision for America that Britain sought to implement, reacted to the critique of colonial history on which it was grounded, and referenced it as a palpable indication of Britain's plans for intrusive control over settled colonies. They saw in the surveys, settlement schemes, and maps a blueprint for their own future subjugation and a model for their displacement as meaningful agents of empire. Few expressions of protest made reference to Britain's development program for the new territories until the controversy over the Quebec Act made the issue of the redefinition of American space central to campaigns of resistance. Colonists objected to the erosion of the practices of a negotiated transatlantic empire. They could see in the maps of the new territories an expression of the "territorial authority" that Britain sought to impose throughout the empire, and they mobilized to overthrow it.[66]

Conclusion

As the British mapped America, they imagined it transformed. The most important meanings contained in the hundreds of hand-drawn maps, plans, and charts that imperial surveyors dispatched to Whitehall appear in the pattern they make together, spread out across eastern North America and the Caribbean Sea, rather than in the particular images framed by the borders of any one of them. Their true authors were not individual mapmakers but rather the Board of Trade—the agency that initiated the surveys and described the role of codified geographic knowledge in a reformed system for planting American colonies. As their central objective, Britain's surveys of the 1760s and 1770s attempted to reconcile scales of spatial representation. Images that compressed feet of terrain into inches of line and color on paper represented space at a resolution that opened colonization to view. Such maps revealed the land's varied terrains as resources to be harvested, grazed, and planted; presented rivers as channels to be navigated; and located structures, fences, fields, and roads that marked where settlers had taken up territory and made it into productive property. With systematic intent, the Board instructed its surveyors to compile and reduce these vivid, high-resolution images into regional overviews designed to give any commander or administrator unmediated access to the places and people they contained. At the broadest scale, metropolitan projectors gathered around maps that viewed the continent and the islands of British America as a whole from an impossible vantage point: high above the surface of the globe.

Britain sought to take command of America by amassing a body of geographic information and recording it in the form of maps that could be produced to suit the task at hand, from military operations to colony-wide settlement schemes. After the Seven Years' War, it undertook purposeful surveys that demonstrated the rise of an imperial state that had grasped the technical methods needed to enforce a more perfect dominion over distant colonies. Maps meant to stand for the places they represented, however, were static glimpses of dynamic spaces. Produced to capture the essential qualities of places, these replicas of the observed world disregarded the adaptive solutions colonists had devised to cultivate land for profit in diverse latitudes as well as the vernacular conceptions of landscape that expressed this working knowledge. Far from functioning as dispassionate scientific records, the maps affirmed metropolitan judgments about what was best for America by supplanting the need to rely on the environmental expertise of resident colonists. By capturing spaces as they appeared in the moment they were surveyed, the maps masked human histories and activities across the landscapes they depicted, enhancing the illusion that they were open to wholesale reconfiguration.

The novelty of Britain's new map of empire reflected the incompatible aims it attempted to reconcile. In 1763, the king charged his Lords Commissioners of Trade and Plantations to demonstrate British sovereignty over vast new territories, but in ways that limited new expenses from an overextended treasury; he urged them to plan new colonies to reap the potential value of these lands, but without provoking armed resistance from the wary Native peoples who inhabited them; and, following a war fought to proclaim the transcendent values of Protestant enterprise and political liberty, he enjoined them to observe the terms of the Treaty of Paris that elevated French Catholics to the standing of fellow subjects. As the commissioners meditated on a map of North America and the West Indies, they made a virtue out of the necessity of honoring these antithetical aims. They drew on memories of their committee's historical mission to impose order on the chaos of unregulated colonization and suggested tantalizing possibilities for the future: new colonies with fewer African slaves and more opportunities for poor white migrants; sincere alliances with recognized indigenous nations; broad toleration for diverse subjects, united under the rule of a virtuous monarch; and an ever-widening circle of peaceful commerce, quelling memories of violence that had made the New World a theater of human misery for the better part of two centuries. They made a case for these noble ends in the authoritative language of contemporary political economy and appealed to ideals of liberality and humanity that resonated with emerging British sensibilities.

In a stunning reversal of colonial, metropolitan, and indigenous expectations, the Board of Trade proposed closing the mainland colonies' open charter boundaries with a line of limitation against settler expansion after the Seven Years' War. When the king gave this recommendation the force of law with his proclamation of October 7, 1763, he challenged what the colonists believed was their collective purpose on the continent: improving a savage wilderness. The idea that new rules for granting, or withholding, land could mold British American populations into new configurations came into direct conflict with entrenched regional settler societies, each powerfully oriented toward territorial expansion. With a few strokes of a paintbrush, the commissioners drew the borders of bounded colonies across a map of North America, although doing so ran counter to a deeply rooted belief, one called repeatedly into service to justify the war's sacrifices: that new territory should be colonized to expand the landed domain of the nation, encourage the reproduction of European inhabitants, increase Britain's capacity for production, and enlarge its markets of consumption. As an aspiration, the Board of Trade's vision for America squared the circle of the king's command to find a way to pursue so many competing goods and attempted to set British America on a new, self-righting course of development. Beyond a coterie of true believers, however, few Britons in power embraced its innovative project or worked to sustain it, and many others undermined it by curbing the power of the Indian superintendents, authorizing the creation of new interior colonies, and extending Quebec's boundaries into the heart of the continent.

In 1696, William III had tasked the Board of Trade to better understand the far-flung colonies that made up England's rising western empire. Commissioned to secure American territories as part of a national domain, Board members made the Plantation Office first into a clearinghouse for gathered statistics, images, and texts that scrutinized these distant places and then into an active agency with its own mandate to comprehend America and redirect its development. After the Seven Years' War, the Board sought to become a "center of all information" to which all intelligence should be sent and from which all appointments, directives, and initiatives should begin. Although its vision for American empire fell apart as the War of Independence raged, its shelves groaned under the accumulated weight of hundreds of maps and thousands of volumes of reports, proposals, letters, and minutes. In 1780, Board of Trade draftsman Francis Assiotti compiled an inventory of 381 sheet maps of Europe, Africa, and North America, almost all of which bore the annotation "M:S" to indicate that they were original manuscripts. Sixty years earlier, the Board had found itself unable to provide a reliable map of the contested Canso Islands; now, the commissioners could sift through sixty-eight separate images of Nova Scotia in addition to the General Survey's thirty-eight charts of the maritime northeast. Envisioned as a dynamic center of calculation, the Board of Trade gathered materials into an inert and disorganized cartographic collection that pictured vital colonial spaces, many lost to Britain after 1776, with new detail and in overwhelming quantity.[1]

Although it failed in its quest to remake America, the Board of Trade shaped a language of empire that framed every serious discussion of American policy around a historical narrative of colonial disorder. Its commissioners advanced the centralization of imperial administration as an imperative, and maintained as a core principle that diverse Atlantic interests could be reconciled through new laws and regulations. Perhaps no one was more attuned to this language of empire than Benjamin Franklin, who framed his *Observations concerning the Increase of Mankind* (1755) to make the case for territorial expansion in the language of political economy by which the Board rationalized its policy initiatives, proposed the Albany Plan of 1754 around the perceived ills of maladministration that the Board had long critiqued, and promoted Vandalia as a model colony that conformed to the standard of governance and development that the Board imposed on Britain's other new colonies. The hints that arrived at the Plantation Office by the score to suggest the best methods for regulating the British Atlantic likewise echoed the Board's touchstones: controlling demography, regulating land, and aligning the interests of inhabitants to the capacities of the places in which they lived. By structuring the way Britons thought about colonies to favor abstract concepts and forms of reasoning that saw them as material systems, participants in this community of reformers found their ideas and roles validated. The Board of Trade's creation of an imperial discourse made it possible for William Johnson, an expert on northern Indian trade and diplomacy, to write extensive commentaries on the colonization of Florida and the sugar islands; it also enabled George Macartney, governor of the Ceded Islands, to write with equal authority about the best way to settle St. John Island in the Gulf of St. Lawrence.

This common language of empire not only framed transatlantic discussions around the Board's presumptions about what was wrong with America and what must be done to fix it but also empowered metropolitan colonizers to see themselves as meaningful agents who should be granted land and appointed to official positions overseeing settlement and trade, even though they lacked American experience. One November evening in 1762, an informal gathering of prominent Scots in London debated the question that preoccupied the Board's commissioners: whether doctrines derived from principle or from direct experience offered the best path to a "proper knowledge of mankind." For these well-placed individuals at the heart of the empire, the problem of reconciling theory and practice was no mere academic diversion. Among them were several who were already, or would soon become, personally involved in Britain's new American territories. Lord Elibank's brother was James Murray, governor of Quebec; Lord Eglinton would gain title to twenty thousand acres in West Florida; and Robert Mylne's brother, William, was to seek his fortune by laying out new estates on ceded Creek Indian land on Georgia's frontiers. Metropolitan Britons with means, education, and connections saw America as a vast proving ground in which to put such knowledge to profitable use as they shaped a new colonial world. With widening access to imperial opportunities, they were no mere armchair philosophers but rather actors striding across a global stage.[2]

A generation of Britons looked west across the Atlantic Ocean to magnify their fortunes after the Seven Years' War. Richard Oswald, a mercantile kingpin and political mover, diversified his investments across vast landholdings in Nova Scotia, East Florida, and West Africa's Guinea Coast. After the Board of Trade rejected the Earl of Egmont's bid to take possession of all of St. John Island as a personal estate, he became master of Amelia Island in East Florida, in addition to securing 122,000 acres in Nova Scotia. Army engineer Harry Gordon established the navigation of the Ohio River, imagined fields of indigo growing on the plains of Natchez, and became a Grenada planter. He also held five patents for 2,129 acres of land in Pennsylvania, two 5,000-acre township grants near Lake Champlain in New York, and a small house and two town lots on Walnut Street in Philadelphia. Members of the Nova Scotia Society and the East Florida Society planned the development of their colonies from London dining rooms and taverns. Over dinner with the Duke of Rutland, a group of these men joked about how many American settlements would be required to set up the nine children of MP Thomas Thoroton, husband of the duke's illegitimate daughter. From 1763, the Board of Trade worked to open America to these and other metropolitan colonizers, who undertook ventures based on their ability to leverage connections in Britain—and speak the Board's language of empire—rather than on any direct relationship to any place overseas.[3]

Edmund Burke railed against the Board of Trade's history of intervention in American colonization before the House of Commons in 1780. Before the Revolutionary War, Burke had dismissed William Knox's claim that new American taxes were needed to relieve the crushing burden of the national debt. He stood up for a view of the empire as a community that demanded acknowledgment of well-worn customary practices, developed to handle particular places and circumstances. During the war, he observed, the Board had receded into such irrelevance that in the "course of our whole dispute with America," it had not laid "so much as a scrap of paper" before Parliament, "respecting the state, condition, and temper of the colonies." He argued in favor of abolishing the Board of Trade as a useless expense of government, concluding that it had served as a "temperate bed of influence" as well as a "gently ripening hothouse" of ill-conceived ideas.[4]

When the Board acted, Burke charged, it either did harm or was "of no use at all." "This Board of Trade and Plantations has not been of any use to the colonies, as colonies," he reasoned, since the "flourishing settlements" on the mainland as well as in the islands took root and grew of their own accord before its existence. "Two colonies alone owe[d] their origin" to the schemes of the commissioners. Georgia made no progress at all "until it had wholly got rid of all the regulations which the Board of Trade had moulded into its original constitutions." The "province of Nova Scotia was the youngest and the favorite child of the Board. Good God! [W]hat sums the nursing of that ill-thriven, hard-visaged, and ill-favored brat has cost to this wittol nation!" According to Burke, only the "overflowings from the exuberant population of New England" had vitalized this moribund province, which had languished under the Board's direction. He ridiculed the twenty-three hundred folio volumes of proceedings the Board had generated since its inception in 1696 as a monument to the folly of a century-long project of imperial restructuring. Burke represented one side of a great debate over the nature of knowledge and the capacities of the state to make use of it. In asserting the Board's insignificance at such length, he protested too much—and, in so doing, revealed his antipathy to his rivals in government, men like Halifax, Grenville, Shelburne, and Hillsborough, who had worked together for years to reform America.[5]

Although Burke dismissed the commissioners' ideas about empire as little more than empty philosophizing, they deflected the charge by devoting unprecedented resources to producing firsthand knowledge about American places after 1763. The principle that colonies should be dependent polities, formed by new land regulations to become stable and productive contributors to Britain's Atlantic economy, applied to all the new territories but was implemented differently in each. The maps, descriptions, and assessments returned by surveyors gave empirical legitimacy to the Board's schemes and helped shape them to fit the conditions of different places without the need to delegate the tasks of colonization to an experienced creole elite. Each American colony possessed wants (for growth, subsistence, protection, and governance) and resources (derived from its climate, geography, inhabitants, and position in networks of trade). To make these needs and capacities match up with the general interests of the nation, the Board initiated a wide-ranging inquiry into the natural characteristics of American regions. Mapmaking had a special place in this quest because it both synthesized different types of knowledge into a single image and put land, the indispensable resource that made colonies worth acquiring in the first place, at the center of this complex of observations about place. Far from attempting to impose uniform strictures on the colonies as part of an indiscriminate "tightening" of imperial control after years of so-called salutary neglect, the Board sought to build a new imperial system fitted to the "nature and situations" of each place that composed it. This meant that even the most high-resolution image of a tract of land as well as the most fine-grained statistical detail about trade mattered only because it could be used to put each colony into a mutually stable relationship to the empire as a whole. To take possession of these new territories and to create a well-defined edge to the North American mainland colonies, it initiated a vast program of surveying and mapmaking designed to reveal the natures of unfamiliar lands.[6]

After Parliament passed the Stamp Act in 1765, protesters challenged its assertions of supreme legislative power in ways that preserved their status as loyal subjects of the king. However, the Crown exercised its prerogative powers to initiate a comprehensive program of reform and regulation, charging its commissioners and privy councilors to formulate new colonial policies, asserting them by royal proclamation, and empowering officials stationed in America to implement them. These officers of the Crown imposed new boundaries, negotiated with Indian nations, and commanded a peacetime standing army that surrounded the settled provinces. When Congress listed grievances against the king to justify American independence in 1776, this shift of focus from Parliament to the Crown was no volte-face: it expressed a generation's experiences with royal power mobilized to take unprecedented action in America.

Americans recognized Britain's effort to take command of colonization in new territories and reacted against the negative presumptions about their societies on which it was premised. As resistance to imperial governance mounted, they looked to the Board's new map of empire as a preview of the future Britain intended for them. They rejected their new status in these schemes as units of political economy rather than moral agents. And they critiqued the general knowledge from which these schemes were conducted, making a vigorous case for the value of local understandings of place. Despite American suspicions of its intentions, the Board of Trade aimed at reconciliation, not subjugation, as it guided the state to exercise new authority over the process of colonization. Its commissioners believed that new boundaries encircling the mainland colonies would prevent disruptive settlement in the west, and that colonial populations would engage more deeply in the Atlantic economy without any threat of coercion simply by following their own interests within these new limits. As Indians knitted their loose polities together to conform to the idea of nations with clear jurisdictions and boundaries, their fears of encroachment and fragmentation would be placated and they would prosper through new opportunities for trade. Reconciliation was the principle at the heart of the philosophy that guided the Board of Trade as it drew its initiating map. Convinced that the mutuality of fair trade offered a key to balancing the large, diverse populations that inhabited British America, the Board sought to match inhabitants with places. In the newly defined colonies, each of the king's subjects was connected by a path or a port to the great network of maritime exchange, which rewarded the capacity to produce with the opportunity to consume.

The circle of metropolitan reformers gathered around the Board of Trade imagined that a reformed America could become a thriving and integral part of a larger British world. With France and Spain expelled from eastern North America, the perpetual state of war by which these rival powers had contested the command of territories and routes of trade could be superseded by a perpetual state of peace. This future would, they hoped, demand no extraordinary mobilization of armies and ships, no costly distribution of Indian presents, and, after infant colonies were settled and brought to maturity in an accelerated process of colonization, no ongoing subsidies to ensure their survival. What could not be purchased from Britain could be exchanged between islands and the continent and among the temperate, subtropical, and tropical latitudes in which these colonies had been planted. This reconciled future for America envisioned a long arc of development that would end, theoretically, at a point of stasis, when the last acre of arable land connected by rivers and sea routes to Britain had been cleared and improved.

The Board of Trade's imperial political economy redefined the social character of the colonial subject. It validated the pursuit of physical comfort as an end in its own sake, and it understood human beings to be naturally driven to seek and enjoy material goods. It stripped the stigma of luxurious consumption out of the equation of economic life, making the pursuit of refinement something that was morally neutral and so fundamentally rooted in the human social condition that to attempt to censure or curtail it was not only unnecessary but impossible. By reducing colonial subjecthood to acts of production and consumption, of pursuing wants without moral meaning, this philosophy invited a response that placed moral discernment at the heart of colonists' participation within the Atlantic economy. British critics feared that Americans would exit spaces of order and civility along the coasts, straying beyond the boundaries the Board drew to contain them. They used a recurring image to embody this fear of disconnected colonists, describing a remote settler family, so lost to imperial commerce that it planted fiber crops and fashioned its own clothing rather than importing the products of British factories. By 1774, when colonists were purchasing more manufactured cloth than they had ever consumed before, the First Continental Congress rebuked Britain by announcing a policy of nonimportation and creating the Continental Association to enforce it. As colonists renounced British goods, they celebrated a unity of purpose and behavior that encouraged them to see their distinctive regions bound together as "one Chain of Colonies extending upward of 1,200 Miles & containing about three Millions of White Inhabitants." Colonists donned coarse, American-made textiles, and this expression of material independence "stamp[ed] on their home-spun all the value, all the pride of ornament." By renouncing imports as unnecessary luxuries, revolutionaries reasserted their power to act in ethical terms, not as a herd of humanity in which individual people were lumped together in populations of inhabitants but as men and women who could pursue political rights as their own reward, regardless of wants and needs.[7]

When the Board of Trade used maps to look down from high above on towns, frontiers, and regions in order to mold the colonies into a reformed Atlantic system, it assumed the position of superiority that political economists routinely attributed to themselves as thinkers uniquely situated to see how particular interests might be arranged to serve a general common good. Its general maps of empire thus invested the abstract idea of geographic scale with moral authority, privileging the view of the general over the particular and associating the former with dispassionate nationalism and the latter with narrow self-interest. Britain's longstanding mission to reform American settlement and agriculture began from a disdain for colonial practices. By viewing the new maps of empire, which pictured an idealized colonial world taking root in Quebec, St. John, East Florida, Grenada, and elsewhere, colonists could not help but see that this vision for the future of British America disregarded the world they had created. In rejecting the Board of Trade's plan for America, mainland colonists reclaimed an ideology of colonization, which was basic to their identity as an independent people. Over and above the desire for land and wealth that moved a thousand expansionist schemes, European Americans united in the national project of extending their provincial societies across space as a demonstration of their expertise, a means of expanding the capacities of their societies, and the highest expression of their civilization.

"Denying freedom," wrote John Locke, deprived men of the capacity to make meaningful choices and, by doing so, reduced them to the status of "bare Machin[e]s." More than any other object, imperial maps stood for the mechanization of human behavior for America's revolutionary generation. These images revealed Britain's intention to take command of American places from a distance and use knowledge acquired by new surveys to turn active colonists into passive inhabitants. New British maps reduced colonists to faceless figures on precisely rendered landscapes, downgrading their aspirations—now calculated as "interests"—and removing their ability to "suspend and deliberate" from the equation of empire. This vision acted on them as "subjects of development," a diminished status compared with their traditional positions as rights-bearing English subjects. By restraining the existing colonies from future expansion and excluding American-born colonists from taking a significant hand in shaping new colonies, it robbed them of a future characterized by grandeur, growth, and the potential for aggrandizement.[8]

William Petty, the Earl of Shelburne, was a key architect of this vision for American empire and presided over the Board of Trade as it drafted a master plan for reformed colonization in 1763. After Shelburne left office in 1770, he helped lead the opposition in the House of Lords for the next twelve years and advocated for reconciliation with America when the War of Independence began. He condemned the "madness, injustice and infatuation of coercing the Americans into a blind and servile submission" rather than regulating their economic lives with a generous regard for their interests. He maintained, however, that he would "never consent that America should be independent" and insisted that "Great Britain should superintend the interests of the whole" empire. Shelburne took pains in public debate to define what he meant when he said that he wished to secure the "dependency of that country upon this." Instead of a "slavish dependency," he envisioned a natural "union" in which the "liberties, properties, and lives of all the subjects of the British empire, would be equally secured." He believed that force would only achieve what those behind the war feared most: a Britain that, once "Deprived of America," would "sink into a petty state," unable to marshal the resources to contend with the "great powers on the continent." In March 1782, Shelburne returned to office in Rockingham's second ministry and succeeded him as prime minister after his death in July. Although Shelburne's tenure at the helm of government lasted less than a year, he oversaw the treaty negotiations that ended the war. He accepted, with great reluctance, the demand that Britain recognize American independence as a precondition to any negotiations. In approving boundaries for the new United States, which critics regarded as overly generous, he preserved the hope that some federal system might yet be put in place to keep American territories within Britain's national domain and, barring that, encouraged a system of commercial reciprocity between the nations that might continue to shore up Britain's stature as a world power. Shortly after Parliament approved the Treaty of Paris in 1783, the forty-five-year-old Shelburne was forced to resign. Since his speech to the Lords in favor of the Peace of Paris in 1762, he had spent two decades working to solve the problem of how to integrate American colonies into a stable imperial system, concluding his career in politics with the singular failure of presiding over the creation of an independent United States.[9]

The lines negotiators marked across copies John Mitchell's map of North America in 1783 were imaginary constructs but ones that gave former colonists an opportunity to imagine their nation by drawing it. Congress dispatched peace commissioners to Paris in 1782 to assert America's sovereignty over the eastern continent. Its instructions, drafted by James Madison, ordered John Adams, Benjamin Franklin, John Jay, and Henry Laurens to repudiate the Board of Trade's vision for America and demand that the legal and geographic artifacts by which that vision was expressed be dismantled in international law. George III's Proclamation of 1763, Madison wrote, was the immediate prelude to the "wicked

& oppressive measures which gave birth to the Revolution" and therefore should be considered a violation of American rights. The Proclamation had in fact never divested the colonies of their claims to the continental interior, Madison asserted; it had merely "shut up the land offices . . . to keep the Indians in peace." The moment Britain recognized the United States as an independent nation, the legal force of its injunction against granting land disappeared, and the original colonial charters resumed unimpeded to define the western limits of the United States. After all, Madison noted, the king's "own Geographer," Emanuel Bowen, had published a map that carried the boundaries of the "States of Georgia, N. Carolina, S. Carolina & Virginia as far as the Mississippi" in 1763. Bowen's *An Accurate Map of North America*—the very map across which the Board of Trade had drawn its invidious lines of limitation—thus affirmed these enduring territorial rights. Likewise, independence invalidated the Quebec Act of 1774 and the extension of the "boundaries of Quebec" into territories first granted to the colonies as part of their "chartered rights." Congress instructed America's negotiators to defend this "Western territory" and stand fast against the "mutilation of our country" if Britain, France, or Spain proposed to detach it. Although one disbelieving French diplomat ridiculed claims to the west as a geographic "extravagance," the Americans refused to consider a boundary that would, as Franklin put it, "coop us up within the Allegheny Mountains" and instead followed the lines of the colonial charters that "extend[ed] the domains of America from the ocean to the South Sea."[10]

In Paris, Spanish ambassador Pedro Pablo Abarca de Bolea, Count d'Aranda, told the Americans that the negotiators should all sit "down together with Maps in our hands, and by that means shall see our Way more clearly." He spread a French edition of Mitchell's map of North America before John Jay, who "pointed with his finger" at the source of the Mississippi and traced it down "almost to New Orleans" to express Congress's demand, based on colonial charters, for that river as a western boundary. The Spanish diplomat countered with his nation's claim to Florida, from which he could draw "parallel lines" that reached toward the north until they crossed the lines of latitude specified in the charters, reducing any map to an abstraction of "lineal squares" that conferred nothing with certainty to either power. He pointed out that these dominions, inhabited as they were by indigenous peoples, were merely "imaginary spaces" to which Spain and the United States had "equal rights, or equally unjust claims." Aranda drew a red line across the continent just beyond the Appalachians, hoping to "force the Americans to adopt a more moderate position, and to counteract Mr. Jay's pretension to the whole Mississippi as a boundary line." Jay informed him that his instructions from Congress left him "without discretion" on the question of the Mississippi boundary, which he and his fellow commissioners maintained as an inviolable point.[11]

Shelburne was First Lord of Trade in 1763, when the Board drew lines across *An Accurate Map of North America* to clarify its vision of a coastal, commercial empire. As prime minister in 1783, Shelburne authorized his peace commissioners to draw lines across a copy of John Mitchell's *A Map of the British Colonies in North America* that nullified this vision. **[256]** Shelburne tapped merchant Richard Oswald, a fellow believer in the harmonizing power of unfettered commerce, to represent the Crown in the treaty talks. If Britain could no longer count America's acres as part of its national domain, Shelburne hoped that large grants of territory combined with enduring cultural sympathies would soften the rupture of independence and keep American land within the orbit of British trade. He believed that the new Treaty of Paris created an America capable of holding up the western end of a free trade zone that spanned the Atlantic. Oswald's red line fractured a continental empire whose unification had been Britain's single most important geopolitical objective in 1763. It bisected Penobscot Bay to separate American Maine from British Nova Scotia, severed the Florida colonies from Georgia and South Carolina before their return to Spain, and split the Great Lakes down their centers.[12]

This red line on Mitchell's map visualized treaty terms that granted the United States "full, complete, and unconditional independence" and expressed a precept of international political economy: that "reciprocal Advantages & mutual Convenience . . . form the only permanent Foundation of Peace" among nations. It described a boundary that would be drawn along high lands, to the edges of lakes, around the headwaters of some rivers, "down a[l]ong the Middle" of others, and in straight courses that followed compass bearings and lines of latitude. Oswald had pored over maps to determine the locations the treaty referenced, but at North America's frontiers these borderlands grew indistinct, and the images that represented them conflicted. The Treaty of Paris thus empowered commissioners to resolve this geographic uncertainty in the future. Their failure to do so—combined with Britain's exclusion of American traders from British ports, its unwillingness to withdraw from western forts, and disputes over the right of navigation along the waterways of the Great Lakes and the Mississippi—tarnished Shelburne's new vision of transatlantic amity with lingering disagreements over the shape of the new nation and the old empire.[13]

The Board of Trade created a formal seal for each new colony featuring a graphic image that captured its essence as a place along with a Latin motto that linked it to a specific imperial desire. Chief engraver Christopher Seaton then inscribed these words and pictures onto silver matrices, which colonial governors used to stamp their wax impressions onto official documents. On Quebec's seal, the king, dressed in the splendor of his coronation robes, points his scepter at a large "chart of that part of America," unscrolled for his examination upon a wooden stand, "thro' which the river of St. Lawrence flows, including the Gulph." The inscription—EXTENSÆ GAUDENT AGNOSCERE METÆ (the extended boundaries rejoice to acknowledge [him])—put forward this idea of geographic grandeur in explicitly cartographic terms: as the king beholds his expansive dominions on a map of the North American northeast, they cannot help but admire his majesty in return, recognizing his natural sovereignty over them. For the other seals, the commissioners drew on *The Aeneid,* a touchstone text for meditations on the meaning of empire. The Ceded Islands seal pictured a "Sugar Mill . . . with Slaves at work." While other nations created great works of art, oratory, and science, Rome distinguished itself by imposing its authority on distant places. "Roman," wrote Virgil, "remember by your strength to rule Earth's peoples—your arts are to be these: / To pacify, to impose the rule of law, / To spare the conquered, and battle down the proud." And so *HÆ TIBI ERUNT ARTES* (your arts are to be these) marked the new islands as places that Britain stood ready to command and transform through slavery, sugar, and trade.[14]

West Florida's seal presented the landscape of a well "cultivated country, interspersed with Vineyards and corn fields." Its motto, *Melioribus utere fatis* (Enjoy a better destiny), was the wish of fellow Trojan Deiphobus when he met Aeneas in the underworld, where Rome's progenitor learned of its glorious imperial future. Such a sentiment expressed an official confidence that the industry of Protestant settlers and their slaves would make this underdeveloped Gulf Coast periphery, once attached to French Louisiana and Spanish Florida, into a productive colonial society. A "fortified Town and harbour" stood for East Florida, signifying historic St. Augustine but also standing for the idea of a new civil order that would expand beyond its bounds. At the beginning of *The Aeneid,* Jupiter assures Venus that Troy's fall will initiate the rise of an expansive Rome that will make war on wild nations, build cities, and impose law, a notion of the civilizing power of colonization captured in East Florida's motto: *Moresque Viris et Menia ponet* (Establish city walls and a way of life). These seals expressed an idea of empire designed to unite Britons at home and abroad in a common cultural project. Yet their symbols and allusions reduced these distinctive places to abstractions, privileging metropolitan fantasies of what they should become before the first British colonist arrived to inhabit and improve them.[15]

The Board of Trade's vision for American empire began from a critique of the history of colonization to establish an official framework for colonial policy in 1763. Implementing this vision depended on generating a new geographic record of British America in the eighteenth century in order to settle and develop the territories acquired at the Peace of Paris. These representations of America, by showing colonists images of their futures within bounded societies, contributed to the discontent that fractured the Atlantic empire. More maps could be drawn into this discussion to describe how links between the mainland and islands were broken and reformed after independence. Other works of history have described how Americans in the early republic expressed a new geographic consciousness and, at times, recapitulated the challenges of empire as they drew new maps of the states, regions, and sections of an expanding nation.[16]

Abel Buell's *A New and Correct Map of the United States of North America* (1784) was the first map of the new nation published in the United

States. [257] Printed in New Haven, Connecticut, it featured the American flag as a decorative image, set its prime meridian to the longitude of Philadelphia, and inscribed international boundaries "agreeable to the Peace of 1783." Buell engraved his copperplate with the names of the states in bold letters, which follow the lines of the old colonial charters west to the Mississippi River. These state boundaries erased the Quebec Act's buffer against settlement north of the Ohio, passing through and negating Britain's negotiated Indian boundary.

Americans searched for icons, shapes, and emblems capable of representing their nascent nation as a coherent continental form while preserving the integrity of the component states to which they owed their allegiance. Like others who described empires, nations, and colonies as suns, planets, and satellites, they used metaphors from celestial observation to describe the natural order of their confederation. Just as astronomers sighted new stars in the night sky through their telescopes, the flag of the United States revealed a "new Constellation" arrayed across a blue field. In contrast to the unitary sun of absolutism, this emblem represented the new nation as a natural fact, novel and yet immutable, interconnected but clearly articulated into separate parts. Just as this image made room for new states as new stars, it rooted the origins of the nation in the thirteen specific colonies that joined to form this union at this particular moment in history. Like geographic forms on a map, the stripes in the flag represented the founding states as bands that did not intersect, overlap, or come to a point of central focus, as they did on the flag of Great Britain. Where Britain's Union Jack had merged the separate national crosses of England, Scotland, and Ireland into a new composite to mark the Act of Union that brought them under a single crown, the American flag kept its stripes as separate as its states. Just as its stars signaled a new political presence in the world, these stripes, as long, horizontal forms, embodied the states' resurgent continental ambitions. This new political order pictured on Buell's map came into focus as a continental space that extended from the Atlantic Ocean to the Mississippi River. It formed "one connected, fertile, wide-spreading country," wrote John Jay, tied together by "navigable waters [that] form a kind of chain round its borders, as if to bind it together." Such images of the republic, seemingly authorized by nature to unite the continent under a unified sovereignty, erased the lines the Board of Trade had drawn across its new map of empire in 1763, when it offered Americans a static, prosperous future confined to Atlantic coasts. Independence dismantled this image of a subordinate western empire and reconfigured the nation into a dynamic union of eastern states, poised to expand into the North American interior.[17]

Abbreviations

Acts of the Privy Council	*Acts of the Privy Council of England,* Colonial Series, vol. 4, *A.D. 1745–1766,* ed. James Munro (Hereford: H. M. Stationery Office, 1911)
Avalon Project	*The Avalon Project: Documents in Law, History and Diplomacy,* Yale Law School
Board of Trade Journals	*Journals of the Commissioners for Trade and Plantations . . . Preserved in the Public Record Office . . . ,* 14 vols. (London: H. M. Stationery Office, 1920–1938)
LC Photostats	British Public Record Office Photostats, Manuscript Division, Library of Congress, Washington, DC
LC Transcripts	British Public Record Office Transcripts, Manuscript Division, Library of Congress, Washington, DC
Calendar of State Papers	*Calendar of State Papers, Colonial Series . . . ,* ed. W. N. Sainsbury [and others], 45 vols. (London: Longmans, H. M. Stationery Office, 1860–1969)
Canadian Documents	Adam Shortt and Arthur G. Doughty, eds., *Documents Relating to the Constitutional History of Canada, 1759–1791* (Ottawa: S.E. Dawson, 1907)
Franklin Papers	*The Papers of Benjamin Franklin, Digital Edition,* Packard Humanities Institute, http://franklinpapers.org/franklin/
Grant Papers	James Grant of Ballindalloch Papers, microfilm copy, Manuscript Division, Library of Congress, Washington, DC
Indian Documents	Alden T. Vaughan, ed., *Early American Indian Documents: Treaties and Laws,*

	1607–1775, 20 vols. (Bethesda, MD: University Publications, 1979–1994)
New York Documents	*Documents Relative to the Colonial History of the State of New-York,* ed. F. B. O'Callaghan, 15 vols. (Albany: Weed, Parsons, 1853)
Oxford DNB	*Oxford Dictionary of National Biography* (Oxford: Oxford University Press, 2004)
State Papers Online	*State Papers Online: Eighteenth Century, 1714–1782,* Gale Cengage Learning
UK Admiralty Library	UK Admiralty Library, Portsmouth
UK Hydrographic Office	Archive of the UK Hydrographic Office of the United Kingdom, Taunton
UK National Archives	The National Archives of the UK, Kew, Richmond, Surrey

Notes

INTRODUCTION

1. Roger Morriss, *The Foundations of British Maritime Ascendancy: Resources, Logistics and the State, 1755–1815* (New York: Cambridge University Press, 2011); Charles M. Andrews, *The Colonial Period of American History,* vol. 4 (New Haven, CT: Yale University Press, 1934); Jack M. Sosin, *Whitehall and the Wilderness: The Middle West in British Colonial Policy, 1760–1775* (Lincoln: University of Nebraska Press, 1961); Bernard Bailyn, *The Ideological Origins of the American Revolution* (Cambridge, MA: Belknap Press of Harvard University Press, 1967).
2. Brendan Simms, *Three Victories and a Defeat: The Rise and Fall of the First British Empire, 1714–1783* (London: Allen Lane, 2007). Works that describe the Board's history and functions include Andrews, *Colonial Period,* vol. 4, chap. 9; Oliver M. Dickerson, *American Colonial Government 1696–1765: A Study of the British Board of Trade in Its Relation to the American Colonies, Political, Industrial, Administrative* (Cleveland, OH: Arthur H. Clark, 1912); Arthur H. Basye, *The Lords Commissioners of Trade and Plantations, Commonly Known as the Board of Trade, 1748–1782* (New Haven, CT: Yale University Press, 1925); Franklin B. Wickwire, *British Subministers and Colonial America, 1763–1783* (Princeton, NJ: Princeton University Press, 1966); Ian K. Steele, *Politics of Colonial Policy: The Board of Trade in Colonial Administration, 1696–1720* (Oxford: Clarendon Press, 1968); Alison G. Olson, "The Board of Trade and London-American Interest Groups in the Eighteenth Century," in *The British Atlantic Empire before the American Revolution,* ed. Peter Marshall and Glyn Williams (London: Frank Cass, 1980), 33–50.
3. John Shy, "Thomas Pownall, Henry Ellis, and the Spectrum of Possibilities, 1763–1775," in *Anglo-American Political Relations, 1675–1775,* ed. Alison Gilbert Olson and Richard Maxwell Brown (New Brunswick, NJ: Rutgers University Press, 1970), 155–186; Andrew Jackson O'Shaughnessy, *The Men Who Lost America: British Leadership, the American Revolution, and the Fate of the Empire* (New Haven, CT: Yale University Press, 2013), chap. 5.
4. James C. Scott, *Seeing Like a State: How Certain Schemes to Improve the Human Condition Have Failed* (New Haven, CT: Yale University Press, 1998), 36 and passim.
5. On the system of negotiated governance in colonial America, see Jack P. Greene, *Negotiated Authorities: Essays in Colonial Political and Constitutional History* (Charlottesville: University Press of Virginia, 1994), chap. 1.
6. Nathan F. Sayre, "Ecological and Geographic Scale: Parallels and Potential for Integration," *Progress in Human Geography* 29 (2005): 276–290;

Sallie A. Marson, John Paul Jones III, and Keith Woodward, "Human Geography without Scale," *Transactions of the Institute of British Geographers* 30 (2005): 416–432; Vernon Meentemeyer, "Geographical Perspectives of Space, Time, and Scale," *Landscape Ecology* 3 (1989): 164; "cross-level fallacy": D. R. Montello, "Scale in Geography," in *International Encyclopedia of the Social and Behavioral Sciences,* ed. N. J. Smelser and P. B. Baltes (Oxford: Pergamon Press, 2001), 13,502; Clark C. Gibson, Elinor Ostrom, and T. K. Ahn, "The Concept of Scale and the Human Dimensions of Global Change: A Survey," *Ecological Economics* 32 (2000): 222; Benjamin Franklin, *The Complete Works of Benjamin Franklin, Volume II, 1744–1757,* ed. John Bigelow (New York: G. P. Putnam's Sons, 1887), 367n.

7. See Edward T. Price, *Dividing the Land: Early American Beginnings of Our Private Property Mosaic* (Chicago: University of Chicago Press, 1995).
8. U.S. National Oceanic and Atmospheric Administration, "Charting and Geodesy," accessed January 12, 2016, http://www.noaa.gov/charts.html.
9. Nicholas Canny and Philip Morgan, "Introduction: The Making and Unmaking of an Atlantic World," in *The Oxford Handbook of the Atlantic World, 1450–1750,* ed. Nicholas Canny and Philip Morgan (Oxford: Oxford University Press, 2011), 13–14; G. F. Heaney, "Rennell and the Surveyors of India," *Geographic Journal* 134 (1968): 318; *Historical Records of the Survey of India,* collected by R. H. Phillimore, 6 vols. (Dehra Dun: Survey of India, 1945), 1:1–5, 13–17, 45, 222–237. See also Matthew H. Edney, *Mapping an Empire: The Geographical Construction of British India, 1765–1843* (Chicago: University of Chicago Press, 1999).
10. See Jack P. Greene, "Social and Cultural Capital in Colonial British America: A Case Study," *Journal of Interdisciplinary History* 29 (1999): 493–495.

CHAPTER ONE: A VISION FOR AMERICAN EMPIRE

1. Charles M. Andrews, *The Colonial Background of the American Revolution: Four Essays in American Colonial History* (New Haven, CT: Yale University Press, 1961), 12–13. On the Board's attempts to constrain William Penn's powers as Pennsylvania's proprietor, for example, see Jean R. Soderlund, *William Penn and the Founding of Pennsylvania: A Documentary History* (Philadelphia: University of Pennsylvania Press, 1983), 33–35.
2. Charles M. Andrews, *British Committees, Commissions, and Councils of Trade and Plantations, 1622–1675* (Baltimore: Johns Hopkins University Press, 1908), 62, 65, 69, 70.
3. Ibid., 108–109; "Instructions for the Council for Foreign Plantations, 1670–1672," in ibid., 118–119, 125, 122. On the Council's interest in checking the rising power of "merchant-councillors," see Nuala Zahedieh, *The Capital and the Colonies: London and the Atlantic Economy, 1660–1700* (New York: Cambridge University Press, 2010), 47.

4. John Evelyn, *Memoirs of John Evelyn, Esq.*, ed. William Bray, 5 vols. (London, 1827), 2:342; Jeannette D. Black, *The Blathwayt Atlas*, vol. 2, *Commentary* (Providence, RI: Brown University Press, 1975), 9, 12–13; see also Ralph Paul Bieber, *The Lords of Trade and Plantations, 1675–1696* (Allentown, PA: H. R. Hass, 1919), 40–41.
5. Black, *Blathwayt Atlas*, 4–6, 58–62, 161–163.
6. April Lee Hatfield, "Intercolonial and Interimperial Relations in the Seventeenth Century," *History Compass* 1 (2003), doi:10.1111 / 1478-0542.059. The forty-eight surviving maps of the atlas can be categorized geographically as follows, with the number of maps within that category in parentheses: world (2), English America (1), northern North America (4), mainland English America (16), West Indies and Atlantic islands (13), Latin America (5), Surinam and Guiana (3), India (1), and Africa (3). Lesley B. Cormack, *Charting an Empire: Geography at the English Universities, 1580–1620* (Chicago: University of Chicago Press, 1997); George Tolias, "Isolarii, Fifteenth to Seventeenth Century," in *The History of Cartography*, vol. 3, *Cartography in the European Renaissance*, ed. David Woodward (Chicago: University of Chicago Press, 2007), 280; George Tolias, "The Politics of the Isolario: Maritime Cosmography and Overseas Expansion during the Renaissance," *Historical Review/La Revue Historique* 9 (2012): 27–52; John E. Crowley, "The Cartographic Origins of the British Empire" (paper presented at the Omohundro Institute of Early American History and Culture 20th Annual Conference, Halifax, Nova Scotia, June 15, 2014).
7. Black, *Blathwayt Atlas*, 186–196, 180–185, 137–140, 149–157, 63–72, 175–179. On Bermuda, see Michael J. Jarvis, *In the Eye of All Trade: Bermuda, Bermudians, and the Maritime Atlantic World, 1680–1783* (Chapel Hill: University of North Carolina Press, 2010), 24.
8. "Commission Establishing a Board of Trade, &c.," in *Documents Relative to the Colonial History of the State of New-York*, ed. F. B. O'Callaghan, 15 vols. (Albany: Weed, Parsons, 1853), 4:145–148 (hereafter cited as *New York Documents*); John Brewer, *The Sinews of Power: War, Money and the English State, 1688–1783* (London: Unwin Hymin, 1989), 221–228; Arthur H. Bayse, *The Lords Commissioners of Trade and Plantations, Commonly Known as the Board of Trade, 1748–1782* (New Haven, CT: Yale University Press, 1925), 4–5. On England's mission to improve empire in the second half of the seventeenth century, see Owen Stanwood, *The Empire Reformed: English America in the Age of the Glorious Revolution* (Philadelphia: University of Pennsylvania Press, 2011).
9. Charles M. Andrews, *The Colonial Period of American History*, vol. 4, *England's Commercial and Colonial Policy* (New Haven, CT: Yale University Press, 1964), 297–298; Andrews, *Guide to the Materials for American History, to 1783, in the Public Record Office of Great Britain*, vol. 1 (Washington, DC: Carnegie Institution, 1912), 83-85; Francis Dodsworth, "*Virtus* on Whitehall: The Politics of

Palladianism in William Kent's Treasury Building, 1733–6," *Journal of Historical Sociology* 18 (2005): 284–285, 295, 302–303.

10. Mary Patterson Clarke, "The Board of Trade at Work," *American Historical Review* 17 (1911): 19–20, 26–30, 34; Bayse, *Board of Trade,* 7–8; Franklin B. Wickwire, *British Subministers and Colonial America, 1763–1783* (Princeton, NJ: Princeton University Press, 1966), 69–71, 87; *Journals of the Commissioners for Trade and Plantations . . . Preserved in the Public Record Office . . . ,* 14 vols. (London: H. M. Stationery Office, 1920–1938), 1768–1775, 401 (hereafter cited as *Board of Trade Journals*).
11. Ian K. Steele, *Politics of Colonial Policy: The Board of Trade in Colonial Administration, 1696–1720* (Oxford: Clarendon Press, 1968), 157, 118–120; Board of Trade to Secretary Stanhope, July 15, 1715, in *Calendar of State Papers, Colonial Series . . . ,* ed. William Noel Sainsbury, J. W. Fortescue, Cecil Headlam, and Arthur Percival Newton, 45 vols. (London: Longmans, H. M. Stationery Office, 1860–1969), 28:232 (hereafter cited as *Calendar of State Papers*). On the importance of maps and geographic ideas to the French, see Dale Miquelon, "Envisioning the French Empire: Utrecht, 1711–1713," *French Historical Studies* 24 (2001): 653–677; Peggy K. Liss, *Atlantic Empires: The Network of Trade and Revolution, 1713–1826* (Baltimore: Johns Hopkins University Press, 1983), 1.
12. Steele, *Politics of Colonial Policy,* 154, 118–120; Board of Trade to Stanhope, July 15, 1715, in *Calendar of State Papers,* 28:232; Board of Trade to Lords Justices, September 15, 1720, in *Calendar of State Papers,* 32:141–142. On early Board of Trade schemes to settle Jamaica, New York, and Nova Scotia, see Steele, *Politics of Colonial Policy,* 161–165; "Circular letter from the Council of Trade and Plantations to Governors of Plantations on the Continent of America," August 7, 1719, in *Calendar of State Papers,* 31:187–188; O. M. Dickerson, *American Colonial Government 1696–1765: A Study of the British Board of Trade in Its Relation to the American Colonies, Political, Industrial, Administrative* (Cleveland, OH: Arthur H. Clark, 1912), 63, 286–290; Steele, *Politics of Colonial Policy,* 165–166.
13. "State of the British Plantations in America, in 1721," September 8, 1721, in *New York Documents,* 5:591–630; Jack P. Greene, "Martin Bladen's Blueprint for a Colonial Union," *William and Mary Quarterly,* 3rd ser., 17 (1960): 517–520. See also Steele, *Politics of Colonial Policy,* 167–170; Warren Hofstra, *The Planting of New Virginia: Settlement and Landscape in the Shenandoah Valley* (Baltimore: Johns Hopkins University Press, 2004), 68–69, 77–81; Rory T. Cornish, "Bladen, Martin (1680–1746)," in *Oxford Dictionary of National Biography* (Oxford: Oxford University Press, 2004) (hereafter cited as *Oxford DNB*); "State of the Plantations," 613–619.
14. What the "continent of America" comprised was a notoriously vague idea in early modern geographic writing. Not only did Europeans lack sufficient knowledge about the far west to see it clearly as part of North America, but

they also viewed global geography without our modern certainty that seven (or by some reckonings, six) continents were inherent and obvious features of it. See Paul W. Mapp, *The Elusive West and the Contest for Empire, 1713–1763* (Chapel Hill: University of North Carolina Press, 2013); W. Lewis and Karen E. Wigan, *The Myth of Continents: A Critique of Metageography* (Berkeley: University of California Press, 1997). See also Paul W. Mapp et al., *H-Diplo Roundtable Review* 14 (2012): 1–24, http://h-diplo.org/roundtables/PDF/Roundtable-XIV-13.pdf. "State of the Plantations," 619–623.

15. Ibid., 623, 621, 627–629; Steele, *Politics of Colonial Policy,* 68–72; "no power": quoted in ibid., 70. "State of the Plantations," 591, 629. The report also advocated appointing a single "Lord Lieutenant, or Captain General," over the colonies. Subsequent plans to unify the governance of the colonies were put forward, to equally little effect, in 1726, 1739, and 1754 (at the Albany Congress). After identifying South Carolina as the mainland's vulnerable "Southern frontier," however, the Board intervened directly in the process of colonization to promote this general vision of an integrated mainland empire by constructing Fort King George in the early 1720s, creating interior townships in the southeastern interior from the 1730s, and authorizing the Georgia Trustees to establish a buffer colony inhabited by yeoman farmers south of the Savannah River in 1732. See S. Max Edelson, "Defining Carolina: Cartography and Colonization in the North American Southeast, 1657–1733," in *Creating and Contesting Carolina: Proprietary Era Histories,* ed. Michelle LeMaster and Bradford W. Wood (Columbia: University of South Carolina Press, 2013), 40–42; see also Edelson, "Visualizing the Southern Frontier: Cartography and Colonization in Eighteenth-Century Georgia," in *Coastal Nature, Coastal Culture: Environmental Histories of the Georgia Coast,* ed. Paul Sutter (Athens: University of Georgia Press, forthcoming).
16. *Some Considerations on the Consequences of the French Settling Colonies on the Mississippi* (London, 1720), 28; "all North America": Duke of Newcastle (1754) quoted in Fred Anderson, *Crucible of War: The Seven Years' War and the Fate of Empire in British North America, 1754–1766* (New York: Knopf, 2001), 67.
17. Matthew H. Edney, "John Mitchell's Map of North America (1755): A Study of the Use and Publication of Official Maps in Eighteenth-Century Britain," *Imago Mundi* 60 (2008): 69; Clarke, "Board at Work," 26; J. B. Harley, "The Bankruptcy of Thomas Jefferys: An Episode in the Economic History of Eighteenth Century Map-Making," *Imago Mundi* 20 (1966): 35–37.
18. Bayse, *Board of Trade,* 50–51; Mary Pedley, "Map Wars: The Role of Maps in the Nova Scotia / Acadia Boundary Disputes of 1750," *Imago Mundi* 50 (1998): 96–104.
19. *Board of Trade Journals,* 1742–1749, 390; Clarke, "Board at Work," 39; "see what room": William Shirley (1748), quoted in Jeffers Lennox, "An Empire on Paper: The Founding of Halifax and Conceptions of Imperial Space, 1744–55,"

Canadian Historical Review 88 (2007): 381–382, 396–399. On Nova Scotia's population, see *Historical Atlas of Canada*, vol. 1, *From the Beginning to 1800*, ed. R. Cole Harris (Toronto: University of Toronto Press, 1987), plate 30; "Aboriginal Peoples," Censuses of Canada, 1665 to 1871, *Statistics Canada*, http://www.statcan.gc.ca/pub/98-187-x/4151278-eng.htm. On the Board's Nova Scotia scheme in general, see Andrew D. M. Beaumont, *Colonial America and the Earl of Halifax, 1748–1761* (New York: Oxford University Press, 2014), chap. 2.

20. Halifax (1754), quoted in Dickerson, *American Colonial Government*, 221; Edney, "John Mitchell's Map."
21. Jack P. Greene, "A Posture of Hostility: A Reconsideration of Some Aspects of the Origins of the American Revolution," *Proceedings of the American Antiquarian Society* 87 (1977): 27–68. On the Board's questionnaires, see George Clinton to Board of Trade, May 23, 1749, in *New York Documents*, 6:507–510. See also John Pownall to John Carteret, April 9, 1755, in *Documenting the American South*, http://docsouth.unc.edu/csr/index.html/document/csr05-0131; Cadwallader Colden (1750), quoted in Greene, "Posture of Hostility," 49; ibid., 51–52, 58; on the packet service, see Ian K. Steele, *The English Atlantic, 1675–1740: An Exploration of Communication and Community* (New York: Oxford University Press, 1986), chap. 9; Liss, *Atlantic Empires*, 16–17.
22. Some sixty-five pamphlets published in Britain between 1759 and 1763 addressed this topic. See Jack M. Sosin, *Whitehall and the Wilderness: The Middle West in British Colonial Policy, 1760–1775* (Lincoln: University of Nebraska Press, 1961), 6–11; see also Paul Mapp, "British Culture and the Changing Character of the Mid-Eighteenth-Century British Empire," in *Cultures in Conflict: The Seven Years' War in North America*, ed. Warren R. Hofstra (Lanham, MD: Rowman & Littlefield, 2007), 44–48. Rodney (1763), quoted in David Syrett, ed., *The Rodney Papers: Selections from the Correspondence of Admiral Lord Rodney*, vol. 2, *1763–1780* (Farnham, UK: Ashgate, 2007), 213.
23. December 9, 1762, *Proceedings and Debates of the British Parliaments Respecting North America, 1754–1783*, ed. R. C. Simmons and P. D. G. Thomas, 5 vols. (Millwood, NY: Kraus International, 1982–1986), 1:422–423.
24. [Shelburne's speech to the House of Lords], December 9, 1762, Shelburne Papers, 165: 309–321, William L. Clements Library, University of Michigan, Ann Arbor; James Boswell, *Boswell's London Journal, 1762–1763* (London: Reprint Society, 1952), 79–82.
25. [Shelburne's speech].
26. Andrew Hamilton, *Trade and Empire in the Eighteenth-Century Atlantic World* (Newcastle upon Tyne, UK: Cambridge Scholars, 2008), 27–33; John Cannon, "Petty, William, second earl of Shelburne and first marquess of Lansdowne (1737–1805)," *Oxford DNB*, Sept. 2013.
27. Nuala Zahedieh, "Trade and Empire," in *Cambridge Economic History of Modern Britain*, vol. 1, *1700–1870*, ed. Roderick Floud, Jane Humphries, and

Paul Johnson (Cambridge: Cambridge University Press, 2014), 392–393; Hamilton, *Trade and Empire,* 27–28. See also Steve Pincus, "Rethinking Mercantilism: Political Economy, the British Empire, and the Atlantic World in the Seventeenth and Eighteenth Centuries," *William and Mary Quarterly,* 3rd ser., 69 (2012): 3–34.

28. Albert O. Hirschman, *The Passions and the Interests: Political Arguments for Capitalism before Its Triumph* (Princeton, NJ: Princeton University Press, 1977); Joyce Oldham Appleby, *Economic Thought and Ideology in Seventeenth-Century England* (Princeton, NJ: Princeton University Press, 1978); Istvan Hont, *Jealousy of Trade: International Competition and the Nation-State in Historical Perspective* (Cambridge, MA: Harvard University Press, 2010); Daniel Carey, *Locke, Shaftesbury, and Hutcheson: Contesting Diversity in the Enlightenment and Beyond* (New York: Cambridge University Press, 2006); T. W. Hutchison, *Before Adam Smith: Emergence of Political Economy, 1662–1776* (New York: Blackwell, 1988); Peter N. Miller, *Defining the Common Good: Empire, Religion, and Philosophy in Eighteenth-Century Britain* (New York: Cambridge University Press, 1994). "[I]mmediate circle": Carey, *Locke, Shaftesbury, and Hutcheson,* 10; "unwieldy empires": Hutcheson (1740), quoted in Liss, *Atlantic Empires,* 5; "great empire": Adam Smith, *An Inquiry into the Nature and Causes of the Wealth of Nations,* vol. 2, ed. Edwin Cannan (London: Methuen, 1904), *Online Library of Liberty,* http://oll.libertyfund.org/titles/119#Smith_0206-02_1006. See also Emma Rothschild, "Adam Smith in the British Empire," in *Empire and Modern Political Thought,* ed. Sankar Muthu (Cambridge: Cambridge University Press, 2012), 185–186.

29. Liss, *Atlantic Empires,* 16; Jack P. Greene, *Creating the British Atlantic: Essays on Transplantation, Adaptation, and Continuity* (Charlottesville: University of Virginia Press, 2013), chap. 5.

30. Charles D'Avenant, "An Essay on the East-India Trade" (1697), *The Avalon Project: Documents in Law, History and Diplomacy,* Yale Law School, http://avalon.law.yale.edu/17th_century/eastindi.asp; Benjamin Franklin to Lord Kames, January 3, 1760, *Papers of Benjamin Franklin, Digital Edition,* http://franklinpapers.org/franklin/framedVolumes.jsp?vol=9&page=005a; "extended / riches": D'Avenant (1698), quoted in Hont, *Jealousy of Trade,* 205. On "ghost acres" and the artificial extension of empire, see Kenneth Pomeranz, *The Great Divergence: China, Europe, and the Making of the Modern World Economy* (Princeton, NJ: Princeton University Press, 2000), 113 and chap. 6; Zahedieh, *Capital and the Colonies,* 234–235, 31–33; E. A. Wrigley, "The Transition to an Advanced Organic Economy: Half a Millennium of English Agriculture," *Economic History Review* 59 (2006): 435–480. On the connections between colonization on European competition, see Sophus A. Reinert, *Translating Empire: Emulation and the Origins of Political Economy* (Cambridge, MA: Harvard University Press, 2011), chaps. 1–2.

31. J. M. Bumsted, "'Things in the Womb of Time': Ideas of American Independence, 1633–1763," *William and Mary Quarterly,* 3rd ser., 31 (1974): 533–564.
32. James Abercromby, *Magna Charta for America: James Abercromby's "An Examination of the Acts of Parliament Relative to the Trade and the Government of Our American Colonies" (1752) and* "De Jure et Gubernatione Coloniarum, *or An Inquiry into the Nature, and the Rights of Colonies, Ancient, and Modern" (1774),* ed. Jack P. Greene, Charles F. Mullett, and Edward C. Papenfuse Jr. (Philadelphia: American Philosophical Society, 1986), 216–217; Thomas Pownall, *Administration of the Colonies* (London, 1764), 6. See Miller, *Defining the Common Good,* 195–202, 207–211.
33. "Treaty of Paris 1763," *Avalon Project,* http://avalon.law.yale.edu/18th_century/paris763.asp; David Armitage, *Ideological Origins of the British Empire* (New York: Cambridge University Press, 2000), 109–113.
34. Egremont to Board of Trade, May 5, 1763, CO 323/15: 101–107, The National Archives of the UK, Kew, Richmond, Surrey (hereafter cited as UK National Archives). This document is published in Adam Shortt and Arthur G. Doughty, eds., in *Documents Relating to the Constitutional History of Canada, 1759–1791* (Ottawa: S. E. Dawson, 1907), 93–96 (hereafter cited as *Canadian Documents*).
35. *Board of Trade Journals,* 1759–1763, 362–363, 368. Historians have variously credited Ellis, Ellis in collaboration with Knox, and Knox alone as the authors of these influential "Hints." See Sosin, *Whitehall and the Wilderness,* 56; Leland J. Bellot, *William Knox: The Life and Thought of an Eighteenth-Century Imperialist* (Austin: University of Texas Press, 1977), 46–47; Edward J. Cashin, *Governor Henry Ellis and the Transformation of British North America* (Athens: University of Georgia Press, 1994), 185; John Shy, *A People Numerous and Armed: Reflections on the Military Struggle for American Independence* (New York: Oxford University Press, 1976), 267n9. In my view, Shy's attributions are the most credible. In this work, Shy also credits Ellis with the composition of the influential military plan for the interior, the "Plan of Forts and Garrisons." Franklin, "Short Hints Towards a Scheme for Uniting the Northern Colonies" (1754), *Franklin Papers,* http://franklinpapers.org/franklin/framedVolumes.jsp?vol=5&page=353a. On the influence of merchants and other interest groups on the Board, see Alison Olson, "The Board of Trade and London-American Interest Groups in the Eighteenth Century," *Journal of Imperial and Commonwealth History* 8 (1980): 33–50. Many more "hints" were housed in volumes titled "America" and "Miscellaneous Papers and Estimates Relative to Indian Trade." *Royal Commission on Historical Manuscripts, Reports, Part I* (London, 1856), 216–218. See also Heather Schwartz, "Re-writing the Empire: Plans for Institutional Reform in British America, 1675–1791" (PhD diss., Binghamton University, State University of New York, 2011).
36. "[O]racle": Francis Maseres to Fowler Walker, August 11, 1768, in *The Maseres Letters 1766–1768* (Toronto: University of Toronto Library, 1919), 99; Henry

Ellis, *To Arthur Dobbs . . . this chart of the coast where a north west passage was attempted* (London, 1748); Henry Ellis, "A New Chart of the parts where a North West Passage was sought in the Years 1746 and 1747," in *Voyage to Hudson's Bay, by the Dobbs Galley and California, in the Years 1746 and 1747, for Discovering a Northwest Passage, etc.* (Dublin, 1749); Henry Ellis, *Considerations on the Great Advantages which would Arise from the Discovery of the North West Passage* (London, 1750); Rory T. Cornish, "Ellis, Henry (1721–1806)," *Oxford DNB*. "[L]arge but weak": J[ohn] Reynolds, "A Representation & Estimate of what is necessary to be done for the Security & Defense of His Majesty's Province of Georgia," [July 23, 1756], Loudon Papers, LO 1332, Huntington Library, San Marino, CA; W. W. Abbot, *The Royal Governors of Georgia, 1754–1777* (Chapel Hill: University of North Carolina Press, 1959), chap. 3; Cashin, *Governor Henry Ellis,* chaps. 5–7; "handwriting": Maseres to Walker, August 11, 1768, in *Maseres Letters,* 99; Jack Stagg, *Anglo-Indian Relations in North America to 1763 and An Analysis of the Royal Proclamation of 7 October 1763* (Ottawa: Research Branch, Indian and Northern Affairs, 1981), 327–328.

37. [Henry Ellis], "Hints Relative to the Division and Government of the Conquered and Newly Acquired Countries in America" [1763], CO 323/16, UK National Archives. For a published version of this text, see Verner W. Crane, "Hints Relative to the Division and Government of the Conquered and Newly Acquired Countries in America," *Mississippi Valley Historical Review* 8 (1922): 367–373.
38. [William Knox], "Hints Respecting the Settlement of our American Provinces" 1763, Shelburne Papers, 46/2: 475–487, Clements Library. For another copy, see Additional MSS, 38,335, fol. 14–33, The British Library, London. It is published in Thomas C. Barrow, "A Project for Imperial Reform: 'Hints Respecting the Settlement for our American Provinces,' 1763," *William and Mary Quarterly,* 3rd ser., 24 (1967): 113–126. On Knox's authorship of this tract, see ibid., 109, 48–49, 227n; for an argument on the possible coauthorship of this tract by Maurice Morgann, see R. A. Humphreys, "Lord Shelburne and the Proclamation of 1763," *English Historical Review* 49, no. 194 (1934): 247–248. Henry Ellis to William Knox, April 30, 1762, in Historical Manuscripts Commission, *Report on Manuscripts in Various Collections,* vol. 6 (Dublin, 1909), 87; Bellot, *William Knox,* 44, 55; William Knox to Governor Lyttleton, March 5, 1760, in HMC, *Report,* 83; on France: Knox to Charles Townshend, August 27, 1763, in ibid., 88; Leland J. Bellot, "Knox, William (1732–1810)," *Oxford DNB*.
39. Benjamin Franklin, "Observations concerning the Increase of Mankind," 1755, in *Franklin Papers,* franklinpapers.org/franklin/framedVolumes.jsp?vol=4&page=225a; [Knox], "Hints Respecting," in Barrow, "Project for Imperial Reform," 113.
40. [Knox], "Hints Respecting," 114–115.

41. Ibid., 113–114, 115, 116, 117, 118, 122; "foreigners": Halifax (1763), quoted in P. J. Marshall, "A Nation Defined by Empire, 1755–1776," in *Uniting the Kingdom? The Making of British History,* ed. Alexander Grant and Keith J. Stringer (London: Routledge, 1995), 220. On negative metropolitan appraisals of American places and people, see Jack P. Greene, *Evaluating Empire and Confronting Colonialism in Eighteenth-Century Britain* (New York: Cambridge University Press, 2013), chap. 2.
42. See Eliga H. Gould, *The Persistence of Empire: British Political Culture in the Age of the American Revolution* (Chapel Hill: University of North Carolina Press, 2000), chaps. 4–5.
43. See Anderson, *Crucible of War,* chap. 58; Wickwire, *British Subministers,* 13; Bellot, "Knox, William"; [Maurice Morgann], "Plan for securing the Future Dependence of the Provinces on the Continent of America," Shelburne Papers, 67:107–110; [Maurice Morgann] to Shelburne, "On American commerce and government in newly acquired territories, in hand of Maurice Morgann," Shelburne Papers, 85: 26–35; "Mr. Pownal's Sketch of a Report concerning the Cessions in Africa and America at the Peace of 1763," in Humphreys, "Shelburne and the Proclamation," 258–264. The original can be found in the Shelburne Papers, 49: 333–364.
44. "Report on Acquisitions in America," June 8, 1763, CO 324/21, fol. 245–291, The National Archives of the UK, Kew, Richmond, Surrey, published in *Canadian Documents,* 97–107; "Report on Acquisitions," in *Canadian Documents,* 100–101.
45. On the publication history of Bowen's map, see Margaret Beck Pritchard and Henry G. Taliaferro, *Degrees of Latitude: Mapping Colonial America* (New York: Colonial Williamsburg/Harry N. Abrams, 2002), 180–183.
46. "Report on Acquisitions," in *Canadian Documents,* 103, 105; Egremont to Bedford, September 7, 1762, in *British Diplomatic Instructions, 1689–1789,* vol. 7, *France, 1745–1789,* ed. L. G. Wickham Legg (London: Offices of the Royal Historical Society, 1934), 65.
47. On "regular" colonies, see Jack P. Greene, *Imperatives, Behaviors, and Identities: Essays in Early American Cultural History* (Charlottesville: University of Virginia Press, 1992), 117–118; "Report on Acquisitions," in *Canadian Documents,* 105, 101, 107, 106, 102; "Republican Mixture": [Ellis], "Hints Relative," in Crane, 372.
48. Shelburne penned this phrase in his "Remarks on the cessions made by France and Spain" [1763], Shelburne Papers, 50: 37–38, and it appeared in Egremont's May 5 charge to the Board and in the "Report on Acquisitions," in *Canadian Documents,* 105. Bute, quoted in William Knox, *Extra-Official State Papers Addressed to the Right Hon. Lord Rawdon* (London, 1789), 29; Rodney quoted in Syrett, *Rodney Papers,* 211–212; Josiah Tucker, "A Letter to Edmund Burke" (1775), in *The Collected Works of Josiah Tucker: Economics and American Colonial Policy,* vol. 5 (London: Routledge/Thoemmes Press, 1993), [8]; *Acts of*

the Privy Council of England, Colonial Series, vol. 4, *A.D. 1745–1766,* ed. James Munro (Hereford: H. M. Stationery Office, 1911), 594–595; "Report on Acquisitions," in *Canadian Documents,* 99–100; Order of the King, October 5, 1763, CO 323 / 16: 341, 342, UK National Archives.

49. December 9, 1762, *Proceedings and Debates,* 1:423.
50. "[E]xact union": Pownall, "Pownal's Sketch," in Humphreys, "Shelburne and the Proclamation," 259; "Report on Acquisitions," in *Canadian Documents,* 103; "Belonging to the Indians": quoted in Stagg, *Anglo-Indian Relations,* 312–313.
51. "Extract from the Minutes of the Proceedings of the Lords Commissioners of Trade and Plantations," December 3, 1762, in *Selections from the Public Documents of the Province of Nova Scotia,* ed. Thomas B. Akins (Halifax, 1869), 337; [Knox], "Hints Respecting," in Barrow, "Project for Imperial Reform," 116; Marshall, "Defined by Empire," 215.
52. "The Royal Proclamation—October 7, 1763," *Avalon Project,* http://avalon.law.yale.edu/18th_century/proc1763.asp; "Report on Acquisitions," in *Canadian Documents,* 104; [Knox], "Hints Respecting," in Barrow, "Project for Imperial Reform," 115, 118, 114; "Report on Acquisitions," in *Canadian Documents,* 104, 106, 99, 101.
53. James Grant to Shelburne, December 22, 1768, CO 5 / 550: 57, British Public Record Office Transcripts, Manuscript Division, Library of Congress, Washington, DC (hereafter cited as LC Transcripts).
54. Board of Trade to the King, December 20, 1763, CO 324 / 21: 379–380, LC Transcripts; "Report on Acquisitions," in *Canadian Documents,* 103. The final boundaries described in the proclamation differed from those detailed in the report, including the decision to change the boundary between Georgia and East Florida from the St. John's River to the St. Mary's River.

CHAPTER TWO: COMMANDING SPACE AFTER THE SEVEN YEARS' WAR

1. Samuel Johnson, *The Works of Samuel Johnson,* 15 vols. (London, 1789), 15:454.
2. Peter Wraxhall (1755), quoted in Stanley Pargellis, ed., *Military Affairs in North America, 1748–1755: Selected Documents from the Cumberland Papers in Windsor Castle* (New York: D. Appleton-Century, 1936), 144. "Allegany": John St. Clair (1755), quoted in ibid., 66. Wraxhall was also secretary for Indian Affairs for New York.
3. On the concept of "centers of calculation," see Bruno Latour, *Science in Action: How to Follow Scientists and Engineers through Society* (Cambridge, MA: Harvard University Press, 1987), chap. 6. See also Stephen Hornsby, *Surveyors of Empire: Samuel Holland, J. F. W. Des Barres, and the Making of the Atlantic Neptune* (Montreal: McGill-Queen's University Press, 2011), 8, 31, 71, 126, 172. On the role of commercial publishing in the creation of English cartography, see Mary Sponberg Pedley, *The Commerce of Cartography: Making and Marketing Maps in Eighteenth-Century France and England* (Chicago: University of Chicago Press, 2005).

4. Of the 144 maps, plans, and charts included within dispatches sent by officials in America to the Board of Trade and the office of the secretary of state for the Southern Department in London, only three date from before 1700, and only twenty-two from the period between 1701 and 1750. The vast majority—118—arrived from America after 1750. S. Max Edelson, "Maps in *Colonial America:* An Introduction," in *Colonial America,* Adam Matthew Digital, accessed March 10, 2016, http://www.colonialamerica.amdigital.co.uk/Explore/Essays/Edelson. See also Rose Mitchell, "From Roanoke to Aden: Colonial Maps in the National Archives of the United Kingdom" (paper presented to the International Symposium on "'Old Worlds-New Worlds': The History of Colonial Cartography 1750–1950," Utrecht University, the Netherlands, August 21–23, 2006), http://www.icahistcarto.org/PDF/Mitchell_Rose_-_From_Roanoke_to_Aden.pdf.
5. See also "Map of New York, New Jersey, Pennsylvania, Maryland and Virginia," 1756, MPG 1/332, The National Archives of the UK, Kew, Richmond, Surrey (hereafter cited as UK National Archives).
6. This image was extracted from a volume of correspondence in which Generals Braddock and Shirley and Admirals Boscawen and Holbourn reported on British military setbacks in the early years of the war. It followed a letter from Shirley to Secretary of State Thomas Robinson discussing the causes and implications of Braddock's defeat at the Battle of the Monongahela on June 9, 1755. Secretary of State: French and Indian War, Jan.–Dec. 1755, CO 5/46 part 1, *Colonial America,* Adam Matthew Digital.
7. Mary Ann Rocque and John Rocque, *A Set of Plans and Forts in America* (London, 1763). See Laurence Worms and Ashley Baynton-Williams, *British Map Engravers: A Dictionary of Engravers, Lithographers and Their Principal Employers to 1850* (London: Rare Book Society, 2011), 559–564.
8. Keith R. Widder, "The Cartography of Dietrich Brehm and Thomas Hutchins and the Establishment of British Authority in the Western Great Lakes Region, 1760–1763," *Cartographica* 36 (1999): 31; Thomas Jefferys, *The West-India atlas; or, A compendious description of the West-Indies* (London, 1775), i, iii; William Laird Clowes, *The Royal Navy: A History from the Earliest Times to the Present,* 5 vols. (London: Sampson Low, Marston & Co., 1898), 3:312–314.
9. Fred Anderson, *The War That Made America: A Short History of the French and Indian War* (New York: Viking, 2005), 61; Eric Hinderaker, "The 'Four Indian Kings' and the Imaginative Construction of the First British Empire," *William and Mary Quarterly,* 3rd ser., 53 (1996): 519; Walker Hovenden, *A Journal: or Full Account of the Late Expedition to Canada* (London, 1720), 67, 88; Dale Miquelon, "Ambiguous Concession: What Diplomatic Archives Reveal about Article 15 of the Treaty of Utrecht and France's North American Policy," *William and Mary Quarterly,* 3rd ser., 67 (2010): 465.
10. Jean Sharpenez, "Plan de la Riviere de Canada," 1755, MSS 368, Admiralty Library, Portsmouth, UK; Rear Admiral G. S. Ritchie, *The Admiralty Chart,*

British Naval Hydrography in the Nineteenth Century, rev. ed. (Edinburgh: Pentland Press, 1995), 25; Donald W. Olson et al., "Perfect Tide, Ideal Moon: An Unappreciated Aspect of Wolfe's Generalship at Québec, 1759," *William and Mary Quarterly,* 3rd ser., 59 (2001): 965–969. Another copy of the Sharpenez chart is "Plan de la Riviere de Canada," 1755, CO 700 / Canada 13, UK National Archives. Other French charts of the St. Lawrence in British hands included "Carte du Detroit de Belle Isle or Embochure du Fleuve de St. Laurent," n.d.; "Carte de la Riuiere du Canada & Golfe de St. Laurant," 1756; M[ssr.] de la Richardiere, "Carte Particuliere de la Riveire du Canada," n.d.; Germain Goynard, "Plane de la grande Riviere de Canada," 1758, all in MS Charts and Maps of America, vol. 1, MSS 368, Admiralty Library.

11. J. B. Harley, "The Bankruptcy of Thomas Jefferys: An Episode in the Economic History of Eighteenth Century Map-Making," *Imago Mundi* 20 (1966): 33–38; Thomas Jefferys, *A general topography of North America and the West Indies. Being a collection of all the maps, charts, plans, and particular surveys, that have been published of that part of the world, either in Europe or America* (London, 1768).
12. Robert Stobo Letter and Plan of Fort Duquesne, July 28, 1754, University of Pittsburgh Digital Research Library, accessed August 12, 2016, http://digital.library.pitt.edu/u/ulsmanuscripts/pdf/31735060225889.pdf; Robert Stobo Letter and Plan of Fort Duquesne, Guide to Archives and Manuscript Collections at the University of Pittsburgh Library System, accessed August 12, 2016, http://digital.library.pitt.edu/cgi-bin/f/findaid/findaid-idx?c=ascead;cc=ascead;q1=DAR.19%2A;view=text;didno=US-PPiU-dar192505.
13. Paul W. Mapp, *The Elusive West and the Contest for Empire, 1713–1745* (Chapel Hill: University of North Carolina Press, 2011), chaps. 9–11.
14. See Douglas Fordham, *British Art and the Seven Years' War: Allegiance and Autonomy* (Philadelphia: University of Pennsylvania Press, 2010).
15. "[S]uffered": Admiralty instructions (1761), quoted in Hornsby, *Surveyors of Empire,* 34; remark book, HMS *Northumberland,* 1758–1762, MS 20 / 3, Admiralty Library; Andrew David, "James Cook's 1762 Survey of St. John's Harbour and Adjacent Parts of Newfoundland," *Terrae Incognitae* 30 (1998): 63–71; remark book, HMS *Norwich,* 1761, MSS 20 / 13, Admiralty Library. For images of a related remark book, see James Cook, "Description of the sea coast of Nova Scotia, 1762," MS 5, National Library of Australia, Trove, accessed March 16, 2016, http://nla.gov.au/nla.obj-229113354.
16. Report of the Board of Trade, March 26, 1763, *Acts of the Privy Council of England,* Colonial Series, vol. 4, *A.D. 1745–1766,* ed. James Munro (Hereford: H. M. Stationery Office, 1911), 588. Other Campbell maps include "Sketch of the Coast Round the Island of Dominique," Maps 188.o.2(6), The British Library, London, and "Sketch of the Island of Dominique," Maps K.Top. 123.93, The British Library. My thanks to Kelvin Smith for his comments on the production of these maps. Other early Admiralty maps and charts of the

Ceded Islands include [Map showing the coastline of Quassyganna Bay], [1764?], MFQ 1 / 1173 / 3; [Map showing part of the coastline and the sites of Fort Royal and other settlements], [1764], MFQ 1 / 1173 / 2; [Map showing the coastline of Privateer's Bay], [1764?], MFQ 1 / 1173 / 4; [Map showing the coastline of Quassyganna Bay], [1764?], MFQ 1 / 1173 / 5, UK National Archives.

17. See, for example, Jacques-Nicolas Bellin, "Les Petites Antilles ou les isles du Vent," in *Le petit atlas maritime* (Paris, 1764); John Stephens Hall, "A Chart of the Caribbee Islands Taken from a French One Published in 1758, by the Sr. Bellin and Improved," 1775, MS 370, Admiralty Library; Hall, [Ceded Islands bay charts], ibid.
18. Carolyn J. Anderson, "Constructing the Military Landscape: The Board of Ordnance Maps and Plans of Scotland, 1689–1815" (PhD diss., University of Edinburgh, 2010); Douglas W. Marshall, "The British Military Engineers in America: 1755–873" (PhD diss., University of Michigan, 1976), 207–208; "General Instructions for the Officers of Engineers Employed in Surveying" (1785), General Roy's Papers, WO 30 / 115 [B], UK National Archives, 63–73. My thanks to C. J. Anderson for this reference.
19. R. A. Skelton, "The Military Surveyor's Contribution to British Cartography in the 16th Century," *Imago Mundi* 24 (1970): 77–78; Mark White, "Manuscript of Navigation" 1752, MS 336, Admiralty Library; Marshal, "Military Engineers," 202–205.
20. Reginald G. Golledge, "Human Wayfinding and Cognitive Maps," in *Wayfinding Behavior: Cognitive Mapping and Other Spatial Processes,* ed. Reginald G. Golledge (Baltimore: Johns Hopkins University Press, 1998), 5–45.
21. Pargellis, *Military Affairs,* 104–109, 176–177, 104n. Hutchins's "Sketch" map may be [Fort Edward to Crown Point], 1755, Geography and Map Division, Library of Congress, Washington, DC; Clarence W. Alvord and Clarence E. Carter, eds., *The New Régime, 1765–1767* (Springfield: Illinois State Historical Library, 1916), 292–296; Amherst (1760), quoted in Widder, "Brehm and Hutchins," 3.
22. Murray (1761), quoted in R. H. Mahon, *Life of General the Hon. James Murray: A Builder of Canada* (London: John Murray, 1921), 288; Hornsby, *Surveyors of Empire,* 27; Jeffrey S. Murray, *Terra Nostra: The Stories behind Canada's Maps, 1550–1950* (Montreal: McGill-Queen's University Press, 2006), chap. 2; see also Nathaniel N. Shipton, "General James Murray's Map of the St. Lawrence," *Canadian Cartographer* 4 (1967): 93–101.
23. Marshall, "Military Engineers," 221–235; Hornsby, *Surveyors of Empire,* 25–29. See also a comparable map of the district from Montreal to Quebec: John Montresor and Samuel Holland, "Plan of the River Saint Lawrence from Montreal to the Parish of Berthier on the North Side of the River, and Sorel on the South," 1762, Maps K.Top. 119.28, The British Library.

24. "Report of the State of the Government of Quebec in Canada," James Murray to the Board of Trade, June 5, 1762, CO 323/15: 62–78, 175–176, UK National Archives; Francis Maseres to Fowler Walker, September 14, 1766, in *The Maseres Letters, 1766–1768* (Toronto: University of Toronto Library, 1919), 42, 43; Maseres to Charles Yorke, May 27, 1768, in ibid., 89.
25. "Report of the State of the Government of Quebec," 62–78, 175–176.
26. "Plan of Forts and Garrisons Propos'd for the Security of North America," [1763?], CO 323/16: 164–85, UK National Archives. For a published version, see Clarence Walworth Alvord and Clarence Edwin Carter, eds., *The Critical Period 1763–1765* (Springfield: Illinois State Historical Library, 1915), 5–11. On the attribution of the plan to Ellis, see John Shy, *A People Numerous and Armed: Reflections on the Military Struggle for American Independence* (New York: Oxford University Press, 1976), 308n10; Richard Brown and Paul E. Cohen, *Revolution: Mapping the Road to American Independence, 1755–1783* (New York: W. W. Norton, 2015), 53–56. See also "Cantonment of the Forces in North America 11th. Octr. 1765," [1765], Library of Congress. A misreading of a notation on Bowen's map led to the mistaken siting of "Apalachie" deep in Creek country rather than in its true location near the coast.
27. "Plan of Forts and Garrisons"; William Johnson, "re. thoughts concerning Florida," n.d., Shelburne Papers, 48: 6, William L. Clements Library, University of Michigan, Ann Arbor.
28. "Plan of Forts and Garrisons."
29. Charles L. Mowat, *East Florida as a British Province, 1763–1784* (Berkeley: University of California Press, 1943), 7, 10; Alan Gallay, ed., *Colonial Wars of North America: An Encyclopedia* (New York: Garland, 1996); 744–745; "Gordon's Journal, May 8, 1766–December 6, 1766," in Alvord and Carter, eds., *The New Régime,* 302–303.
30. "Plan of Forts and Garrisons"; Jack M. Sosin, *Whitehall and the Wilderness: The Middle West in British Colonial Policy, 1760–1775* (Lincoln: University of Nebraska Press, 1961), 18–19; Thomas Gage to Halifax, May 21, 1764, in Alvord and Carter, eds., *The Critical Period,* 249. For reproductions of maps that show the Iberville passage, see Alfred E. Lemmon, John T. Magill, and Jason R. Wiese, eds., *Charting Louisiana: Five Hundred Years of Maps* (New Orleans: Historic New Orleans Collection, 2003), 56–57, 64–68, 73–74, 78–79, 80–81. On British concerns for the free navigation of the Mississippi, see "Instructions for . . . John, Duke of Bedford, September 4, 1762," in *British Diplomatic Instructions, 1689–1789,* vol. 7, *France, 1745–1789,* ed. L. G. Wickham Legg (London: Offices of the Royal Historical Society, 1934), 57, 74. Douglas Stewart Brown, "The Iberville Canal Project: Its Relation to Anglo-French Commercial Rivalry in the Mississippi Valley, 1763–1775," *Mississippi Valley Historical Review* 32 (1946): 499–500. See Philip Pittman, *The Present State of the European Settlements on the Missis[s]ippi* (London, 1770), 29–32. See also Pittman, "Copy of a Draught of Massiac River, Lake Maurepas, part of the River Amitt

and the River Iberville," [1765], CO 700 / Florida 27, UK National Archives; Pittman, *Draught of the R. Ibbeville* (London, 1770), originally published in Pittman, *State of the Settlements*. Both the copy and the published version of this map alter the original manuscript's designation that feet were used to measure the river and fathoms used to measure the lake to show the opposite. I have assumed the original manuscript to be correct.

31. "Gordon's Journal," 302–303; Pittman, *State of the Settlements,* 31; Brown, "Iberville Canal Project," 511–512. See also Thomas Hutchins, *An Historical Narrative and Topographical Description of Louisiana and West-Florida* (Gainesville: University of Florida Press, 1968), xix, 40–44. These Iberville maps include Elias Durnford, "Map of part of the River Mississippi and Ibberville, showing the proposed new cut," [1771], CO 700 / Florida 46, UK National Archives; "The Communication Between the Iberville & the River Mississippi," 1770, MPG 1 / 359 / 1, UK National Archives; Durnford, "The Communication between the Iberville and the Mississippi," 1770, Clements Library; Durnford, "An actual Survey of part of the River Ibberville," 1771, CO 700 / Florida 44, UK National Archives; "Sketch Map of part of the River Iberville," [1771?], CO 700 / Florida 48, UK National Archives; Tho[mas] Hutchins, "A plan of the Lakes Ponchartrain and Maurepas and the River Ibberville," [ca. 1772?], MR 1 / 107, UK National Archives; Hutchins, "Horizontal Line from Manchac to the forks of the Iberville," 1773, Clements Library.
32. "Reference to the Land surveyed on the River Mississippi since the Establishment of the Civil Government of West Florida," CO 700 / Florida 45, UK National Archives. The West Florida Council issued warrants for these grants in 1766, before Fort Panmure was abandoned. Douglas Brymner, Arthur George Doughty, Edouard Richard, eds., *Documents Relating to the Constitutional History of Canada, 1759–1791, 1791–1818* (Ottawa: Maclean, Roger, 1886), 161; Robert V. Haynes, *The Natchez District and the American Revolution* (Jackson: University Press of Mississippi, 2008), 11–18.
33. "Relative to Fortifications in America," CO 5 / 232, British Public Record Office Transcripts, Manuscript Division, Library of Congress (hereafter cited as LC Transcripts); "Journal of an Officer's [Lord Adam Gordon's] Travels in America and the West Indies, 1764–1765," in *Travels in the American Colonies,* ed. Newton D. Mereness (New York: Macmillan, 1916), 387–388; "not worth repairing": T. Sower (1772), quoted in Peter J. Hamilton, *Colonial Mobile: An Historical Study* (Boston: Houghton Mifflin, 1897), 212.
34. Gallay, ed., *Colonial Wars,* 744–745; James W. Parker, "Archaeological Test Investigations at 1Su7: The Fort Tombecbe Site," *Journal of Alabama Archaeology* 28 (1982): 1–104; letter of Lieutenant Ford, November 24, 1763, in *Mississippi Provincial Archives, 1763–1766: English Dominion,* ed. Dunbar Roland (Nashville: Brandon Printing, 1911), 25.

35. Letter of George Johnstone and John Stuart, June 12, 1765, in *Mississippi Provincial Archives,* 186; Major Robert Farmar to Secretary of War Welbore Ellis, January 24, 1764, in ibid., 12.
36. Louis De Vorsey Jr., *The Indian Boundary in the Southern Colonies, 1763–1775* (Chapel Hill: University of North Carolina Press, 1966), 207, 210, 212.
37. William Forbes to Secretary of State, June 20, 1768, in *Mississippi Provincial Archives,* 141–142; "Relative to Fortifications in America"; "Journal of an Officer," 381–384; "horrid place": John Stuart to James Grant, October 12, 1764, James Grant of Ballindalloch Papers, microfilm, Manuscript Division, Library of Congress.
38. "Journal of an Officer," 382; James Grant to Richard Oswald, September 20, 1764, Grant Papers; Forbes to Secretary of State, January 29, 1764, *Mississippi Provincial Archives,* 142; John D. Ware and Robert R. Rea, *George Gauld: Surveyor and Cartographer of the Gulf Coast* (Gainesville: University Presses of Florida, 1982), 36; "A Proposed plan of a new House and Offices for the Governor of West Florida," MPG 1/611, UK National Archives; Martha Pollak, "Military Architecture and Cartography in the Design of the Early Modern City," in *Envisioning the City: Six Studies in Urban Cartography,* ed. David Buisseret (Chicago: University of Chicago Press, 1998), 118. See also Elias Durnford, "Plan of the New Town of Panzacola," [1765], CO 700/Florida 13, UK National Archives; Elias Durnford, "Plan of the New Town of Pensacola and country adjacent," 1765, CO 700/Florida 20/1, UK National Archives. Other plans documented British construction within the original stockade. See Durnford, "Plan of the Stockade fort at Pensacola," [1768], MPG 1/349, UK National Archives; Durnford, "Plan of the Stockade Fort," 1771, MPG 1/527, UK National Archives; Durnford, "Plan of the fort," 1778, MPG 1/358, UK National Archives.
39. Forbes to Secretary of State, June 20, 1768, in *Mississippi Provincial Archives,* 142; Governor Johnstone and John Stuart, June 12, 1765, in ibid., 187; John Stuart to Hillsborough, December 29, 1771, CO 5/73: 41–42, LC Transcripts.
40. "[U]nknown coast": Alexander Colville (1763), quoted in Ware, *George Gauld,* xx; ibid., 40–43.
41. William Roberts, *An Account of the First Discovery and Natural History of Florida* (London, 1763), 97; Mark F. Boyd, "The Fortifications at San Marcos de Apalache (St. Marks, Wakulla Co., Florida)," *Florida Historical Society Quarterly* 15 (1936): 1–14; Grant to Lieut. Pompellonne, August 24, 1765, Grant Papers; James Grant to [Shelburne], September 28, 1768, ibid., 335.
42. Boyd, "Fortifications at Apalache," 1–14; Grant to Lieut. Pompellonne, August 24, 1765, Grant Papers; James Grant to [Shelburne], September 28, 1768, ibid.; Clifton Paisley, *The Red Hills of Florida, 1528–1865* (Tuscaloosa: University of Alabama Press, 1989), 35–39; see also Ware, *George Gauld,* 86–91.

43. Johnstone (1766), quoted in Ware and Rea, *George Gauld,* 37; Johnson, "thoughts concerning Florida"; William Knox, "Hints respecting the Settlement of Florida," Shelburne Papers, William Clements Library, 48: 5; "Journal of an Officer," 392; John Stuart to Hillsborough, December 29, 1771, CO 5 / 73: 37, LC Transcripts.
44. See Robert Paulette, *An Empire of Small Places: Mapping the Southeastern Anglo-Indian Trade, 1732–1795* (Athens: University of Georgia Press, 2012).
45. James Grant to [Henry] Laurens, November 18, 1764, Grant Papers; Grant to Richard Oswald, September 20, 1764, ibid.; John Stuart to Hillsborough, December 29, 1771, CO 5 / 73, LC Transcripts. See also Bernard Bailyn, *Voyagers to the West: A Passage in the Peopling of America on the Eve of the Revolution* (New York: Vintage, 1986), chap. 13.
46. John Lindsay to James Grant, November 1, 1764, Grant Papers.
47. Egremont (1763), quoted in Jack Stagg, *Anglo-Indian Relations in North America to 1763 and an Analysis of the Royal Proclamation of 7 October 1763* (Ottawa: Research Branch, Indian and Northern Affairs, 1981), 301; Marshall, "Military Engineers," 203; [Maurice Morgann], "Plan for securing the Future Dependence of the Provinces on the Continent of North America," Shelburne Papers, 67: 107–110.
48. "Mr. Pownal's Sketch of a Report concerning the Cessions in Africa and America at the Peace of 1763," in R. A. Humphreys, "Lord Shelburne and the Proclamation of 1763," *English Historical Review* 49 (1934): 262.

CHAPTER THREE: SECURING THE MARITIME NORTHEAST

1. "Report on Acquisitions in America," June 8, 1763, in *Documents Relating to the Constitutional History of Canada,* vol. 1, *1759–1791* (Ottawa, 1907), 100, 142, 137, 134. On the general extent and uncertain boundaries of New France, see Gage to Amherst, March 20, 1762, CO 323 / 16: 40–42, The National Archives of the UK, Kew, Richmond, Surrey (hereafter cited as UK National Archives). The text of the report suggests "Cape Roziere" on the Gaspé Peninsula as the northernmost limit of Nova Scotia; on the Board's annotated map annexed to the report, however, the border with Quebec follows a more southerly course, along the Restigouche River to Chaleur Bay.
2. D. C. Harvey, *Holland's Description of Cape Breton Island and other Documents* (Halifax, NS: Public Archives of Nova Scotia, 1935), 31–32; See also [Samuel Holland], "Map of Quebec and the environs, with plan of military operations of 1759," CO 700 / Canada 19, UK National Archives; Board of Trade to the King, December 20, 1763, CO 324 / 21: 318–21, British Public Record Office Transcripts, Manuscript Division, Library of Congress, Washington, DC (hereafter cited as LC Transcripts).
3. Holland (1763), quoted in Harvey, *Holland's Description of Cape Breton,* 36; Board of Trade to the King, December 20, 1763, CO 324 / 21: 317–319, LC Transcripts.

4. John Pownall to William De Brahm, August 15, 1764, CO 324/17: 170–171, LC Transcripts; Pownall to Samuel Holland, May 19, 1766, CO 324/18: 8–9, LC Transcripts; Pownall to Holland, April 17, 1764, CO 32/417: 392–396, LC Transcripts; Pownall to De Brahm, August 15, 1764, CO 324/17: 425–426, LC Transcripts.
5. John D. Ware and Robert R. Rea, *George Gauld: Surveyor and Cartographer of the Gulf Coast* (Gainesville: University Presses of Florida, 1982), 16; [Samuel] Holland to [Dartmouth], May 2, 1773, CO 5/228: 247–248, LC Transcripts; [John] Pownall to Samuel Holland, May 23, 1771, CO 324/18: 86, LC Transcripts.
6. Elizabeth Mancke, "Spaces of Power in the Early Modern Northeast," in *New England and the Maritime Provinces: Connections and Comparisons,* ed. Stephen J. Hornsby and John G. Reid (Montreal: McGill-Queen's University Press, 2005), 33–42.
7. See Peggy K. Liss, *Atlantic Empires: The Network of Trade and Revolution, 1713–1826* (Baltimore: Johns Hopkins University Press, 1982), 1–20.
8. Judith Tulloch, "The New England Fishery and Trade and Canso, 1720–1744," in *How Deep Is the Ocean? Historical Essays on Canada's Atlantic Fishery,* ed. James E. Candow and Carol Corbin (Sydney, NS: University College of Cape Breton Press, 1997), 69, 66; Mancke, "Spaces of Power," 38–45; see Jeffers Lennox, "*L'Acadie Trouvée:* Mapping, Geographic Knowledge, and Imagining Northeastern North America, 1710–1763" (PhD diss., Dalhousie University, 2010), chap. 2.
9. Quoted in Harriet Hart, "History of Canso, Guysborough County, N.S.," in *Collections of the Nova Scotia Historical Society* (Halifax, NS: Wm. Macnab & Son, 1927), 21:4; "State of His Majesty's Plantations on the Continent of America," September 8, 1721, in *Documents Relative to the Colonial History of the State of New-York,* ed. F. B. O'Callaghan, 15 vols. (Albany, NY: Weed, Parsons, 1853), 5:592; Mr. Delafay to the Board of Trade, May 28, 1719, in *Calendar of State Papers Colonial, America and West Indies,* vol. 31, *1719–1720,* ed. Cecil Headlam (London: H. M. Stationery Office, 1933): 98–100, no. 208; see Lennox, "*L'Acadie Trouvée,*" 87–102. See also Jeffers Lennox, "A Time and a Place: The Geography of British, French, and Aboriginal Interactions in Early Nova Scotia, 1726–44," *William and Mary Quarterly,* 3rd ser., 72 (2015): 423–460.
10. Council of Trade and Plantations to the King, September 8, 1721, 32 (1720–1721): 408–449, in *Calendar of State Papers, Colonial Series, America and West Indies, 1574–1739,* CD-ROM, consultant editors Karen Ordahl Kupperman, John C. Appleby, and Mandy Banton (London: Routledge, 2000); Ian K. Steele, *The Politics of Colonial Policy: The Board of Trade in Colonial Administration 1696–1720* (New York: Oxford University Press, 1968), 154, 118–120; Board of Trade to Secretary Stanhope, July 15, 1715, *Calendar of State Papers, Colonial Series, America and the West Indies, August, 1714–December 1715,* ed. Cecil

Headlam (London: H. M. Stationery Office, 1928), 232, no. 518; Board of Trade to Lords Justices, September 15, 1720, *Calendar of State Papers, Colonial Series, America and the West Indies, March 1720–December 1721* (London: H. M. Stationery Office, 1933), 141–142, no. 231.

11. British maps of the area made after this directive included [Cyprian] Southack, [Copy of Chart of Fishing Limits], 1718, Geography and Map Division, Library of Congress; Cyprian Southack, "The Harbour & Islands of Canso part of the boundaries of Nova Scotia," 1718, MS Charts and Maps of America, vol. 1, MSS 368, Admiralty Library, Portsmouth, UK; Tho[ma]s Durell, "A Draught of the Harbour & Islands of Canso in the Government of Nova Scotia," 1732, MS Charts and Maps of America, vol. 2, MS 369, Admiralty Library; for another copy, see CO 700 / Nova Scotia 9, UK National Archives. William Welch, "A Further Note on Captain Thomas Durell's Charts of Nova Scotia," *Journal of the Royal Nova Scotia Historical Society* 14 (2001): 170–173; Jerry Lockett, *Captain James Cook in Atlantic Canada: The Adventurer and Map Maker's Formative Years* (Halifax, NS: Formac, 2010), 94; Clara Egli LeGear, "The New England Coasting Pilot of Cyprian Southack," *Imago Mundi* 11 (1954): 137–144. See also Lennox, "A Time and a Place."
12. David Flaherty, "Envisioning the British Atlantic: Strategies for Settlement and Sovereignty on the North American and Caribbean Frontiers, 1700–1763" (PhD diss., University of Virginia, forthcoming), chap. 3; Alan Taylor, *Liberty Men and Great Proprietors: The Revolutionary Settlement on the Maine Frontier, 1760–1820* (Chapel Hill: University of North Carolina Press, 1990), 11–14; "Board of Trade report and accompanying letter on the proposed settlement of Irish and Palatine Germans in New England and Nova Scotia," 1729, in *Colonial America,* accessed November 12, 2015, http://www.colonialamerica.amdigital.co.uk/Documents/Details/CO_5_4_Part2_008; D. W. Meinig, *The Shaping of America,* vol. 1, *Atlantic America, 1492–1800* (New Haven, CT: Yale University Press, 1986), 107–109.
13. "Report on the settlement in New England and Nova Scotia."
14. F. W. Maitland, *Township and Borough* (Cambridge: The University Press, 1898), 8, 22–23, 30, 33; Robert L. Meriwether, *The Expansion of South Carolina, 1729–1765* (Kingsport, TN: Southern Publishers, 1940), chap. 2.
15. W. Bollan, "The Course of the War bet.n the English & French & th.r Sevl. Int.s on the Continent of America & the Isl.ds of C Breton & Newfoundl.d Consid.d with Proposals for Prosecuting the War there to the Advantage of this Kingdom," Grenville Papers, Americana, STG Box 12 (7), Huntington Library, San Marino, CA.
16. William Bollan, *The Importance and Advantage of Cape Breton . . .* (London, [1746]), Eighteenth-Century Collections Online, 83, 104–106, 139–140, 120, 118–119, 47, 95, 154–155.
17. Ibid., 104–105; Mancke, "Spaces of Power," 45.

18. "[E]xtended boundaries": Newcastle (1750), quoted in T. R. Clayton, "The Duke of Newcastle, the Earl of Halifax, and the American Origins of the Seven Years' War," *Historical Journal* 24 (1981): 23; J. B. Harley, "The Bankruptcy of Thomas Jefferys: An Episode in the Economic History of Eighteenth Century Map-Making," *Imago Mundi* 20 (1966): 27–37; Mary Pedley, "Map Wars: The Role of Maps in the Nova Scotia / Acadia Boundary Disputes of 1750," *Imago Mundi* 50 (1998): 96–104; Thomas Jefferys, *Explanation for the new map of Nova Scotia and Cape Britain* (London, 1755); [Anonymous], *An Account of the Present State of Nova-Scotia, in Two Letters to a Noble Lord* (London, 1756), 5.
19. T[homas] Pownall, "[Plan for new Settlements on the Lakes] & annexed Chart [which] represents . . . the several English Colonies & the British Territories up to the River St. Laurence & the Great Lakes," 1755, Loudoun Papers, LO 740, Huntington Library; Herman Friis, "A Series of Population Maps of the Colonies and the United States, 1625–1790," *Geographical Review* 30 (1940): 463–470; Carville Earle, "Place Your Bets: Rates of Frontier Expansion in American History, 1650–1890," in *Cultural Encounters with the Environment: Enduring and Evolving Geographic Themes,* ed. Alexander B. Murphy and Douglas L. Johnson (Lanham, MD: Rowman and Littlefield, 2000), 93. For a visualization of the maps, see S. Max Edelson, "The Territorial Pattern of Settler Populations in North America, 1625–1790," a MapScholar Digital Atlas, accessed December 8, 2015, http://mapscholar.org/population/.
20. Pownall, "[Plan] and Chart"; Pownall cited Franklin's plan for western expansion: Benjamin Franklin, "A Plan for Settling Two Western Colonies, 1754," in *Founders Online,* accessed December 8, 2015, http://founders.archives.gov/documents/Franklin/01-05-02-0132.
21. [Proceedings of the Albany Congress, July 3 and 9, 1754], in *Early American Indian Documents: Treaties and Laws, 1607–1789,* vol. 10, *New York and New Jersey Treaties, 1754–1775,* ed. Barbara Graymont (Bethesda, MD: University Publications of America, 2001), 50, 38. On Tonge's political career in Nova Scotia, see Ronald H. Mcdonald, "Tonge, Winckworth," in *Dictionary of Canadian Biography,* vol. 4, *1771–1800* (Toronto: University of Toronto Press, 1979), http://www.biographi.ca/en/bio/tonge_winckworth_4E.html. See also "A Map of the Surveyed Parts of Nova Scotia," 1756, Maps K.Top. 119.60, The British Library, from the Norman B. Leventhal Map Center at the Boston Public Library, http://maps.bpl.org/id/n51607.
22. Charles Morris, "A Description of the Several Towns in the Province of Nova Scotia with the Lands comprehended in, and bordering upon said Towns," [1763], Shelburne Papers, William L. Clements Library, University of Michigan, Ann Arbor; Graeme Wynn, "A Province Too Much Dependent on New England," *Canadian Geographer* 31 (1987): 98–112; Michael Franklin (1766), quoted in John B. Brebner, *The Neutral Yankees of Nova Scotia: A Marginal*

Colony during the Revolutionary Years (Toronto: McClelland and Stewart, 1969), 124; "rich & fertile": *Account of the Present State,* 26. See Margaret Conrad, ed., *They Planted Well: New England Planters in Maritime Canada* (Fredericton, NB: Acadiensis Press, 1988).

23. Joyce E. Chaplin, *Subject Matter: Technology, the Body, and Science on the Anglo-American Frontier, 1500–1676* (Cambridge, MA: Harvard University Press, 2003), 116–117.
24. "Treaty of Paris 1763," *The Avalon Project: Documents in Law, History, and Diplomacy,* Yale Law School, accessed December 8, 2015, http://avalon.law.yale.edu/18th_century/paris763.asp; "Report on Acquisitions," 134.
25. "[G]enius": quoted in Lockett, *Captain James Cook,* 114; William H. Whiteley, "James Cook and British Policy in the Newfoundland Fisheries, 1763–7," *Canadian Historical Review* 54 (1973): 246–249; "surrender[ed]": Thomas Graves (1763), quoted in *Canadian Historical Review* 54 (1973): 249; "Treaty of Paris 1763."
26. "Report on Acquisitions,"135; remark book, HMS *Tweed,* 1763, MS 20 / 14, Admiralty Library; Whiteley, "James Cook," 249–250 (another version of this chart is James Cook, "Plan of the Islands St Peters, Langly and Miquelong," ADM 352 / 61, UK National Archives); Stephen Hornsby, *Surveyors of Empire: Samuel Holland, J. F. W. Des Barres, and the Making of the Atlantic Neptune* (Montreal: McGill-Queen's University Press, 2011), 92; Governor Wilmot to Halifax, December 18, 1764, Thomas B. Akins, ed., *Selections from the Public Documents of Nova Scotia* (Halifax, NS: C. Annand, 1869), 350–351. Later reports confirmed the island's poor conditions for settlement. Deposition of Capt. Farnham, February 1, 1786, Box 6, Vol. 3, Records of the Great Britain Board of Trade, Manuscript Division, Library of Congress, 116.
27. Andrew David, "James Cook's 1763–4 Survey of Newfoundland's Northern Peninsula Reassessed," *Northern Mariner / le marin du nord* 19 (2009): 393–403; Whiteley, "James Cook," 256–257.
28. Leonard Smelt, "Report on the state of the Fortifications at Placentia," November 22, 1751, CO 5 / 232, LC Transcripts. See also Capt. Ruthuen, "A Sketch of the Coast of Newfoundland Between Canada Head and Cape St. Antoine," 1763, Admiralty Library; Thomas Graves (1763), quoted in Lockett, *Captain James Cook,* 126.
29. R. A. Skelton, "Captain James Cook as a Hydrographer," *Mariner's Mirror* 40 (1954): 94; Whiteley, "James Cook," 265–266.
30. The manuscript originals on which this chart is based include "Chart of the Coasts, Bays and Harbours in Newfoundland between Griquet and Point Ferolle," 1764, C 54 / 7, Press 49c, Archive of the Hydrographic Office of the United Kingdom, Taunton, UK (hereafter cited as UK Hydrographic Office); "Chart of the seacoast, bays, harbours and islands in Newfoundland between the Bay of Despair and the harbours of St. Laurence," 1765, C 58 / 71, UK

Hydrographic Office; "Chart of the seacoast, bays, harbours and islands in Newfoundland between Cape Anguille and the harbor of Great Jervis," 1766, C 54/5, UK Hydrographic Office; "West coast of Newfoundland," 1767, C 54/1, UK Hydrographic Office. The complete corpus that included these charts was published as Thomas Jefferys, *A Collection of Charts of the Coasts of Newfoundland and Labradore, &c.* (London, 1770). For a facsimile edition of this work, see *James Cook, Surveyor of Newfoundland . . . with an Introductory Essay by R. A. Skelton* (San Francisco: David Magee, 1965). Two subsequent published atlases included these charts: Thomas Jefferys, *The Newfoundland Pilot* (London, 1769), and R. Sayer and J. Bennett, *Sailing Directions for the North-American Pilot* (London, 1775). Palliser (1765), quoted in Whiteley, "James Cook," 260–261; ibid., 262–263, 272.

31. Hornsby, *Surveyors of Empire,* 46–49, 101; John Pownall to Samuel Holland, April 17, 1764, CO 324/17: 144–145, LC Transcripts; Alexander Johnson, "Charting the Imperial Will: Colonial Administration & the General Survey of British North America 1764–1775" (PhD diss., University of Exeter, 2011), 92–94. For a detailed history of the island, see J. M. Bumsted, *Land, Settlement, and Politics on Eighteenth-Century Prince Edward Island* (Kingston, ON: McGill-Queen's University Press, 1987). See also Andrew Hill Clark, *Three Centuries and the Island: A Historical Geography of Settlement and Agriculture in Prince Edward Island, Canada* (Toronto: University of Toronto Press, 1959).
32. Col. Franquet (1751), quoted in Hill, *Three Centuries,* 33; ibid., 32–35, 39.
33. Hornsby, *Surveyors of Empire,* 50–54. For another medium-scale coastal chart, see Samuel Holland, "The Three Rivers, Island of St. John's by Captain Holland," 1764, Uh B5301, UK Hydrographic Office. For a composite coastal chart, see [Chart of St. John Island], [ca. 1772?], ADM 352/144, UK National Archives.
34. Hornsby, *Surveyors of Empire,* fig. 2.3, 53; Thomas Wright Field Book, Public Archives and Record Office of Prince Edward Island, Charlottetown, Prince Edward Island, Canada.
35. Hornsby, *Surveyors of Empire,* 54. Other, reduced versions of this map include ADM 352/143 and MR 1/1785, UK National Archives; Hornsby, *Surveyors of Empire,* 39–40; John Pownall to Samuel Holland, April 17, 1764, CO 324/17: 392–395, LC Transcripts.
36. "Land Grants in Prince Edward Island, 1767," in Public Archives of Canada, *Report concerning Canadian Archives for the Year 1905,* 3 vols. (Ottawa: S. E. Dawson, 1906), 1:3.
37. Spence, Mure, and Burns received two townships outright, as did Mill, Cathcart, and Higgens, since both firms had been hard at work "establishing and Fishery & making Improvements on this Island" since 1764. Ibid., 7; Jack M. Sosin, *Whitehall and the Wilderness: The Middle West in British Colonial Policy, 1760–1775* (Lincoln: University of Nebraska Press, 1961), 124–126.

38. "Land Grants in Prince Edward Island," 8. On the Feast of St. Michael, September 29, 1767, the clock began to tick on this quitrent liability, rising gradually over a ten-year period. After five years, the proprietors must pay their rents for half their land; after ten, the full annual quitrent was due every year. Stevens was liable for the discounted rate of 2 shillings for every hundred acres of land, which meant he owed the Crown £23 sterling per year from 1772 for his less-than-ideal township and the full £46 sterling per year from 1777. By that time, his revenue from land sales, produce, and fishing receipts had to generate at least that amount annually above expenses to avoid losing money on the venture. John Pownall, by contrast, owed an annual quitrent of £30 from 1772 and £60 from 1777 for his township.
39. "Committee Report for Granting to Sundry persons sixty-six Townships in this Island in the Gulph of St. Lawrence," August 24, 1767, State Papers: Registers of the Privy Council, PC 2 / 112, fo. 419, *State Papers Online: Eighteenth Century, 1714–1782,* accessed March 24, 2016; Hornsby, *Surveyors of Empire,* 39–40; John Pownall to Samuel Holland, April 17, 1764, CO 324 / 17: 144–145, LC Transcripts.
40. John Fry (1739), quoted in Clara Egli LeGear, "The New England Coasting Pilot of Cyprian Southack," *Imago Mundi* 11 (1954): 141; Mr. [Thomas] Knight to [Dartmouth], October 10, 1773, CO 5 / 228: 268–269, LC Transcripts; Harry Baglole, "Patterson, Walter," in *Dictionary of Canadian Biography,* vol. 4, http://www.biographi.ca/en/bio/patterson_ walter_4E.html.
41. Fred Anderson, *The War That Made America: A Short History of the French and Indian War* (New York: Penguin, 2005), 144; Hornsby, *Surveyors of Empire,* 41. On Cape Breton's geology as well as its strategic importance, see John Robert McNeill, *Atlantic Empires of France and Spain: Louisbourg and Havana, 1700–1763* (Chapel Hill: University of North Carolina Press, 1985), 10–11.
42. Samuel Holland to [Secretary of State?], November 24, 1765, in Harvey, *Holland's Description of Cape Breton,* 39–40; Holland, quoted in Hornsby, *Surveyors of Empire,* 144.
43. Samuel Holland, "An Extract from the Discription of the Island of Cape Breton," Stowe-Grenville Correspondence, Americana, Box 12, folder 31, Huntington Library. My thanks to David Flaherty for this reference.
44. John Pownall to Samuel Holland, April 17, 1764, CO 324 / 17: 149, LC Transcripts; May 23, 1764, *Acts of the Privy Council of England,* Colonial Series, vol. 4, *A.D. 1745–1766,* ed. James Munro (Hereford: H. M. Stationery Office, 1911), 660; December 3, 1766, ibid., 19–20; Dartmouth (1774), quoted in Harvey, *Holland's Description of Cape Breton,* 29–30; Morris (1774), quoted in Beamish Murdoch, *A History of Nova-Scotia or Acadie,* 2 vols. (Halifax, NS: James Barnes, 1866), 2:527.
45. See Johnson, "Imperial Will," 174–179. For another map of the western Gulf, see "Sketch Map of Miramich Bay," [1770], CO 700 / New Brunswick 6, UK National Archives.

46. Copy of a letter from Samuel Holland to Lauchlin Macleane, July 31, 1767, Shelburne Papers, 51: 451–453. Wright (1766), quoted in Johnson, "Imperial Will," 185; Francis Maseres to Fowler Walker, August 11, 1768, in Francis Maseres, *The Maseres Letters 1766–1768* (Toronto: University of Toronto Library, 1919), 100. The Admiralty mapped the Seven Islands at the close of the war in "A Draft of the Seven Islands," 1761, MS 20, Admiralty Library; Assistant General Surveyor George Sproule mapped Mingan in 1769 on his "Survey of the Mingan Islands with the adjacent Coast of Labrador, forming the Channel of Anticost[i] to the Northward," 1769, CO 700/Canada 30, UK National Archives. See also Johnson, "Imperial Will," 182–184, 152.
47. Samuel Holland to Hillsborough, December 19, 1770, in Willis Chipman, "The Life and Times of Major Samuel Holland, Surveyor-General, 1764–1801," *Papers and Records—Ontario Historical Society* 21 (1924): 27–28. See also Hornsby, *Surveyors of Empire,* 75–78. William Knox revived the idea to settle loyal subjects in a Penobscot/"New Ireland" colony in 1778. See Leland J. Bellot, *William Knox: The Life & Thought of an Eighteenth-Century Imperialist* (Austin: University of Texas Press, 1977), 166–167, 176–177.
48. Stephen J. Hornsby and Richard W. Judd, eds., *Historical Atlas of Maine* (Orono: University of Maine Press, 2015), plate 13; [Samuel] Holland to [Hillsborough], November 10, 1772, CO 5/228: 200, LC Transcripts; [Samuel] Holland to Hillsborough, December 19, 1770, CO 5/229: 251–252, LC Transcripts; Holland to Hillsborough, June 15, 1772, CO 5/73: 483, LC Transcripts.
49. [Samuel] Holland to [Dartmouth], May 2, 1773, CO 5/228: 249–250, LC Transcripts. Dartmouth denied the surveyors' memorial because of Massachusetts's claims under its charter, but he urged them to resubmit their request, which they did for lands in New Hampshire. Dartmouth to Samuel Holland, August 4, 1773, CO 5/74: 181, LC Transcripts; [Samuel] Holland to [Dartmouth], October 28, 1773, CO 5/228: 288, LC Transcripts; Holland (1766), quoted in Hornsby, *Surveyors of Empire,* 99; "Trade of Plantations, Commissioners of," in William Berry, *Encyclopædia Heraldica: or, Complete Dictionary of Heraldry, Vol. 1* (London: Sherwood, Gilbert, and Piper, [1828]).
50. Samuel Holland to Hillsborough, June 20, 1768, CO 6/69: 211, LC Transcripts; [John] Wentworth to [Hillsborough], October 22, 1770, CO 5/227, LC Transcripts; Holland to Hillsborough, December 19, 1770; Holland to Hillsborough, June 15, 1772, CO 5/73: 483–484, LC Transcripts. The Privy Council approved a 1769 Treasury order to map the location of mast trees in the king's woods to better produce them for the Admiralty. *Acts of the Privy Council,* 22–23.
51. Hillsborough to Samuel Holland, March 8, 1768, CO 6/69: 100, LC Transcripts; Dartmouth to Samuel Holland, November 4, 1772, CO 5/73: 785; [Samuel] Holland to [Dartmouth], May 2, 1773, CO 5/228: 247–248. Holland did not complete this general map, in part because he was waiting for Admiralty

surveyor J. F. W. Des Barres to supply him with charts of Nova Scotia. Holland to [John] Pownall, December 20, 1774, in *Documents Relating to the Colonial History of the State of New Jersey,* vol. 10, *Administration of Governor William Franklin, 1767–1776* (Newark, NJ: Daily Advertiser, 1886), 518–521; [Samuel] Holland to [Dartmouth], April 14, 1774, CO 5/228: 323, LC Transcripts.

52. Holland to [Dartmouth], October 28, 1773, CO 5/228: 288, LC Transcripts.
53. Macartney (1773), quoted in *The Oxford History of the British Empire,* vol. 2, *The Eighteenth Century,* ed. P. J. Marshall (New York: Oxford University Press, 1998), 263; George Macartney, "Observations on the Island of St. John in the Gulf of St. Lawrence," 1775, George Macartney Common Book, HM 686, Huntington Library. On political turmoil in St. John/Prince Edward Island, see Bumsted, *Land, Settlement, and Politics,* as well as Rusty Bittermann, *Rural Protest on Prince Edward Island: From British Colonization to the Escheat Movement* (Toronto: University of Toronto Press, 2006).
54. Macartney, "Observations on St. John"; Hill, *Three Centuries,* 59.
55. Joseph Gridley to John Adams, [August 28, 1780], *The Adams Papers Digital Edition,* ed. C. James Taylor (Charlottesville: University of Virginia Press, Rotunda, 2008); F. J. Thorpe, "Holland, Samuel Johannes," in *Dictionary of Canadian Biography,* vol. 5, *1801–1820* (Toronto: University of Toronto Press, 1983), http://www.biographi.ca/en/bio/holland_samuel_johannes_5E.html. On Holland's landholdings, see Hornsby, *Surveyors of Empire,* 154–156.
56. [Joseph Robinson], *To the Farmers in the Island of St. John, in the Gulf of St. Lawrence* (n.p., [ca. 1796?]), Public Archives and Records Office of Prince Edward Island; J. M. Bumsted, "Robinson, Joseph," in *Dictionary of Canadian Biography,* vol. 5, http://www.biographi.ca/en/bio/robinson_joseph_5E.html. On the enduring differences between British and American societies in this region, see Elizabeth Mancke, *The Fault Lines of Empire: Political Differentiation in Massachusetts and Nova Scotia, ca. 1760–1830* (New York: Routledge, 2005), especially chap. 3.
57. Clark, *Three Centuries,* 45; Louis Maltais, Manager of Hydrographic Operations, Canadian Hydrographic Service, email to the author, December 16, 2015.

CHAPTER FOUR: MARKING THE INDIAN BOUNDARY

1. Patrick Griffin, *American Leviathan: Empire, Nation, and Revolutionary Frontier* (New York: Hill and Wang, 2007); Jack M. Sosin, *Whitehall and the Wilderness: The Middle West in British Colonial Policy, 1760–1775* (Lincoln: University of Nebraska Press, 1961); Francis Jennings, *Empire of Fortune: Crowns, Colonies, and Tribes in the Seven Years War in America* (New York: W. W. Norton, 1988), 461–463; Alan Taylor, *The Divided Ground: Indians, Settlers, and the Northern Borderland of the American Revolution* (New York: Vintage, 2007), 40–48; Daniel K. Richter, *Before the Revolution: America's Ancient Pasts* (Cambridge, MA: Harvard University Press, 2011), 408–412; Colin G. Calloway, *The Scratch*

of a Pen: 1763 and the Transformation of North America (New York: Oxford University Press, 2006), chap. 4; Gary Nash, *The Unknown American Revolution: The Unruly Birth of Democracy and the Struggle to Create America* (New York: Viking, 2005), 71–72; "did not exist": Griffin, *American Leviathan,* 50.

2. Dorothy V. Jones, *License for Empire: Colonialism by Treaty in Early America* (Chicago: University of Chicago Press, 1982), 46.
3. *The Charter of the Colony of Connecticut 1662* ([Hartford, CT], 1900), 19; "A Copy of the Queries relating to the Province of South Carolina prepounded by the Lords of Trade," August 23, 1720, Records of the Great Britain Board of Trade, Manuscript Division, Library of Congress, Washington, DC, 88–89; *Acts of the Privy Council of England,* Colonial Series, vol. 4, *A.D. 1745–1766,* ed. James Munro (Hereford: H. M. Stationery Office, 1911), 496–497.
4. Board of Trade to the King, September 8, 1721, in *Documents Relative to the Colonial History of the State of New York,* 15 vols. (Albany: Weed, Parsons and Company, 1856), 5:623–627 (hereafter cited as *New York Documents*); see Warren R. Hofstra, *The Planting of New Virginia: Settlements and Landscape in the Shenandoah Valley* (Baltimore, MD: Johns Hopkins University Press, 2004), 77–81.
5. G. Malcolm Lewis, ed., *Cartographic Encounters: Perspectives on Native American Mapmaking and Map Use* (Chicago: University of Chicago Press, 1998); Gregory A. Waselkov, "Indian Maps of the Colonial Southeast," in *Powhatan's Mantle: Indians in the Colonial Southeast,* ed. Peter H. Wood, Gregory A. Waselkov, and M. Thomas Hatley (Lincoln: University of Nebraska Press, 1989), 296–306, 320–324; James H. Merrell, *The Indians' New World: Catawbas and Their Neighbors from European Contact through the Era of Removal* (New York: W. W. Norton, 1989), 92–99. For an argument that this map was a Cherokee creation, see Ian Chambers, "A Cherokee Origin for the 'Catawba' Deerskin Map (c. 1721)," *Imago Mundi* 65 (2013): 207–216.
6. *New York Documents,* 5:670; James H. Merrell, *Into the American Woods: Negotiators on the Pennsylvania Frontier* (New York: W. W. Norton, 1999), 19–41; Allan Greer, "Commons and Enclosure in the Colonization of North America," *American Historical Review* 117 (2012): 365–386.
7. "Albany Plan of Union 1754," in *The Avalon Project: Documents in Law, History, and Diplomacy,* Yale Law School, http://avalon.law.yale.edu/18th_century/albany.asp; Jack Stagg, *Anglo-Indian Relations in North America to 1763 and an Analysis of the Royal Proclamation of 7 October 1763* (Ottawa: Research Branch, Indian and Northern Affairs, 1981), 107; *Early American Indian Documents: Treaties and Laws, 1607–1789,* vol. 10, *New York and New Jersey Treaties, 1754–1775,* ed. Barbara Graymont (Bethesda, MD: University Publications of America, 2001), 10:78 (hereafter cited as *Indian Documents*); see Timothy J. Shannon, *Indians and Colonists at the Crossroads of Empire: The Albany Congress of 1754* (Ithaca, NY: Cornell University Press, 2000); Wilbur R. Jacobs, ed., *The Appalachian Indian Frontier: The Edmond*

Atkin Report and Plan of 1755 (Lincoln: University of Nebraska Press, 1967), xxxv, 3.

8. Andrew D. M. Beaumont, *Colonial America and the Earl of Halifax, 1748–1761* (New York: Oxford University Press, 2015), 117–119; *Indian Documents,* 10:78; Francis Fauquier to the Board of Trade, March 13, 1760, in *The Official Papers of Francis Fauquier, Lieutenant Governor of Virginia, 1758–1768,* ed. George Reese (Charlottesville: University Press of Virginia, 1980), 1:331–332, 333n.
9. *Indian Documents,* 10:330–331, 335; Lewis Evans, *Geographical, Historical, Political, Philosophical and Mechanical Essays* (Philadelphia, 1755), 6–7, 9; Charles Thompson, *An Enquiry into the Causes of the Alienation of the Delaware and Shawanese Indians from the British Interest* (London, [1759]).
10. James H. Merrell, ed., *The Lancaster Treaty of 1744 with Related Documents* (Boston: Bedford, 2008), 30; "No Savage": Shingas (1756), quoted in Jennings, *Empire of Fortune,* 154; Tamaqua (Beaver) (1758), quoted in Calloway, *Scratch of a Pen,* 55; "Brethren": *Indian Documents,* vol. 3 *Pennsylvania Treaties, 1756–1775,* ed. Alison Duncan Hirsch (Bethesda, MD: University Publications of America, 2004), 569.
11. *Indian Documents,* 10:320, 248, 335. On the psychology of human territoriality, see William J. Szlemko et al., "Territorial Markings as a Predictor of Driver Aggression and Road Rage," *Journal of Applied Psychology* 38 (2008): 1667–1671; Irwin Altman, *The Environment and Social Behavior: Privacy, Personal Space, Territory, and Crowding* (Monterey, CA: Brooks/Cole, 1975), 112–118, 122–138.
12. Jennings, *Empire of Fortune,* 25–28; [Thomson], *Alienation of the Delaware,* 115–116. On the use of maps in this and previous negotiations, see ibid., 43, 64, 120. Anthony F. C. Wallace, *King of the Delawares: Teedyuscung, 1700–1763* (Freeport, NY: Books for Libraries, 1970), 21–26; Stephen F. Auth, *The Ten Years' War: Indian-White Relations in Pennsylvania, 1755–1765* (New York: Garland, 1989), 83, 103, 106.
13. *Indian Documents,* 10:337, 248, 336, 349; James Hamilton, "Narrative," [ca. September 1761], in *Fauquier Papers,* 2:604.
14. Fauquier to Board of Trade, December 1, 1759, in *Fauquier Papers,* 1:276–277; Board of Trade to Fauquier, June 13, 1760, in ibid., 1:376–377; *Indian Documents,* 10:330–331, 320; *Acts of the Privy Council,* 4:500, 496–497.
15. William Johnson to Board of Trade, November 11, 1763, in *New York Documents,* 7:578; *Acts of the Privy Council,* 4:496–497; *New York Documents,* 7:478.
16. [John] Pownall, "Mr. Pownal's Sketch of a Report concerning the Cessions in Africa and America at the Peace of 1763," in R. A. Humphreys, "Shelburne and the Proclamation of 1763," *English Historical Review* 49 (1934): 259; Egremont to Jeffrey Amherst, January 27, 1763, in *Correspondence and Documents during Thomas Fitch's Governorship of the Colony of Connecticut, 1754–1766* (Hartford: Connecticut Historical Society, 1920), 224; Ward H. Goodenough, "Moral Outrage: Territoriality in Human Guise," *Zygon* 32 (1997): 17–18; "Ld.

Barrington's Plan relative to the Out Posts, Indian Trade &c with [Gage's] remarks," Shelburne Papers, 50: 51, William L. Clements Library, University of Michigan, Ann Arbor.

17. Shelburne to James Grant, February 19, 1767, CO 5 / 548: 172, British Public Record Office Transcripts, Manuscript Division, Library of Congress (hereafter cited as LC Transcripts); *Acts of the Privy Council,* 4:587, 610; "Some thoughts on the Settlement and Government of our Colonies in North America," March 10, 1763, Shelburne Papers, 48: 44; see especially Peter Silver, *Our Savage Neighbors: How Indian War Transformed Early America* (New York: W. W. Norton, 2008). See also, for example, Adam Smith, *The Theory of Moral Sentiments* (Edinburgh, 1759), Library of Economics and Liberty, I.I.7, accessed April 7, 2016, http://www.econlib.org/library/Smith/smMS1.html; "merciless": Sosin, *Whitehall and the Wilderness,* 46n.
18. Verner W. Crane, "Hints Relative to the Division and Government of the Conquered and Newly Acquired Countries in America," *Mississippi Valley Historical Review* 8 (1922): 371; "Report on Acquisitions in America," June 8, 1763, in *Documents Relating to the Constitutional History of Canada, 1759–1791,* ed. Adam Shortt and Arthur G. Doughty (Ottawa: S. E. Dawson, 1907), 139, 141; [Pownall], "Report concerning the Cessions," 259–260; "Barrington's Plan," 45; "Report on Acquisitions," 140; "The Royal Proclamation—October 7, 1763," *Avalon Project,* http://avalon.law.yale.edu/18th_century/proc1763.asp; "Plan for the future Management of Indian Affairs" (1764), in *New York Documents,* 7:641; CO 324 / 21: 233, 235, LC transcripts; Louis De Vorsey Jr., *The Indian Boundary in the Southern Colonies, 1763–1775* (Chapel Hill: University of North Carolina Press, 1966), 32–34.
19. Jack Williams, *East 40 Degrees: An Interpretive Atlas* (Charlottesville: University of Virginia Press, 2007), 1; Lawrence Henry Gipson, *Lewis Evans* (Philadelphia: Historical Society of Pennsylvania, 1939), 11–12; [Pownall], "Report concerning the Cessions," 260.
20. "Pennsylvania Assembly Committee: Report on the Western Bounds" [March 7, 1754], in *The Papers of Benjamin Franklin, Digital Edition,* Packard Humanities Institute (hereafter cited as *Franklin Papers*); Margaret Beck Pritchard and Henry G. Taliaferro, *Degrees of Latitude: Mapping Colonial America* (New York: Colonial Williamsburg and Harry N. Abrams, 2002), 183; Jacobs, ed., *Appalachian Indian Frontier,* 49; François De Bussy (1761), quoted in Paul W. Mapp, *The Elusive West and the Contest for Empire, 1713–1763* (Chapel Hill: University of North Carolina Press, 2011), 371; Pritchard and Taliaferro, *Degrees of Latitude,* 174.
21. "A Report from the Lords of Trade concerning that part of North America which lies to the West of the Old Colonies and which was acquired at the Peace in 1763," August 3, 1763, Shelburne Papers, 49: 22; "The Royal Proclamation"; "Draught of a Proclamation," October 4, 1763, CO 324 / 21: 336, 334, LC Transcripts.

22. De Vorsey, *Indian Boundary,* 36–38; see also Barbara McCorkle, *New England in Early Printed Maps, 1513 to 1800* (Providence, RI: Brown University Press, 2001), 157; Charles James Fox (1763), quoted in Linda Colley, *Britons: Forging the Nation, 1707–1837* (New Haven, CT: Yale University Press, 1992), 101.
23. Fauquier (1767), quoted in Sosin, *Whitehall and the Wilderness,* 170.
24. De Vorsey, *Indian Boundary,* 41–42; Board of Trade to Shelburne, December 23, 1766, *New York Documents,* 7:1004.
25. John R. Alden, *John Stuart and the Southern Colonial Frontier* (New York: Gordian Press, 1966), 115–122; J. Russell Snapp, *John Stuart and the Struggle for Empire on the Southern Frontier* (Baton Rouge: Louisiana State University Press, 1996), 56–57.
26. Alden, *John Stuart,* 176–185; [Henry Ellis], "On the methods to prevent giving any alarm to the Indians by taking possession of Florida and Louisiana," Shelburne Papers, 60: 131–134; Copy of a Circular Letter to the Southern Governors and John Stuart, March 16, 1763, CO 323/16: 101–104, The National Archives of the UK, Kew, Richmond, Surrey (hereafter cited as UK National Archives); *Journal of the congress of the four southern governors, and the superintendent of that district, with the Five Nations of Indians, at Augusta, 1763* (Charlestown, 1764), Early American Imprints, Series I: Evans, 1639–1800, no. 9706, Readex Digital Collections, 12–14.
27. *Journal of the congress,* 23–24; *Indian Documents,* vol. 14, *North and South Carolina Treaties, 1756–1775,* ed. W. Stitt Robinson (Bethesda, MD: University Publications of America, 2003), 258–259.
28. *Acts of the Privy Council,* 552–553; Hagler (Nopkehe) (1762), quoted in Merrell, *The Indians' New World,* 199, 195–201.
29. Merrell, *Indians' New World,* 205; *Journal of the congress,* 30; Stuart quoted in Merrell, *Indians' New World,* 205.
30. "Proceedings of Sir William Johnson with the Indians," [April 29–May 22, 1765], in *New York Documents,* 7:737; *Indian Documents,* 10:558, 44–45.
31. [Thomas] Pownall, Appendix to "Considerations towards a general Plan," July 11, 1764, in *Indian Documents,* 10:61; Isaac Nakhimovsky, "Vattel's Theory of the International Order: Commerce and the Balance of Power in the *Law of Nations,*" *History of European Ideas* 33 (2007): 157–173.
32. Jacobs, ed., *Appalachian Indian Frontier,* 38; Christopher Tomlins, *Freedom Bound: Law, Labor, and Civil Identity in Colonization English America, 1580–1865* (New York: Cambridge University Press, 2010), 132; P. G. McHugh, *Aboriginal Societies and the Common Law: A History of Sovereignty, Status, and Self-Determination* (New York: Oxford University Press, 2004), 62, 103, 110; [George] Johnstone to John Stuart, June 12, 1765, in *Mississippi Provincial Archives 1763–1766: English Dominion,* ed. Dunbar Roland (Nashville: Brandon Printing, 1911), 186; Daniel K. Richter, "Native Americans, the Plan of 1764, and a British Empire That Never Was," in *Cultures and Identities in Colonial British*

America, ed. Robert Olwell and Alan Tully (Baltimore, MD: Johns Hopkins University Press, 2006), 287.

33. *Journal of the congress,* 10; Board of Trade to Francis Fauquier, June 13, 1760, in *Fauquier Papers,* 1:377; Tomlins, *Freedom Bound,* 134–156; Egremont to Jeffrey Amherst, January 27, 1763, in *The Fitch Papers: Correspondence and Documents during Thomas Fitch's Governorship of the Colony of Connecticut, 1754–1766,* 2 vols. (Hartford: Connecticut Historical Society, 1918–1920), 2:224; Egremont to Board of Trade, May 5, 1763, CO 323/15: 101–107, UK National Archives; "Barrington's Plan," 46.
34. Lauren Benton, *Law and Colonial Cultures: Legal Regimes in World History, 1400–1900* (New York: Cambridge University Press, 2002), 168–169; William De Brahm, *De Brahm's Report of the General Survey in the Southern District of North America,* ed. Louis De Vorsey Jr. (Columbia: University of South Carolina Press, 1971), 253; James Grant to Board of Trade, August 30, 1766, CO 5/541: 121, LC Transcripts.
35. *Journal of the congress,* 34; "Report of the Congress [Picolata]," in James W. Covington, *The British Meet the Seminoles: Negotiations between British Authorities in East Florida and the Indians: 1763–68* (Gainesville: University Press of Florida, 1761), 37; Kathryn E. Holland Braund, "'The Congress Held in a Pavilion': John Bartram and the Indian Congress at Fort Picolata, East Florida," in *America's Curious Botanist: A Tercentennial Reappraisal of John Bartram (1699–1777),* ed. Nancy E. Hoffmann and John C. Van Horne (Philadelphia: American Philosophical Society, 2004), 79–96; Holland Braund, "'Like a Stone Wall Never to Be Broke': The British-Indian Boundary Line with the Creek Indians, 1763–1773," in *Britain and the American South: From Colonialism to Rock and Roll,* ed. Joseph P. Ward (Jackson: University Press of Mississippi, 2003), 53–79.
36. Louis De Vorsey Jr., "Maps in Colonial Promotion: James Edward Oglethorpe's Use of Maps in 'Selling' the Georgia Scheme," *Imago Mundi* 38 (1986): 35–45.
37. De Brahm, *Report of the General Survey,* 31. For a similar map, see Henry Yonge, "A Map of the Sea Coast of Georgia," 1763, Add MSS 14,036, fol. g., The British Library, London.
38. *Journal of the congress,* 31.
39. De Vorsey, *Indian Boundary,* 131–132; *Indian Documents,* 14:255.
40. De Vorsey, *Indian Boundary,* 128; quoted in ibid., 117; De Brahm, *Report of the General Survey,* 72; *Indian Documents,* vol. 12, *Georgia and Florida Treaties, 1763–1776,* ed. John T. Juricek (Bethesda, MD: University Publications of America, 2002), 1.
41. Jones, *License for Empire,* 49, 54. See *New York Documents,* 8:22.
42. *New York Documents,* 8:22; Board of Trade to the King, March 7, 1768, in *New York Documents,* 8:22, 23.

43. Jones, *License for Empire,* 94. See also Woody Holton, *Forced Founders: Indians, Debtors, Slaves, and the Making of the American Revolution in Virginia* (Chapel Hill: University of North Carolina Press, 1999), 7–32.
44. *Indian Documents,* 10:550–551, 534.
45. Jon Parmenter, *The Edge of the Woods: Iroquoia, 1534–1701* (East Lansing: Michigan State University Press, 2010).
46. In 1760, Johnson used his influence to secure a personal 92,000-acre grant of Mohawk land, known as the Canajoharie Patent, a tract shown in "Draft of a tract granted by the whole Conajohare Indians to Sir William Johnson, Baronet, in 1760," 1766, CO 700 / New York 33, UK National Archives. Jones, *License for Empire,* 77; Sosin, *Whitehall and the Wilderness,* 174–177; Taylor, *Divided Ground,* 44.
47. *Indian Documents,* 10:551; Philip Vickers Fithian, *Philip Vickers Fithian: Journal, 1775–1776, Written on the Virginia-Pennsylvania Frontier and in the Army around New York,* ed. Robert Greenhalgh Albion and Leonidas Dodson (Princeton, NJ: Princeton University Press, 1934), 46–52. See also Peter C. Mancall, *Valley of Opportunity: Economic Culture along the Upper Susquehanna, 1700–1800* (Ithaca, NY: Cornell University Press, 1991), chap. 5.
48. Thomas Gage (1770), quoted in Jones, *License for Empire,* 103; Jones, quoted in ibid., 107; ibid., 108.
49. *Indian Documents,* 12:81–85; De Vorsey, *Indian Boundary,* 161–172, 123. Other maps documented the southern New Purchase process, including [Map of an area to the south of the Savannah River, showing the Ogeechee and Oconee Rivers, Briar Creek, Creek and Cherokee lands, lands assigned by the Cherokee for the payment of debts], [1772], MR 1 / 18, UK National Archives; Andrew Way, [Map showing the boundary line between Georgia and the territory of the Creek Indians running from the Altamaha to the Ogeechee River], 1773, MPG 1 / 357 / 2, UK National Archives; "Map of Georgia showing towns, rivers, forts, Indian boundary lines, paths, and hunting lands," [1774], MPG 1 / 20, UK National Archives.
50. Alan Gallay, *The Formation of a Planter Elite: Jonathan Bryan and the Southern Colonial Frontier* (Athens: University of Georgia Press, 1989), 139–145; *Indian Documents,* 14:297–301; Merrell, *The Indians' New World,* 200–201; "Established Boundary": *Indian Documents,* vol. 5, *Virginia Treaties, 1723–1775,* ed. W. Stitt Robinson (Bethesda, MD: University Publications of America, 1983), 372; *Indian Documents,* 12:80.
51. Delf Norona, "Joshua Fry's Report on the Back Settlements of Virginia (May 8, 1751)," *Virginia Magazine of History and Biography* 56 (1948): 37.
52. John Stuart to President Blair, October 17, 1768, in *Indian Documents,* 5:331; Governor Botetourt to Colonel Lewis and Dr. Walker, December 20, 1768, in *Indian Documents,* 5:338; "remain to be purchased": quoted in De Vorsey, *Indian Boundary,* 70; "The Royal Proclamation."

53. Benjamin Franklin and Samuel Wharton, Memorial to Congress, February 26, 1780, in *Franklin Papers;* De Vorsey, *Indian Boundary,* 83.
54. John Stuart to Hillsborough, November 28, 1770, CO 5/72: 29, LC Transcripts; "Treaty with the Cherokees at Lochabor, S.C., 1770," *Virginia Magazine of History and Biography* 9 (1902): 360–363; Alexander Cameron to John Stuart, March 19, 1771, CO 5/72: 439, LC Transcripts.
55. Peter H. Wood, "The Changing Population of the Colonial South: An Overview by Race and Region, 1685–1790," in *Powhatan's Mantle,* 61–65; De Vorsey, *Indian Boundary,* 48; quoted in ibid., 75; *Indian Documents,* 12:120; *Indian Documents,* 5:364, 357; John Stuart to Hillsborough, November 28, 1770, CO 5/72: 51, LC Transcripts; Alexander Cameron to John Stuart, March 19, 1771, CO 5/72: 440, LC Transcripts.
56. *Indian Documents,* 5:344; John Stuart to Hillsborough, April 27, 1771, CO 5/72: 423, LC Transcripts. On Dunmore's War, see Richter, *Facing East,* 213. See also James Corbett David, *Dunmore's New World: The Extraordinary Life of a Royal Governor in Revolutionary America* (Charlottesville: University of Virginia Press, 2013), chap. 3.
57. Benjamin Franklin, *The Interest of Great Britain Considered, With Regard to her Colonies . . .* (London, 1760), in *Franklin Papers;* Gage, quoted in Sosin, *Whitehall and the Wilderness,* 109; Stuart (c. 1765), quoted in De Vorsey, *Indian Boundary,* 128, 169.
58. Habersham, quoted in De Vorsey, *Indian Boundary,* 168; *Indian Documents,* 12:5; Deposition of William Frazier, 1768, in *Indian Documents,* 12:42; John Stuart to John Blair, October 17, 1768, in *Indian Documents,* 5:332; James Grant to Richard Oswald, November 21, 1764, James Grant of Ballindalloch Papers, microfilm copy, Manuscript Division, Library of Congress; *Indian Documents,* 12:54.
59. Sosin, *Whitehall and the Wilderness,* 107; William Johnson to [Henry] Conway, June 28, 1766, Shelburne Papers, 51: 200–201; [Richard Jackson], "Remarks on Ld. Barrington's Plan" [1766], Shelburne Papers, 50: 78; William P. Courtney and Jean-Marc Alter, "Jackson, Richard (1721/2–1787)," *Oxford Dictionary of National Biography* (Oxford: Oxford University Press, 2004); Lord Barrington, "Ld. Barrington's Plan relative to the Out Posts, Indian Trade &c with [Gage's] remarks," Shelburne Papers, 50: 45; "Report on Acquisitions," 102; "The Royal Proclamation"; Egremont to Board of Trade, July 14, 1763, CO 323/16: 320, UK National Archives.
60. Jeffrey Amherst, Copy of "Remarks on Ld. Barrington's Plan," [1766], Shelburne Papers, 50: 65; Henry Cavendish, *Debates of the House of Commons in the Year 1774* (London, 1839), 134; Sian E. Rees, "The Political Career of Wills Hill, Earl of Hillsborough (1718–1793) with Particular Reference to His American Policy" (PhD diss., University of Wales, 1976), 317–319; see also Thomas Gage and Clarence Edwin Carter, *The Correspondence of General Thomas*

Gage, 2 vols. (New Haven, CT: Yale University Press, 1931), 1:175–179, 2:138; Humphreys, "Shelburne and Colonial Policy," 263.

61. "Extracts of the Journal of Captain Harry Gordon," *Journal of the Illinois State Historical Society* 2 (1909): 56–59; T[homas] Hutchins, *The Courses of the Ohio River taken by Lt. T. Hutchins Anno 1766 and Two Accompanying Maps*, ed. Verly W. Bond Jr. (Cincinnati: Historical and Philosophical Society of Ohio, 1942); Thomas Hutchins, *A Topographical Description of Virginia, Pennsylvania, Maryland, and North Carolina* (London, 1778); Hutchins, *Courses of the Ohio*, 14.
62. Andrew Hamilton, *Trade and Empire in the Eighteenth-Century Atlantic World* (Newcastle upon Tyne: Cambridge Scholars, 2008), 35; Richter, "Plan of 1764."
63. "Plan Proposed by Gen[era]l Phineas Lyman, for settling Louisiana, and for erecting New Colonies between West Florida and the Falls of St. Anthony," Shelburne Papers, 48: 3. Other schemes included William Franklin's "project of a colony in the Illinois country," which his father, Benjamin Franklin, presented to Shelburne with a copy of Evans's map of the middle colonies. Benjamin Franklin to William Franklin, May 10, 1766, Benjamin Franklin to William Franklin, September 12, 1766, in *Franklin Papers*. On West Florida lieutenant governor Montfort Browne's ambitions to establish a Mississippi colony (and his use of a George Gauld chart to make his case to Dartmouth), see Robin F. A. Fabel, "An Eighteenth Colony: Dreams for Mississippi on the Eve of the Revolution," *Journal of Southern History* 59 (1993): 658–666. See also David Narrett, *Adventurism and Empire: The Struggle for Mastery in the Louisiana-Florida Borderlands, 1762–1803* (Chapel Hill: University of North Carolina Press, 2015).
64. *Report of the Lords Commissioners for Trade and Plantations on the Petition of the Honourable Thomas Walpole, Benjamin Franklin, John Sargent, and Samuel Wharton, Esquire* (London, 1772), 6–7, 13, 17–18, 21.
65. Ibid., 67–68, 96, 73–76, 80–81. Hillsborough accused Benjamin Franklin of "'Writing that Pamphlet against his [report] about the Ohio,'" but Franklin noted that it had in fact been written by three other Grand Ohio Company associates: Henry Dagge, Samuel Wharton, and Thomas Walpole. Franklin to William Franklin, July 14, 1773, in *Franklin Papers*.
66. Peter Marshall, "Lord Hillsborough, Samuel Wharton and the Ohio Grant, 1769–1775," *English Historical Review* 80 (1965): 717–739; Benjamin Franklin and Samuel Wharton, Memorial to Congress, February 26, 1780, in *Franklin Papers*. See Merrill Jensen, *The Founding of a Nation: A History of the American Revolution, 1763–1776* (Indianapolis: Hackett, 1968), 389–391; Francis Jennings, *The Creation of America: Through Revolution to Empire* (New York: Cambridge University Press, 2000), 122–124.
67. John Locke, *The Works of John Locke in Nine Volumes*, rev. ed., 9 vols. (London: Rivington, 1824), vol. 4, chapter 4, section 108, in *Online Library of Liberty*, http://oll.libertyfund.org/title/763/65497. See Harold M. Baer, "An Early Plan for the Development of the West," *American Historical Review* 30 (1925):

537–543; see also Robert E. Toohey, *Liberty and Empire: British Radical Solutions to the American Problem, 1774–1776* (Lexington: University Press of Kentucky, 1978), 36–52; see also Craig Yirush, *Settlers, Liberty, and Empire: The Roots of Early American Political Theory, 1675–1775* (New York: Cambridge University Press, 2011), 129–138; John Cartwright, *American Independence: The Interest and Glory of Great-Britain; A New Edition [with] Letter to Edmund Burke, Esq; Controverting the Principle of American Government, Laid down in his lately published Speech on American Taxation . . .* (London, 1775), "Postscript," 44, 63, 35. Report from the Committee for the Western Territory to the United States Congress, March 1, 1784, *Envisaging the West: Thomas Jefferson and the Roots of Lewis and Clark,* University of Nebraska at Lincoln / University of Virginia, http://jeffersonswest.unl.edu/archive/view_doc.php?id=jef.00155, accessed August 22, 2013. See Peter S. Onuf, *Statehood and Union: A History of the Northwest Ordinance* (Bloomington: Indiana University Press, 1987); see also Peter J. Kastor, *William Clark's World: Describing America in an Age of Unknowns* (New Haven, CT: Yale University Press, 2011), 31–36.

68. Cartwright, *American Independence,* "Postscript," 44.
69. George Washington to William Crawford, September [17], 1767, in *Founders Online,* National Archives, http://founders.archives.gov/documents/Washington/02-08-02-0020; Edward Redmond, "George Washington: Surveyor and Mapmaker," in *American Memory,* Library of Congress, accessed August 21, 2016, https://memory.loc.gov/ammem/gmdhtml/gwmaps.html.

CHAPTER FIVE: CHARTING CONTESTED CARIBBEAN SPACE

1. "Report on Acquisitions in America," June 8, 1763, in *Documents Relating to the Constitutional History of Canada, 1759–1791,* ed. Adam Shortt and Arthur G. Doughty (Ottawa: S. E. Dawson, 1907), 106.
2. Herman Moll, *The Compleat geographer; or, The chorography and typography of all the known parts of the earth,* 2 vols. (London, 1709), 2:487.
3. Braithwaite (1726), quoted in Christopher Taylor, *The Black Carib Wars: Freedom, Survival, and the Making of Garifuna* (Jackson: University Press of Mississippi, 2012), 27; Barth[olomew] Candler, "Barbados on The W. Side" and [Chart of Prince Rupert's Bay, Dominica], in "Observations and Surveys taken in Severall parts of the West Indies, in the Years 1717 & 18," [1718], MSS 362, Admiralty Library, Portsmouth, UK. See also P[eter] Kenna[r], "Chart of the Caribe Islands," [1715?], Vz 10/7, UK Admiralty Library.
4. Bedford to Albemarle, November 22, 1750, in *British Diplomatic Instructions 1689–1789,* vol. 7, *France, Part IV, 1745–1789,* ed. L. G. Wickham Legg (London: Offices of the Royal Historical Society, 1934), 14–15; Sir Thomas Robinson to Albemarle, October 3, 1754, in ibid., 49.
5. Kenneth Morgan, "Robert Dinwiddie's Reports on the British American Colonies," *William and Mary Quarterly* 65 (2008): 27; "advantage": Josiah Tucker

(1749), quoted in Lawrence Henry Gipson, *The British Empire before the American Revolution*, vol. 2, *The British Isles and the American Colonies: The Southern Plantations, 1748–1754* (New York: Knopf, 1958), 253; Address of the Jamaica Assembly (1752), quoted in Gipson, *Southern Plantations*, 258; Gipson, *Southern Plantations*, 263; Richard Pares, *War and Trade in the West Indies, 1739–1763* (London: F. Cass, 1763), 83; Gipson, *Southern Plantations*, 257–259; "fair trader": Charles Knowles (1752), quoted in Gipson, *Southern Plantations*, 256. See also S. D. Smith, *Slavery, Family and Gentry Capitalism in the British Atlantic: The World of the Lascelles, 1648–1834* (New York: Cambridge University Press, 2006), 132–133.

6. [John Campbell], *Candid and Impartial Considerations on the Nature of the Sugar Trade* (London, 1763), 22–26.
7. D. L. Niddrie, "Eighteenth-Century Settlement in the British Caribbean," *Transactions of the Institute of British Geographers* 40 (1966): 67; *Acts of the Privy Council of England*, Colonial Series, vol. 4, *A.D. 1745–1766*, ed. James Munro (Hereford: H. M. Stationery Office, 1911), 580–581; D. H. Murdoch, "Land Policy in the Eighteenth-Century British Empire: The Sale of Crown Lands in the Ceded Islands, 1763–1783," *Historical Journal* 27 (1984): 556; Edward T. Price, *Dividing the Land: Early American Beginnings of Our Private Property Mosaic* (Chicago: University of Chicago Press, 1995), 106–119.
8. "Considerations on Settling the Granadoes and the other Islands Ceded to the Crown of Great Brittain by the Definitive Treaty," Stowe-Grenville Correspondence, Americana, 12–28, Huntington Library, San Marino, CA; Campbell, *Candid and Impartial Considerations*, 88; "Abstract of several Informations & Plans relative to the Settling of Granada, Tobago, St. Vincent & Dominico," Grenville Americana, 12–12; "Some Hints for the Better Settlement of the ceded Islands" (1763), Shelburne Papers, 48: 47, William L. Clements Library, University of Michigan, Ann Arbor; David Watts, *The West Indies: Patterns of Development, Culture and Environmental Change since 1492* (New York: Cambridge University Press, 1987), 294, 296. In Grenada's well-developed sugar economy, British planters built on top of preexisting French landholdings toward the upper end of the 500-acre limit; in Dominica, thought to be "better adapted to the Cultivation of Cotton, Coffee & Cocoa than of Sugar," lots of 50–100 acres were more common; the new sugar islands of St. Vincent and Tobago featured middling landholdings, most of which were between 100 and 300 acres. Board of Trade Report, March 26, 1763, in *Acts of the Privy Council*, 583, 588, 604.
9. *Acts of the Privy Council*, 583–584; Thomas Whately to Mr. Sedgwick, April 3, 1764, SP 37/3 f.53, *State Papers Online: Eighteenth Century, 1714–1782*, Gale Cengage Learning, 2013, document number: MC4589183048; William Young, *Some Observations; which May Contribute to Afford a Just Idea of the Nature, Importance, and Settlement, of our New West-India colonies* (London, 1764), 38–39; *Acts of the Privy Council*, 593–594, 598–599; "Some Hints about selling

the Lands in our New Sugar Islands," Grenville Americana, 12–19; "Considerations on Settling the Granadoes."

10. "Sketch of a plan"; *Acts of the Privy Council,* 592–593, 600; "Some Hints about selling the Lands." See David Buisseret, ed., *Rural Images: Estate Maps in the Old and New Worlds* (Chicago: University of Chicago Press, 1996). The preprinted land grant documents left spaces in which to enter the name of the surveyor who completed the plan of the property that was to be annexed to each official grant. T 1/475, The National Archives of the UK, Kew, Richmond, Surrey (hereafter cited as UK National Archives).
11. Comparable maps that feature cadastral boundaries include William De Brahm's *A map of South Carolina and a part of Georgia* (London, 1757) (at a scale of 1:310,000, or 4.89 miles per inch) and Samuel Holland's *A topographical map of the Province of New Hampshire* (London, 1784) (at a scale of 1:260,000, or 4.1 miles per inch). Byres's St. Vincent map, by contrast, is drawn at a scale of 1:32,000.
12. John Byres, *References to the Plan of the Island of Dominica, as Surveyed from the Year 1765 to 1773; By John Byres, Chief Surveyor* (London, 1777); *Acts of the Privy Council,* 582. [Grey] Cooper to the Ceded Islands Land Commission, January 10, 1766, in William Hewitt Papers (Ms. 522) 1756–1790, microfilm edition, University of London, 370.
13. "Relative to Fortifications in America," [ca. 1763], CO5/232, British Public Record Office Transcripts, Manuscript Division, Library of Congress, Washington, DC (hereafter cited as LC Transcripts); *Acts of the Privy Council,* 588; "Abstract of . . . Plans relative to the Settling of Granada, Tobago, St. Vincent & Dominico"; Winthrop Pickard Bell, *The Foreign Protestants and the Settlement of Nova Scotia: The History of a Piece of Arrested British Colonial Policy in the Eighteenth Century* ([Toronto]: University of Toronto Press, [1961]), 432, 476–477; see also Lunenburg County Land Grants, Lunenberg Garden Lots, accessed June 19, 2013, http://www.seawhy.com/lggdnlot.html.
14. Copy of a Letter to the Earl of Hertford, "The Regulations for the Settlement of the Conquered Sugar Islands," [1765], Shelburne Papers, 49: 30; Watts, *West Indies,* 313–315; "Hints for the Better Settlement," 47; Hugh Græme, "Some Hints Concerning the new Islands Humbly offered to the Right Hon[ora]ble George Greenvile Esqr.," Grenville Americana, 12–16.
15. Townshend to [Dartmouth], March 8, 1773, *State Papers Online,* document number: MC4589400100.
16. "Memorial of the . . . Proprietors of Plantations in the Islands of Grenada, Tobago and the Granadines," 1777, CO 101/21: 147, British Public Record Office Photostats, Manuscript Division, Library of Congress, Washington, DC (hereafter cited as LC Photostats); "Persons interested in Grenada" to George Germain, December 24, 1778, CO 101/22: 181–182.
17. Bryan Edwards, *The history, civil and commercial, of the British colonies in the West Indies,* 2nd ed., 2 vols. (London, [1794]), 1: 378. On naming American places

for patrons, see Stephen Hornsby, *Surveyors of Empire: Samuel Holland, J. F. W. Des Barres, and the Making of the Atlantic Neptune* (Montreal: McGill-Queen's University Press, 2011), 127–139.

18. Watts, *West Indies,* 249, 346; "A General Account of the Island of Dominique," Grenville Americana, 12–22.
19. These values, and all others noted in this chapter, are denominated in British pounds sterling. John Byres, *References to the Plan of the Island of Dominica* (London, 1777); Report to the Treasury from the Commissioner[']s Office, Dominica, August 1, 1777, MPD 3278, UK National Archives. For detailed backgrounds on the partners in the Rosalie Company and surveyor Isaac Werden, see Anthony Mullan, "A Web of Imperial Connections: Surveyors and Planters in Eighteenth-Century Dominica," *Terrae Incognitae* 48 (2016): 183–205.
20. See Geoff Quilley, "Pastoral Plantations: The Slave Trade and the Representation of British Colonial Landscape in the Late Eighteenth Century," in *An Economy of Colour: Visual Culture and the Atlantic World, 1660–1830,* ed. Geoff Quilley and Kay Dian Kriz (Manchester: Manchester University Press, 2003), 106–128.
21. "Some Observations on the Nature, Importance, and Settlement of Our New West-India Colonies," Grenville Americana, 12–12.
22. [Petition of James Clark for Charles O Hara, William Stuart, Roberts & Philip Browne and Self to William Hewitt Esquire], [ca. 1777], T 1 / 535 / 343, UK National Archives.
23. George Scott to Board of Trade, July 17, 1762, CO 323 / 16, UK National Archives; Scott to Board of Trade, January 19, 1763, CO 323 / 16, UK National Archives; Board of Trade Report, March 26, 1763, 584–585.
24. Josef Konvitz, *Cartography in France, 1660–1848: Science, Engineering, and Statecraft* (Chicago: University of Chicago Press, 1987), chap. 1. A Dutch map first published in 1735 laid a matrix of property lines that showed how lands along the Commewigne and Suriname Rivers had been claimed as private property and how these individual plantation tracts constituted the Suriname colony, southeast of the British Windwards on the coast of South America. See Alesandre Lavaux, *Algemeene kaart van de Colonie of Provintie van Suriname* (Amsterdam, [ca. 1758]), JCB Archive of Early American Images, http://jcb.lunaimaging.com/luna/servlet/detail/JCBMAPS~1~1~1129~109580002. Bellin's *Carte de l'isle de la Grenade* (Paris, 1760) offered Pinel a specific precursor image from which to draw as he adapted this precise new cartographic style to the features of the island. See a copy at University of Florida Map and Image Library, http://ufdc.ufl.edu/UF00077084/00001. George Scott to the Board of Trade, July 17, 1762.
25. Watts, *West Indies,* 296, 246, 229, 242. Scott to Board of Trade, July 17, 1762. On plans to improve the harbor, see George Macartney to George Germain, October 25, 1778, CO 101 / 21: 154, LC Photostats.

26. Daniel Paterson, *A Topographical Description of the Island of Grenada* (1780; repr., St. George's, Grenada: Carenage Press, 1972). Paterson's text draws liberally from the publication that accompanied Pinel's map, George Scott, *Explication du plan de l'Isle de La Grenade* (London, 1763). George Macartney to George Germain, July 7, 1778, CO 101/22: 54–55, LC Photostats; Watts, *West Indies*, 410; Paterson, *Topographical Description*, 6–7; George Macartney to George Germain, July 22, 1778, CO 101/22: 54–55, LC Photostats; "Memorial of the . . . Proprietors of Plantations in the Islands of Grenada, Tobago and the Granadines" 1777, CO 101/21: 147–48. On the extent of British control over land, see Smith, *Slavery, Family and Gentry Capitalism*, 211–213.
27. George Scott to the Board of Trade, July 17, 1762.
28. Ibid.; George Scott to Egremont, January 19, 1763. Grants made by Scott prior to the commission's sales were later confirmed by the Treasury. [Grey] Cooper to the Ceded Islands Land Commission, January 10, 1766, in William Hewitt Papers, 370. A report on revenues from Grenada under the French is included in Grenville Americana, 12–17.
29. Græme, "Some Hints Concerning the new Islands"; Report of the Board of Trade, March 26, 1763, 589–590.
30. Report of the Board of Trade, March 26, 1763, 589–590; "Treaty of Paris 1763," *The Avalon Project: Documents in Law, History and Diplomacy*, Yale Law School, accessed April 14, 2016, http://avalon.law.yale.edu/18th_century/paris763.asp; "Considerations on Settling the Granadoes"; [Grey] Cooper to the Ceded Islands Land Commission, January 10, 1766, in William Hewitt Papers, 370; Earl of Halifax to the Lords of Trade, September 27, 1763, SP 44/138: 102, *State Papers Online*, document number: MC4589182283; Report of the Board of Trade, March 26, 1763, 590–591; "Some Hints for the better Settlement."
31. Scott to [Board of Trade], July 17, 1762; Scott to Egremont, January 19, 1763, CO 323/16; Report of the Board of Trade, March 26, 1763, 585; Ceded Islands Land Commission to the Treasury, July 19, 1773, T 1/499, UK National Archives.
32. [George] Macartney to William Hewitt, January 28, 1778, William Hewitt Papers, 388; Macartney to George Germain, February 7, 1778, CO 101/21: 156–157, LC Photostats; David Beck Ryden, "'One of the Finest and Most Fruitful Spots in *America*': An Analysis of Eighteenth-Century Carriacou," *Journal of Interdisciplinary History* 43 (2013): 539–570; Francis Kay Brinkley, "An Analysis of the 1750 Carriacou Census," *Caribbean Quarterly* 24 (1978): 44–60; H. Gordon Slade, "Craigston and Meldrum Estates, Carriacou, 1769–1841," *Proceedings of the Society of Antiquaries of Scotland* 114 (1984): 481–537; Thomas Campbell (1790), quoted in Ryden, "Fruitful Spots," 547.
33. John Byres, *References to the Plan of the Island of St. Vincent* (London, 1776); Joseph Spinelli, "Land Use and Population in St. Vincent, 1763–1960: A Contribution to the Study of the Patterns of Economic and Demographic

Change in a Small West Indian Island" (PhD diss., University of Florida, 1973), 53.

34. "[J]ealous": quoted in Murdoch, "Land Policy," 586; "Names of the different Quarters Actually possessed by the French in the Island of St. Vincent," [ca. 1764], Grenville Americana, 12–11; "State of the Island of St. Vincent's," May 19, 1764, MFQ 1173/23, UK National Archives; Richard Maitland to Hillsborough, n.d., in [William Young], *Authentic Papers Relative to the Expedition against the Charibbs, and the Sale of Lands in the Island of St. Vincent* (London, 1773), 25.

35. Jacques-Nicolas Bellin, "Carte de I'isle de la Grenade," in *Le Petit Atlas Maritime Recueil De Cartes et Plans Des Quatre Parties Du Monde* (Paris, 1764), David Rumsey Historical Map Collection, http://www.davidrumsey.com/luna/servlet/detail/RUMSEY~8~1~232805~5509427; Report of the Board of Trade, March 26, 1763, 583; James Grant to the Board of Trade, received January 29, 1772, CO 5/545: 115–116, LC Transcripts; Samuel Holland, *Holland's Description of Cape Breton Island and Other Documents* (Halifax: Public Archives of Nova Scotia, 1935), 23; "State of the Island of St. Vincent's."

36. Campbell, *Candid and Impartial Considerations,* 103–104, 105, 199; William Young (1764), quoted in Jack P. Greene, *Evaluating Empire and Confronting Colonialism in Eighteenth-Century Britain* (New York: Cambridge University Press, 2013), 3.

37. Memorial of William Young, April 11, 1767, "Papers Relative to the Expedition against the Caribbs," in *The Parliamentary History of England,* vol. 17, *A.D. 1771–1774* (London, 1813), 579, 576–578, 581.

38. Campbell, *Candid and Impartial Considerations,* 198–199.

39. Ibid., 200.

40. Board of Trade report, March 26, 1763, 586–87; "no survey": quoted in Taylor, *Black Caribs,* 53; Taylor, *Black Caribs,* 54–55, 60–61; T 1/475, UK National Archives; [Grey] Cooper to the Ceded Islands Land Commission, January 10, 1766, William Hewitt Papers, 370.

41. William Young, *An Account of the Black Charaibs in the Island of St. Vincent's . . .* (London, 1795), 44–47; Harry Alexander to Lieutenant Governor Fitzmaurice, May 3, 1769, in Young, *Authentic Papers,* 22; Fitzmaurice to Hillsborough, June 10, 1769, in Young, *Authentic Papers,* 31–32; "no roads": Report of Richard Maitland [1771], in Young, *Authentic Papers,* 43; William Young to Harry Alexander, May 1, 1769, in Young, *Authentic Papers,* 21; Harry Alexander to Lieutenant Governor Fitzmaurice, May 3, 1769, in Young, *Authentic Papers,* 24; Memorial of Richard Maitland, in Young, *Authentic Papers,* 25.

42. Hilary McD. Beckles, "Kalingo (Carib) Resistance to European Colonization of the Caribbean," *Caribbean Quarterly* 38 (1992): 6; Memorial of Richard Maitland, in Young, *Authentic Papers,* 26; Address of the Council and Assembly of St. Vincent to the King, May 1773, in Young, *Authentic Papers,* 27; "Consid-

erations on Settling the Granadoes"; Ulysses Fitzmaurice to Hillsborough, June 10, 1769, in Young, *Authentic Papers,* 34.

43. Memorial of William Young, 579; "A Proclamation," June 10, 1769, in Young, *Authentic Papers,* 35; "hunting the Caribs": quoted in Taylor, *Black Caribs,* 61; Fitzmaurice to Hillsborough, June 10, 1769, in Young, *Authentic Papers,* 29.
44. Records of the Ceded Islands Land Commission, August 12, 1771, CO 106 / 12, UK National Archives; Fitzmaurice to Hillsborough, June 10, 1769, in Young, *Authentic Papers,* 30; "Report of the Commissioners for the Sale of Lands in the Ceded Islands to the Lords of the Treasury," October 16, 1771, in Young, *Authentic Papers,* 37–38.
45. "Report of the Commissioners," October 16, 1771, in Young, *Authentic Papers,* 39–40.
46. Report of Richard Maitland, in Young, *Authentic Papers,* 41–46; Memorial of the Assembly and Council of St. Vincent, in Young, *Authentic Papers,* 51.
47. Board of Trade to the King, March 29, 1770, in Young, *Authentic Papers,* 61; Hillsborough to Governor Leyborne, April 18, 1772, in Young, *Authentic Papers,* 73; Leyborne to Hillsborough, August 25, 1772, in Young, *Authentic Papers,* 77; J. Pownall to W. Knox, October 3, 1772, in Historical Manuscripts Commission, *Report on Manuscripts in Various Collections,* vol. 6 (Dublin, 1909), 109; Taylor, *Black Caribs,* 71–75; Christopher Leslie Brown, *Moral Capital: Foundations of British Abolitionism* (Chapel Hill: University of North Carolina Press, 2006), 156; Greene, *Evaluating Empire,* 16–17, 348–351; Young, *Black Charaibs,* 1.
48. Cha[rles] Winston to William Hewitt, March 13, 1773, William Hewitt Papers; Young, *Black Charaibs,* 90–92.
49. Petition of John Byres to Treasury, August 8, 1776, and August 10, 1776, T 1 / 526: 310–311, UK National Archives; Taylor, *Black Caribs,* 90.
50. *Acts of the Privy Council,* 581–582; Lawrence Henry Gipson, *The British Empire before the American Revolution,* vol. 5, *Zones of International Friction: The Great Lakes, Frontier, Canada, the West Indies, India, 1748–1754* (New York: Knopf, 1958), 218–223; *British Diplomatic Instructions,* 3–4.
51. These revised copies of the 1749 chart include "A plan of the Island of Tobago," 1762, CO 700 / Tobago 1, UK National Archives, and "A Plan of the Island of Tobago," 1763, Vz 10 / 34, UK Admiralty Library; "The Neutral Islands," Grenville Americana, 12–23.
52. "Some Hints about selling the Lands."
53. James Simpson to the Treasury, n.d., T 1 / 500, UK National Archives; "Some Hints about selling the Lands."
54. "Some Hints for the better Settlement."
55. *Acts of the Privy Council,* 582–583; Richard H. Grove, *Green Imperialism: Colonial Expansion, Tropical Island Edens and the Origins of Environmentalism, 1600–1860* (New York: Cambridge University Press, 1996), chap. 6. On

the environmental impact of sugar agriculture, see Watts, *West Indies,* especially 395–397.

56. John Fowler, *A Summary Account of the Present Flourishing State of the Respectable Colony of Tobago, in the British West Indies* (London, 1774), 24. See "A Map of the Island and A Plan of the Settlement of Tobago in the West-Indies Humbly Dedicated to all the Proprietors and Friends of the very respectable and flourishing Colony MDCCLXXIV," in Fowler, *Summary Account*; Land Commissioners (1768), quoted in Murdoch, "Land Policy," 561.
57. Extract from the Minutes of the Board of Treasury, August 9, 1776, William Hewitt Papers, 416 / 2; Memorial of Robert Walton, n.d., William Hewitt Papers, 416; William Pulteney to Grey Cooper, n.d., William Hewitt Papers, 416 / 4; Memorial of the Council and Assembly of Tobago to the Treasury, n.d., William Hewitt Papers, 416 / 8.
58. "The Neutral Islands"; "Some Hints for the better Settlement"; "Copy of a Letter to a French Gentleman in Grenada from an Adventurer in the New Spanish Settlement of Trinidada," January 11, 1777, and April 17, 1777, George MacCartney Common Book, HM 686, Huntington Library, trans. Sage Morghan. On the size of Trinidadian *quarrés,* see Robert Montgomery Martin, *History of the Colonies of the British Empire* (London: Wm. H. Allen, 1843), 32.
59. "Letter to a French Gentleman." A rare version of a later map of the island was published: E. H. Columbine, *A Survey of the Island of Trinidad. By Capt. E. H. Columbine, 1803; with additions from Capt*[n] *F. Mallet.* ([London], 1816). See also Kit Candlin, *The Last Caribbean Frontier, 1795–1815* (New York: Palgrave Macmillan, 2012).
60. Niddrie, "Eighteenth-century Settlement," 80n; *London Gazette,* 10403, March 24–March 27, 1764.
61. Abstract of Dispatches from the Governor of Grenada, [1767], Shelburne Papers, 52: 443; Watts, *West Indies,* Table 7.7, 316.
62. Isaac Werden to William Hewitt, May 17, 1777, William Hewitt Papers, 329 / 1.
63. Aaron Willis, "The Standing of New Subjects: Grenada and the Protestant Constitution after the Treaty of Paris (1763)," *Journal of Imperial and Commonwealth History* 42 (2014): 11; Jack P. Greene, ed., *Exclusionary Empire: English Liberty Overseas, 1600–1900* (New York: Cambridge University Press, 2010), 70–72; "Memorial of the . . . Proprietors of Plantations in the Islands of Grenada, Tobago and the Granadines," 1777, CO 101 / 21: 147–148, UK National Archives.
64. See Craig Yirush, *Settlers, Liberty, and Empire: The Roots of Early American Political Theory* (New York: Cambridge, 2011); Willis, "Standing of New Subjects," 6–7; *Acts of the Privy Council,* 742–743.
65. Quoted in Willis, "Standing of New Subjects," 9, 7; "Regulations for the Sugar Islands," 1765. See also Hannah Weiss Muller, "An Empire of Subjects: Unities and Disunities in the British Empire, 1760–1790" (PhD diss., Princeton University, 2010), 93.

66. Henry Cowper, *Reports of cases adjudged in the Court of King's Bench: from Hilary term, the 14th of George III. 1774, to Trinity term, the 18th of George III. 1778,* 2 vols. (London, 1800), 212–213; Edwards, *The history . . . of the . . . West Indies,* 1: 359.
67. Valentine Morris to the Treasury, August 8, 1776, and August 10, 1776, John Byres to the Treasury, August 8, 1776, T 1/526/310–311, UK National Archives. My thanks to Gillian Hutchinson for citations to these documents. Bruce B. Solnick and Christopher M. Klein, "Cartography and Colonization: The British in the West Indies after 1763," in *Imago et Mensura Mundi: Atti Del IX Congresso Internazionale di Storia della Cartographia,* ed. Carla Clivio Marzoli (Roma: Istituto della Enciclopedia Italiana, 1985), 410; K. F. H. Smith, "Located Interconnections: The Social Construction of Place, Identity and Community in the Carib Territory, Dominica" (PhD diss., University of Essex, 2007), 51–55, 78–81.
68. *Acts of the Privy Council,* 589; Murdoch, "Land Policy," 565–566. Buyers who eyed particular tracts could employ their own surveyors as long as they produced a plat at a scale of a quarter mile to the inch and submitted copies of their field notes to the chief surveyor. Such hired hands were to do double duty as they laid out tracts for purchase, producing a survey of up to a mile of riverside or coastline to contribute to the general project of mapping each island. Solnick and Klein, "Cartography and Colonization," 410. Alexander Forbes to William Hewitt, June 23, 1777, William Hewitt Papers, 384; John Robinson to Ceded Islands Land Commission, October 25, 1776, William Hewitt Papers, 380.
69. "Abstract of Informations & Plans"; "Heads of Enquiry relating to the State of the Islands of Dominico, St. Vincents, Tobago, Grenada," March 5, 1763, CO 324/21: 229, 231, LC Transcripts.
70. Murdoch, "Crown Land Sales," Tables 1–3, 5, 562–565, 569, 571; Watts, *West Indies,* 294; Smith, *Slavery, Family, and Gentry Capitalism,* 192, 211–212, 220; "Private information of the present State of the Island of Grenada," c. 1770–1779, quoted in Smith, *Slavery,* 213; Watts, *West Indies,* Tables 7.2 and 7.4, 286, 288; May 31, 1778, CO 101/21: 217–219, LC Photostats.

CHAPTER SIX: DEFINING EAST FLORIDA

1. Amy Turner Bushnell and Jack P. Greene, "Peripheries, Centers, and the Construction of Early Modern American Empires: An Introduction," in *Negotiated Empires: Centers and Peripheries in the Americas, 1500–1820,* ed. Christine Daniels and Michael Kennedy (New York: Routledge, 2002), 8.
2. Pierre Crignon (1545), quoted in *American Beginnings: Exploration, Culture, and Cartography in the Land of Norembega,* ed. Emerson W. Baker et al. (Lincoln: University of Nebraska Press, 1994), xxv.
3. Michele Currie Navakas, "Island Nation: Mapping Florida, Revising America," *Early American Studies* 11 (2013): 248–256; William P. Cumming, *The Southeast in Early Maps,* ed. Louis De Vorsey Jr., 3rd ed. (Chapel Hill: University of

North Carolina Press, 1998), 254–255. See [Map of the coast of Florida from Fort William to Musketae River] [north sheet], 1743, Maps K.Top. 122.83, The British Library, from Norman B. Leventhal Map Center at the Boston Public Library, Boston, MA, http://maps.bpl.org/id/n51594; see also Justly Watson, "A Survey of the Coast from Fort William near St. Iuans River to Mosquito River," 1743, Maps K.Top. 122.82, The British Library, from Norman B. Leventhal Map Center at the Boston Public Library, http://maps.bpl.org/id/n51593. R. Baldwin, *The Present State of the West Indies* (London, 1778), 43.

4. George Gauld, *An Account of the Surveys of Florida, &c* (London, 1790), 4; Jack D. L. Holmes and John D. Ware, "Juan Baptista Franco and Tampa Bay, 1756," *Tequesta* 28 (1968): 91–97; Charles W. Arnade, "Celi's Expedition to Tampa Bay: A Historical Analysis," *Florida Historical Quarterly* 47 (1968): 1–7.
5. Jefferys's map also appeared as the frontispiece of William Roberts, *An Account of the First Discovery, and Natural History of Florida* (London, 1763); ibid., vii; "Report on Acquisitions in America," June 8, 1763, in *Documents Relating to the Constitutional History of Canada, 1759–1791,* ed. Adam Shortt and Arthur G. Doughty (Ottawa: S. E. Dawson, 1907), 138, 140, 143; [John] Pownall, "Mr. Pownal's Sketch of a Report concerning the Cessions in Africa and America at the Peace of 1763," in R. A. Humphreys, "Lord Shelburne and the Proclamation of 1763," *English Historical Review* 49 (1934): 262; "Hints relative to the Division and Government of the conquered and newly acquired Countries in America" (1763), CO 323/16, The National Archives of the UK, Kew, Richmond, Surrey (hereafter cited as UK National Archives); "Report on Acquisition," 144.
6. Gauld, *Surveys of Florida,* 4; "Report on Acquisitions," 144; James Grant to Richard Oswald, September 20, 1764, James Grant of Ballindalloch Papers, microfilm copy, Manuscript Division, Library of Congress, Washington, DC (hereafter cited as Grant Papers); James Grant to Brigadier Bouguett, August 11, 1765, Grant Papers; Bernard Romans, *A Concise Natural History of East and West Florida,* ed. Kathryn E. Holland Braund (Tuscaloosa: University of Alabama Press, 1999), 257.
7. John Savage to Francis Johns and Corn[eliu]s Hinson, December 14, 1764, CO 5/540: 221, British Public Record Office Transcripts, Manuscript Division, Library of Congress (hereafter cited as LC Transcripts); William Stork, *A Description of East-Florida, with a Journal, kept by John Bartram of Philadelphia . . .* (London, 1774), ii; "Report on Acquisitions," 138; William Stork, *An Account of East-Florida. With Remarks on its Future Importance to Trade and Commerce* (London, [1766]), 43; Romans, *Concise Natural History,* 257; Alexander Johnson, "Charting the Imperial Will: Colonial Administration and the General Survey of British North America, 1764–1775" (PhD diss., University of Exeter, 2011), 215. On the publication history of Bowen's map, see Margaret Beck Pritchard and Henry G. Taliaferro, *Degrees of Latitude:*

Mapping Colonial America (New York: Colonial Williamsburg/Harry N. Abrams, 2002), 180–183.

8. William Johnson, "Re. thoughts concerning Florida," [1763], Shelburne Papers, William L. Clements Library, University of Michigan, Ann Arbor, 48: 6.
9. "Journal of an Officer's [Lord Adam Gordon's] Travels in America and the West Indies, 1764–1765," in *Travels in the American Colonies,* ed. Newton D. Mereness (New York: Macmillan, 1916), 392–393, 387; "congealed shells": "Sketch of the Castle of St. Augustin," [1763], CO 700/Florida 6, UK National Archives.
10. William De Brahm, *De Brahm's Report of the General Survey in the Southern District of North America,* ed. Louis De Vorsey Jr. (Columbia: University of South Carolina Press, 1971), 205, 225; "Journal of an Officer," 393; James Grant to Colonel Robertson, December 8, 1764, Grant Papers. On the importance of hearths, see John E. Crowley, *The Invention of Comfort: Sensibilities and Design in Early Modern Britain and Early America* (Baltimore, MD: Johns Hopkins University Press, 2001), chaps. 1–3; John Bartram to Peter Collinson, August 26, 1766, *The Correspondence of John Bartram, 1734–1777,* ed. Edmund Berkeley Jr. and Dorothy Smith Berkeley (Gainesville: University Press of Florida, 1992), 675–676; James Grant to William Knox, May 6, 1765, Grant Papers.
11. Augustin Prevost (1763), quoted in John D. Ware and Robert R. Rea, *George Gauld: Surveyor and Cartographer of the Gulf Coast* (Gainesville: University Presses of Florida, 1982), 26; Francis Ogilvi[e] to Board of Trade, January 26, 1764, CO 5/540: 63–64, LC Transcripts.
12. James Grant to Colonel Robinson, October 2, 1764, Grant Papers.
13. John Greg to Board of Trade, February 22, 1764, CO 5/540: 27–28, LC Transcripts; "divided": Maps and Plans, c. 1700s, Grant Papers; James Grant to Colonel Robinson, October 2, 1764, Grant Papers; Jane Landers, *Black Society in Spanish Florida* (Urbana: University of Illinois Press, 1999), 61; see also Charles L. Mowat, *East Florida as a British Province, 1763–1784* (Berkeley: University of California Press, 1943), 53–54.
14. James Grant to Board of Trade, March 1, 1765, CO 5/540: 214, LC Transcripts. Other St. Augustine maps include James Moncrief, "A Plan and Section of Fort St. Mark's, St. Augustine," [1765], CO 700/Florida 15, UK National Archives, and [James Moncrief], "A Plan of the City, Harbour, Fortifications and Environs of Saint Augustine," [1765], CO 700/Florida 20/2, UK National Archives; James Grant to Board of Trade, March 1, 1765, CO 5/540: 207–209, LC Transcripts; John Savage to Francis Johns and Corn[eliu]s Hinson, December 14, 1764, CO 5/540: 221, LC Transcripts; James Grant to Gilbert Elliot, December 6, 1764, Grant Papers; James Grant to [Henry] Laurens, December 29, 1764, Grant Papers.

15. Hunting ground: James Grant to Board of Trade, December 10, 1767, CO 5/549: 26, UK National Archives; "any thing": James Grant to [Henry] Laurens, February 6, 1765, Grant Papers.
16. James Grant to Board of Trade, March 1, 1765, CO 5/540: 207–208, LC Transcripts. Moncrief also completed an unfinished sketch, "A Plan of St. Mary's Harbour in the province of East Florida," [1765], Maps 6-K-6, Clements Library.
17. "Journal of an Officer," 388–391; "Report on Acquisitions," 143; James Grant to Halifax, December 23, 1763, Grant Papers; Mr. Greene to the Treasury, June 6, 1764, CO 5/540: 115, LC Transcripts; James Grant to Brigadier Bouguett, August 11, 1765, Grant Papers.
18. De Brahm, *De Brahm's Report,* 7–33.
19. Ibid., 6, 33. See also Mart A. Stewart, "William Gerard de Brahm's 1757 Map of South Carolina and Georgia," *Environmental History* 16 (2011): 524–535; see also Cumming, *Southeast in Early Maps,* 280–281.
20. John Pownall to William De Brahm, August 15, 1764, CO 324/17: 392–393, UK National Archives; Pownall to Samuel Holland, May 19, 1766, CO 324/18: 8–9, UK National Archives; Pownall to Holland, April 17, 1764, CO 324/17: 392–396, UK National Archives.
21. De Brahm, *De Brahm's Report,* 215, 35–36; see Johnson, "Charting the Imperial Will," 196–215.
22. James Grant to Hillsborough, November 26, 1763, Grant Papers; "parallel": Stork, *Description of East-Florida,* 10.
23. The Middle Inlet and River were renamed the Rio Blanco Inlet and the Rio Blanco River on the general map of 1770. The total arable acreage for the lands represented in charts 11 and 12 was 97,821. This total includes acres in the 12th section along the Middle River, served by the Middle Inlet.
24. "Proclamation Published by Governor Grant in the severall provinces of North America," November 22, 1764, CO 5/540: 173, LC Transcripts.
25. Johnson, "Charting the Imperial Will," 210; De Brahm to Board of Trade, April 4, 1765, CO 323/18: 141–153, UK National Archives; De Brahm to Board of Trade, March 24, 1765, CO 5/548: 136–140, UK National Archives.
26. James Grant to John Pownall, April 4, 1765, Grant Papers; Grant to Board of Trade, March 1, 1765, CO 5/540: 207–208, LC Transcripts; Grant to [Board of Trade], May 8, 1765, CO 5/548: 65, LC Transcripts; Grant to Governor Wright, April 1, 1765, Grant Papers; Grant to [James] Wright, April 1, 1765, Grant Papers; Grant to John Savage, April 4, 1765, Grant Papers; H. S. Conway to [James] Grant, September 12, 1765, CO 5/548: 71, LC Transcripts. A note on De Brahm's Mosquito Inlet chart indicates that although it was surveyed and drafted in 1765, it was not received until January 29, 1767.
27. De Brahm, *De Brahm's Report,* 63.
28. James Grant to John Savage, April 4, 1765, Grant Papers; Grant to [Henry] Laurens, November 18, 1764, Grant Papers; Michael J. Jarvis, *In the Eye of All*

Trade: Bermuda, Bermudians, and the Maritime Atlantic World, 1680–1783 (Chapel Hill: University of North Carolina Press, 2010), 339–343; John Savage, "Proposal Made for Settling a Tract of Land in East Florida," CO 4 / 540: 177–179, LC Transcripts; James Grant to Richard Oswald, November 21, 1764, Grant Papers; Grant to John Savage, July 12, 1765, Grant Papers.

29. Messrs. Ephraim and John Gilbert to John Savage, January 25, 1765, CO 5 / 540: 233, LC Transcripts; James Grant to Jonathan Bryan, July 4, 1765, Grant Papers; Grant to John Savage, July 12, 1765, Grant Papers; Grant to the Board of Trade, August 5, 1766, CO 5 / 541: 41, LC Transcripts.
30. Bernard Bailyn, *Voyagers to the West: A Passage in the Peopling of America on the Eve of the Revolution* (New York: Vintage, 1986), 453 and chap. 12.
31. *Acts of the Privy Council of England,* Colonial Series, vol. 4, *A.D. 1745–1766,* ed. James Munro (Hereford: H. M. Stationery Office, 1911), 610; Mowat, *East Florida,* 58–61.
32. Bailyn, *Voyagers to the West,* 440; George C. Rogers Jr., "The East Florida Society of London, 1766–1767," *Florida Historical Quarterly* 54 (1976): 479–484; Nigel Ramsay, "Astle, Thomas (1735–1803)," *Oxford Dictionary of National Biography* (Oxford: Oxford University Press, 2004); Thomas Astle to Board of Trade, February 3, 1769, CO 5 / 544: 17, LC Transcripts; "Representation of Board of Trade for granting 5,000 acres to Thomas Astle," 1769, PC 1 / 59 / 6 / 2, UK National Archives.
33. "English Grantees": James Grant (1766), quoted in Rogers, "East Florida Society," 494; Charles L. Mowat, "The Land Policy in British East Florida," *Agricultural History* 14 (1940): 76; Grant, "Proclamation"; James Grant to Denys Rolle, September 24, 1764, Grant Papers; Grant to Denys Rolle, January 16, 1765, Grant Papers; Grant to Board of Trade, November 22, 1764, CO 5 / 540: 135, LC Transcripts.
34. De Brahm's manuscript map of a portion of this coast reveals the density of information generated by the St. Johns River survey. [William De Brahm], [A Survey of the Part of the Eastern Coast of East Florida from St. Mary's Inlet to Mount Halifax. Showing the Ascertained Boundary Between East Florida and the Creek Indians], [1766], CO 700 / Florida 53, UK National Archives; Johnson, "Charting the Imperial Will," 303n840; James Grant to Richard Oswald, September 20, 1764, Grant Papers; "pointed out": Grant (1768), quoted in [Daniel L. Schafer], "New World in a State of Nature: British Plantations on the St. Johns River, East Florida, 1763–1784," *Florida History Online,* accessed October 27, 2015, http://www.unf.edu/floridahistoryonline /Plantations/plantations/John_Rawdon.htm.
35. James Grant to William Knox, May 6, 1765, Grant Papers; Grant to Richard Oswald, September 20, 1764, Grant Papers; De Brahm, *De Brahm's Report,* 202. Historian Daniel L. Schafer has located many of these sites settled before the American Revolution. He has documented how roughly a dozen English grantees made credible attempts to settle portions of their vast lands along

with two dozen more for whom there is evidence that they developed land obtained by provincial grants. Daniel L. Schafer, *William Bartram and the Ghost Plantations of British East Florida* (Gainesville: University Press of Florida, 2010), chaps. 4–6; see also [Schafer], "New World in a State of Nature."

36. On Rolle: Bailyn, *Voyagers to the West,* 434–436, 447–451; James Grant to Denys Rolle, September 14, 1764, Grant Papers; Grant to Richard Oswald, November 21, 1764, Grant Papers; James Grant to Lachlin Macleane, Feburary 13, 1767, CO 5 / 541: 98–99, 101, LC Transcripts; Grant to [Shelburne], July 2, 1768, CO 5 / 549: 95–96, LC Transcripts. On Turnbull: Bailyn, *Voyagers to the West,* 447–461; Romans, *Concise Natural History,* 247; James Grant to Board of Trade, March 4, 1769, CO 5 / 550: 53, LC Transcripts; Grant to Board of Trade, September 1, 1770, CO 5 / 551: 87–88, LC Transcripts; Mowat, *East Florida,* 7.
37. Henry Laurens to James Grant, January 28, 1768, in *The Papers of Henry Laurens,* ed. Philip M. Hamer et al., 16 vols. (Columbia: University of South Carolina Press, 1968–2003), 5:576 (hereafter cited as *Laurens Papers*); Romans, *Concise Natural History,* 3; [Schafer], "New World in a State of Nature," accessed October 27, 2015, http://www.unf.edu/floridahistoryonline/Plantations/plantations/Doctors_Lake_Plantations.htm; Daniel L. Schafer, "Plantation Development in British East Florida: A Case Study of the Earl of Egmont," *Florida Historical Quarterly* 63 (1984): 172–183.
38. Grant (1766), quoted in Thomas W. Taylor, "'Settling a Colony over a Bottle of Claret': Richard Oswald and the British Settlement of Florida" (MA thesis, University of North Carolina at Greensboro, 1984), 28; ibid., 38–55, 75; *Laurens Papers,* 7:344n, 4:465–66; David Hancock, *Citizens of the World: London Merchants and the Integration of the British Atlantic Community, 1735–1785* (New York: Cambridge University Press, 1995), 153–171; Schafer, *Ghost Plantations.*
39. *Laurens Papers,* 8:433; Romans, *Concise Natural History,* 108, 159.
40. [Schafer], "New World in a State of Nature," accessed October 27, 2015, http://www.unf.edu/floridahistoryonline/Plantations/plantations/New_Switzerland.htm.
41. "A List of the Inhabitants of East Florida . . . from 1763 to 1771," in De Brahm, *De Brahm's Report,* 180–186. Planters who had died or departed the province were not included among these 73. James Grant to Shelburne, December 25, 1767, CO 5 / 549: 41, LC Transcripts; Mowat, *East Florida,* 77; John J. McCusker, "Table Eg1027–1032—Indigo and silk exported from South Carolina and Georgia: 1747–1788," *Historical Statistics of the United States: Millennial Edition Online,* Cambridge University Press, accessed October 28, 2015.
42. S. Max Edelson, "The Characters of Commodities: The Reputations of South Carolina Rice and Indigo in the Atlantic World," in *The Atlantic Economy during the Seventeenth and Eighteenth Centuries: Organization, Operation, Practice, and Personnel,* ed. Peter A. Coclanis (Columbia: University of South

Carolina Press, 2005), 344–345; John Moultrie to Board of Trade, May 23, 1771, CO 5/552: 53, LC Transcripts; James Grant to Board of Trade, December 14, 1770, CO 5/552: 13–14, LC Transcripts; Henry Laurens to James Grant, October 27, 1769, in *Laurens Papers,* 7:176; ibid., 171n; Joyce E. Chaplin, *An Anxious Pursuit: Agricultural Innovation and Modernity in the Lower South, 1730–1815* (Chapel Hill: University of North Carolina Press, 1993), 204–205.

43. "Journal of an Officer," 393; James Grant to Richard Oswald, November 21, 1764, Grant Papers; slave driver: [Schafer], "New World in a State of Nature," accessed October 27, 2015, http://www.unf.edu/floridahistoryonline/Plantations/plantations/San_Marco-Jericho-Chichester.htm; Daniel L. Schafer, "Governor Grant's Villa: A British East Florida Indigo Plantation," *El Escribano* 37 (2000): 3–5, 23; James Grant to [Henry] Laurens, February 6, 1765; between April 4, 1765, and June 12, 1765; and July 16, 1765, Grant Papers; Schafer, "Grant's Villa," 18–21.
44. "Puff General": Adam Gordon to James Grant, February 14, 1767, in [Letters of Andrew Turnbull], *Florida History Online,* accessed October 29, 2015, https://www.unf.edu/floridahistoryonline/Turnbull/letters/2.htm.
45. John Moulture to Hillsborough, October 16, 1771, CO 5/552: 89, LC Transcripts.
46. Another unfinished manuscript map was intended as a general map at a higher scale of resolution: William De Brahm, "East Florida, East of the 82nd degree of Longitude from the Meridian of London," [1773], CO 700/Florida 3, UK National Archives; see Johnson, "Charting the Imperial Will," 287–288.
47. James Grant to Board of Trade, December 9, 1765, "Treaty for Settling Limits between his Majesty's Province of East Florida, and the Upper and Lower Creek Nations, concluded the 16th Novr. 1765," CO 5/548: 106, 90, 93, LC Transcripts; "Text of the Treaty of Picolata," in James W. Covington, *The British Meet the Seminoles: Negotiations Between British Authorities in East Florida and the Indians: 1763–68* (Gainesville: University Press of Florida, 1961), 37; "fine concession": John Bartram (1765–1766), quoted in Kathryn E. Holland Braund, "'The Congress Held in a Pavilion': John Bartram and the Indian Congress at Fort Picolata, East Florida," in *America's Curious Botanist: A Tercentennial Reappraisal of John Bartram, 1699–1777,* ed. Nancy Hoffman and John C. Van Horne (Philadelphia: American Philosophical Society, 2004), 93.
48. "Report of the Congress [Picolata]," in Covington, *British Meet the Seminoles,* 32; James Grant to John Stuart, February 1, 1769, CO 5/550: 67–68, LC Transcripts.
49. De Brahm, *De Brahm's Report,* 253–257.
50. James Grant (1766), quoted in Taylor, "Settling a Colony," 24.
51. De Brahm, *De Brahm's Report,* 39–57; James Grant to Board of Trade, April 23, 1770, CO 5/551: 42–43, LC Transcripts; Hillsborough to William De Brahm, July 3, 1771, CO 5/72: 373, LC Transcripts.

52. William De Brahm to Hillsborough, January 6, 1771, CO 5/228: 76, UK National Archives; De Brahm to James Grant, April 10, 1764, Grant Papers; Grant to Benjamin Barton, July 4, 1765, Grant Papers; Grant to Board of Trade, April 26, 1766, CO 5/541: 13, LC Transcripts.
53. James Grant to Board of Trade, April 26, 1766, CO 5/541: 7, LC Transcripts; Grant to Hillsborough, November 26, 1763, Grant Papers; James Grant to Henry Laurens, July 16, 1765, Grant Papers; James L. Hill, "'Bring them what they lack': Spanish-Creek Exchange and Alliance Making in a Maritime Borderland, 1763–1783," *Early American Studies* 12 (2014): 36–67.
54. Shelburne to James Grant, October 25, 1766, CO 5/548: 157, LC Transcripts. Emphasis in the original manuscript.
55. William De Brahm to Hillsborough, March 15, 1771, CO 5/72: 281–282, LC Transcripts; Louis De Vorsey Jr., "Pioneer Charting of the Gulf Stream: The Contributions of Benjamin Franklin and William Gerard De Brahm," *Imago Mundi* 28 (1976): 113; W[illiam] [D]e Brahm to Hillsborough, March 15, 1771, CO 5/72: 281, LC Transcripts.
56. De Brahm, *De Brahm's Report,* 214; De Brahm, *The Atlantic Pilot. A Facsimilie Reproduction of the 1772 edition* (Gainesville: University Presses of Florida, 1974), v, 1; for published versions of manuscript maps discussed above, see William De Brahm, "The Ancient Tegesta, now Promontory of East Florida," in *The Atlantic Pilot* (London, 1772), from Norman B. Leventhal Map Center at the Boston Public Library, http://maps.bpl.org/id/12056; William De Brahm, "Chart of the south end of east Florida, and Martiers," in *The Atlantic Pilot* (London, 1772), from Norman B. Leventhal Map Center at the Boston Public Library, http://maps.bpl.org/id/12057. De Vorsey, "Pioneer Charting of the Gulf Stream," 105–120.
57. William Stork (1768), quoted in [Schafer], "New World in a State of Nature," *Florida History Online,* accessed October 27, 2015, http://www.unf.edu/floridahistoryonline/Plantations/plantations/William_Beresford.htm.

CHAPTER SEVEN: ATLASES OF EMPIRE

1. Linda Colley, *Britons: Forging the Nation, 1707–1837* (New Haven, CT: Yale University Press, 2009), 86; Gary B. Nash and T. H. Breen, "The American Revolution: Social or Ideological?" in *Interpretations of American History,* vol. 1, *Patterns and Perspectives,* ed. Francis G. Couvares et al., 7th ed. (New York: Free Press, 2000), 169; T. H. Breen, *The Marketplace of Revolution: How Consumer Politics Shaped American Independence* (New York: Oxford University Press, 2004), 82; Benjamin Franklin to Isaac Norris, March 19, 1759, in *The Papers of Benjamin Franklin, Digital Edition,* Packard Humanities Institute, http://franklinpapers.org/franklin/framedVolumes.jsp?vol=8&page=291a; T. H. Breen, "Ideology and Nationalism on the Eve of the American Revolution: Revisions Once More in Need of Revising," *Journal of American History*

84 (1997): 32–33; see also Jack P. Greene, *Evaluating Empire and Confronting Colonialism in Eighteenth-Century Britain* (New York: Cambridge University Press, 2013).

2. Dora Mae Clark, *The Rise of the British Treasury: Colonial Administration in the Eighteenth Century* (Hamden, CT: Archon Books, 1969), 115–124; "general good": [Thomas Whately], *Regulations Lately Made concerning the Colonies and the Taxes Imposed upon Them Considered* (London, 1765), 43; "defending": "The Sugar Act: 1764," *The Avalon Project: Documents in Law, History and Diplomacy,* Yale Law School, http://avalon.law.yale.edu/18th_century/sugar_act_1764.asp; "The Stamp Act," March 22, 1765, *Avalon Project,* http://avalon.law.yale.edu/18th_century/stamp_act_1765.asp. See Ian R. Christie, "A Vision of Empire: Thomas Whately and the Regulations Lately Made concerning the Colonies," *English Historical Review* 113 (1998): 300–320.
3. *America Vindicated from the High Charge of Ingratitude and Rebellion* (Devizes, UK, 1774), 12; James Otis, "The Rights of the British Colonies Asserted and Proved" (1764), in *Pamphlets of the American Revolution,* vol. 1, *1750–1775,* ed. Bernard Bailyn (Cambridge, MA: Harvard University Press, 1965), 447; "viewing the numbers"/"interest of a part": *The Constitutional Right of the Legislature of Great Britain to Tax the British Colonies in America* (London, 1768), 3, 55; "enormous debt"/"Assistance": [Soame Jenyns], *The Objection to the Taxation of our American Colonists* (London, 1765), 22, 13. On the war debt, see John Brewer, *The Sinews of Power: War, Money, and the English State, 1688–1783* (Boston: Unwin Hyman, 1989), 30.
4. "Cato," *Thoughts on a Question of Importance, Proposed to the Public* (London, 1765), 11, 14, 17–23, 29–35.
5. *The Late Occurrences in North America, and Policy of Great Britain Considered* (London, 1766), 31; "custom-house books" /"one third": *America Vindicated,* 9; "essential": *America Vindicated,* 14; "imperious ambition": *Constitutional Right,* 28.
6. [Jenyns], *Objection,* 15; Martin Howard Jr., "A Letter From a Gentleman at Halifax to his Friend in Rhode Island" (Newport, 1765), in *Tracts of the American Revolution 1763–1776,* ed. Merrill Jensen (Indianapolis: Bobbs-Merrill, 1967), 76; *Colonising, or a Plain Investigation of that Subject: With a Legislative, Political and Commercial View of Our Colonies* (London, [1774]), 13.
7. John Shy, *A People Numerous and Armed: Reflections on the Military Struggle for American Independence* (New York: Oxford University Press, 1976), chap. 3; "insolent": Ellis (1774), quoted in Shy, *Numerous and Armed,* 43; "dictators": Joseph Craddock (1826), quoted in Shy, *Numerous and Armed,* 39; Thomas Pownall, *The Administration of the British Colonies . . .* 2nd ed. (London, 1774), appendix, 17–18.
8. Richard Bland, "The Colonel Dismounted: or the Rector Vindicated," in Bailyn, ed., *Pamphlets of the American Revolution,* 320; Barré quoted in

Michael G. Kammen, *Empire and Interest: the American Colonies and the Politics of Mercantilism* (Philadelphia: J. B. Lippincott, 1970), 133; James Otis, "Rights of the Colonies," 435–436; Bland, "Colonel Dismounted," 324; Otis, "Rights of the Colonies," 448–449; "Examination before the Committee of the Whole of the House of Commons," February 13, 1766, in *Franklin Papers,* http://franklinpapers.org/franklin//framedVolumes.jsp?vol=13&page=124a.

9. *The General Opposition of the Colonies to the Payment of the Stamp Duty* (London, 1766), 39; "Janus," *The Critical Moment, on which the Salvation or Destruction of the British Empire Depends* (London, 1776), 39, 42, 58; *A Short and Friendly Caution to the Good People of England* (London, 1766), 5; Richard Bland, "An Inquiry into the Rights of the British Colonies" (1766), in Jensen, ed., *Tracts of the American Revolution,* 114.
10. Otis, "Rights of the Colonies," 445, 440.
11. *Critical Moment,* 34–35; *Late Occurrences,* 6; Howard, "Letter from Halifax," 70; [Jenyns], *Objection to the Taxation,* 18; *Late Occurrences,* 4.
12. *Late Occurrences,* 29; *Constitutional Right,* 455.
13. Otis, "Rights of the Colonies," 461; *Late Occurrences,* 35–37; *Constitutional Right,* 21; Thomas Hopkins, "Adam Smith on American Economic Development and the Future of the European Atlantic Empires," in *The Political Economy of Empire in the Early Modern World,* ed. Sophus Reinert and Pernille R. Røge (Houndmills, UK: Palgrave Macmillan, 2013), 53–75.
14. *Colonising,* 6, 8, 13.
15. Otis, "The Rights of the Colonies," 435–436, 444; *Late Occurrences,* 13.
16. "Pre-eminent rights": Elizabeth Mancke, "Early Modern Imperial Governance and the Origins of Canadian Political Culture," *Canadian Journal of Political Science / Revue canadienne de science politique* 32 (1999): 15; Bland, "Colonel Dismounted," 31; Otis, "Rights of the Colonies," 453; [William Goddard?], "The Constitutional Courant: Containing Matters Interesting to Liberty, and No Wise Repugnant to Loyalty" (1765), in Jensen, ed., *Tracts of the American Revolution,* 81; Bland, "Inquiry," 117–118.
17. Otis, "Rights of the Colonies," 435; *General Opposition,* 29; Paul A. Varg, "The Advent of Nationalism, 1758–1776," *American Quarterly* 16 (1964): 180–181, 175–176. See also Richard L. Merritt, *Symbols of American Community, 1735–1775* (New Haven, CT: Yale University Press, 1966), 58–59, 130–131.
18. *Constitutional Right,* 14, 21; *America Vindicated,* 35–43; William Knox, *The Controversy Between Great Britain and her Colonies Reviewed* (London, 1769), 13.
19. *The Political Balance, in which the Principles and Conduct of the Two Parties are Weighed* (London, 1765), 40; Bland, *Inquiry,* 114–116. On the rise of American identity, see Breen and Nash, "American Revolution," 29–33. On the connection between occupation and natural rights claims, see Craig Yirush, *Settlers, Liberty, and Empire: The Roots of Early American Political Theory, 1675–1775*

(New York: Cambridge University Press, 2011). See James D. Drake, *The Nation's Nature: How Continental Presumptions Gave Rise to the United States of America* (Charlottesville: University of Virginia Press, 2011).

20. *Constitutional Right,* 28; *General Opposition,* title page; *Late Occurrences,* 7; *Constitutional Right,* 24, 27, 60.
21. [Maurice Morgann], "Plan for securing the Future Dependence of the Provinces on the Continent of North America," Shelburne Papers, 67: 107, William L. Clements Library, University of Michigan, Ann Arbor; "On American Commerce and Government," Shelburne Papers, 85:26; Thomas C. Barrow, "A Project for Imperial Reform: 'Hints Respecting the Settlement for our American Provinces,' 1763," *William and Mary Quarterly,* 3rd ser., 24 (1967): 114–116, 118, 123; "A Sketch of a plan for the disposal of lands in the newly acquired islands," [1763?], Shelburne Papers, 74: 95–109; [Morgann], "Plan for securing Dependence," 107–110; "Plan of Forts and Garrisons propos'd for the Security of North America," [1763?], CO 323 / 16: 164–185, The National Archives of the UK, Kew, Richmond, Surrey (hereafter cited as UK National Archives).
22. Howard, "Letter from Halifax," 77; *Critical Moment,* 14, 46.
23. "Report on Acquisitions in America," June 8, 1763, in *Documents Relating to the Constitutional History of Canada, 1759–1791,* ed. Adam Shortt and Arthur G. Doughty (Ottawa: S. E. Dawson, 1907), 149; Bailyn, ed., *Pamphlets of the American Revolution,* 524–530; S. Max Edelson, *Plantation Enterprise in Colonial South Carolina* (Cambridge, MA: Harvard University Press, 2006), 192–196; Bland, *Inquiry,* 110, 114–116.
24. William Nelson (1770), quoted in Peter Onuf, *Origins of the Federal Republic: Jurisdictional Controversies in the United States, 1775–1787* (Philadelphia: University of Pennsylvania Press, 1983), 80.
25. Egremont (1763), quoted in Jack M. Sosin, *Whitehall and the Wilderness: The Middle West in British Colonial Policy, 1760–1775* (Lincoln: University of Nebraska Press, 1961), 25.
26. Shelburne to Gage, December 11, 1766, in *Collections of the Illinois State Historical Library,* vol. 11, *The New Régime, 1765–1767,* ed. Clarence Walworth Alvord and Clarence Edwin Carter (Springfield: Illinois State Historical Library, 1916), 457.
27. Thomas Gage to Samuel Holland, July 27, 1767, Gage Papers, Clements Library; St. George L. Sioussat, "The Breakdown of the Royal Management of Lands in the Southern Provinces, 1773–1775," *Agricultural History* 3 (1929): 68–70; Clarence W. Alvord, *The Mississippi Valley in British Politics: A Study of the Trade, Land Speculation and Experiments in Imperialism Culminating in the American Revolution,* 2 vols. (Cleveland: Arthur H. Clark, 1917), 2:212.
28. Earl of Dartmouth to the Governors in America, February 5, 1774, in *Documents Relative to the Colonial History of the State of New-York,* ed. F. B. O'Callaghan, 15 vols. (Albany: Weed, Parsons, 1853), 8:409–413; D. H. Murdoch, "Land Policy in the Eighteenth-Century British Empire: The Sale

of Crown Lands in the Ceded Islands, 1763–1783," *Historical Journal* 27 (1984): 573; Charles L. Mowat, "The Land Policy in British East Florida," *Agricultural History* 14 (1940): 75–77.

29. Sioussat, "Breakdown of Royal Management," 78–81; James Wright (1774), quoted in ibid., 79. This figure was denominated in British pounds sterling.
30. Preston (1775), quoted in Sioussat, "Breakdown of Royal Management," 91; Resolution of the Darien Committee, January 12, 1775, quoted in ibid., 79; ibid., 82–83; Dunmore (1775), quoted in ibid., 92; ibid., 97.
31. Thomas Jefferson, "A Summary View of the Rights of British America" (1774), *Avalon Project,* http://avalon.law.yale.edu/18th_century/jeffsumm.asp.
32. Ibid.; see Yirush, *Settlers, Liberty, and Empire,* 248–252; see also Onuf, *Origins of the Federal Republic,* 81–82.
33. Aziz Rana, *The Two Faces of American Freedom* (Cambridge, MA: Harvard University Press, 2011), 62.
34. "Report on Acquisitions," 103.
35. Ibid., 102–103; Egremont to the Board of Trade, July 14, 1763, in Shortt and Doughty, eds., *Canadian Documents,* 147.
36. Egremont to the Board of Trade, July 16, 1763, CO 323/16: 319–327, UK National Archives; Board of Trade to Egremont, August 5, 1763, in Shortt and Doughty, eds., *Canadian Documents,* 110–111; Halifax to the Board of Trade, September 19, 1763, in Shortt and Doughty, eds., *Canadian Documents,* 112.
37. R. A. Humphreys, "Lord Shelburne and British Colonial Policy, 1766–1768," *English Historical Review* 50 (1935): 263. On North, see Andrew Jackson O'Shaughnessy, *The Men Who Lost America: British Leadership, the American Revolution, and the Fate of Empire* (New Haven, CT: Yale University Press, 2013), chap. 2; R. C. Simmons and P. D. G. Thomas, eds., *Proceedings and Debates of the British Parliaments Respecting North America, 1754–1783* (Millwood, NY: Kraus International, 1982), 446.
38. Shortt and Doughty, eds., *Constitutional Documents,* 339; Hilda Neatby, *The Quebec Act: Protest and Policy* (Scarborough, ON: Prentice-Hall, 1972), 37–38; see Sosin, *Whitehall and the Wilderness,* 241–243.
39. Sian E. Rees, "The Political Career of Wills Hill, Earl of Hillsborough (1718-1793) with Particular Reference to his American Policy" (PhD diss., University of Wales, 1976), 319–320; Philip Lawson, *The Imperial Challenge: Quebec and Britain in the Age of the American Revolution* (Montreal: McGill-Queen's University Press, 1989), 132; Neatby, *Quebec Act,* 37. See also Rees, "Hillsborough," 315–320.
40. Neatby, *Quebec Act,* 23, 39; Wedderburn, quoted in Neatby, *Quebec Act,* 40; Dartmouth (1775), quoted in Neatby, *Quebec Act,* 59.
41. Dunning, quoted in Lawson, *Imperial Challenge,* 128; Ezra Stiles, *The Literary Diary of Ezra Stiles,* vol. 1, *Jan. 1, 1769–Mar. 13, 1766* (New York: Scribner's, 1901), 455.

42. "Journal of the Continental Congress—The Articles of Association; October 20, 1774," *Avalon Project,* http://avalon.law.yale.edu/18th_century/contcong_10-20-74.asp. See also Lawson, *Imperial Challenge,* 141; Mark R. Anderson, *The Battle for the Fourteenth Colony: America's War of Liberation in Canada, 1774–1776* (Hanover, NH: University Press of New England, 2013), 14–15; Livingston, quoted in Rana, *Two Faces of American Freedom,* 74; Elizabeth Mancke, "Spaces of Power in the Early Modern Northeast," in *New England and the Maritime Provinces: Connections and Comparisons,* ed. Stephen J. Hornsby and John G. Reid (Montreal: McGill-Queen's University Press, 2005), 49–50, 73.
43. "The Declaration of Independence: A Transcription," National Archives and Record Administration, http://www.archives.gov/exhibits/charters/declaration_transcript.html; John Shy, *Toward Lexington: The Role of the British Army in the Coming of the American Revolution* (Princeton, NJ: Princeton University Press, 1965), 269–278; Michael N. McConnell, *Army and Empire: British Soldiers on the American Frontier, 1758–1775* (Lincoln: University of Nebraska Press, 2004), 51; Shy, *Toward Lexington,* chap. 5; Stiles, *Literary Diary,* 1:458.
44. "Declaration of Independence"; Jefferson, "Summary View"; Alvord, *Mississippi Valley,* 2:215–216; see also Yirush, *Settlers, Liberty, and Empire,* 250–255; Pauline Maier, *American Scripture: Making the Declaration of Independence* (New York: Vintage, 1998), 140–141.
45. Jefferson, "Summary View"; Meier, *American Scripture,* 138.
46. Gabriel de Dartine to the Commissioners, July 14, 1778, and July 16, 1778, in *The Adams Papers, Digital Edition,* ed. C. James Taylor (Charlottesville: University of Virginia Press, Rotunda, 2008).
47. On the publication history of Bowen's map, see Margaret Beck Pritchard and Henry G. Taliaferro, *Degrees of Latitude: Mapping Colonial America* (New York: Colonial Williamsburg / Harry N. Abrams, 2002), 180–183.
48. Mary Sponberg Pedley, *The Commerce of Cartography: Making and Marketing Maps in Eighteenth-Century France and England* (Chicago: University of Chicago Press, 2005), chap. 5.
49. Richard Brown and Paul E. Cohen, *Revolution: Mapping the Road to American Independence, 1755–1783* (New York: W. W. Norton, 2015), 63–137.
50. Thomas Jefferys, *The West-India atlas; or, A compendious description of the West-Indies* (London, 1775), 20, ii.
51. Ibid., i–ii, 26.
52. Ibid., i–iv, 5.
53. *Critical Moment,* 55–56; Letter of Charles Winstone, December 22, 1777, Charles Winstone "Dominica" Letterbook, 1777–1786, Clements Library.
54. Captain Morse, "Report of the Defences with a state of the ordnance department in Tobago," 1777, CO 101/21, British Public Record Office Photostats, Manuscript Division, Library of Congress, Washington, DC (hereafter cited

as LC Photostats); "Memorial of the Proprietors and Merchants Concerned in the Island of Tobago," [January 1778], CO 101 / 21: 120–121, LC Photostats; [George] Macartney to George Germain, October 24, 1777, CO 101 / 21: 68–69, LC Photostats; David Watts, *The West Indies: Patterns of Development, Culture and Environmental Change since 1492* (New York: Cambridge University Press, 1987), 347.

55. Charles Winstone to David Chollet, October 26, 1778, "Dominica" Letterbook; Winstone to John Rae, October 26, 1778, "Dominica" Letterbook; Winstone to Rae, September 10, 1779, "Dominica" Letterbook; Winstone to [ill.], January 13, 1780, "Dominica" Letterbook; Winstone to Joseph Smith, April 5, 1780, "Dominica" Letterbook; Winstone to Rae, July 25, 1784, "Dominica" Letterbook; Winstone to Samuel Chollet and [ill.], May 7, 1784, "Dominica" Letterbook; Winstone to William Stuart, January 12, 1780, "Dominica" Letterbook; Thomas Atwood, *The History of the Island of Dominica* (London, 1791), 228–229, 72; Winstone to John Gregg, July 25, 1784, "Dominica" Letterbook.

56. *Critical Moment*, 64, 52, 62.

57. Great Britain, *Royal Commission on Historical Manuscripts, The Manuscripts of the Earl of Dartmouth* (London: H. M. Stationery Office, 1895), 2:340; William B. Wilcox, "The Clinton-Parker Controversy over British Failure at Charleston and Rhode Island," in *Sources of American Independence: Selected Manuscripts from the Collections of the William L. Clements Library*, ed. Howard H. Peckham, 2 vols. (Chicago: University of Chicago Press, 1978), 1:189–191; Thomas Anburey, *Travels through the Interior Parts of America*, 2 vols. (Cambridge, MA: Riverside Press, 1923), 2:3.

58. John Adams to William Tudor, August 24, 1770, *Founding Families: Digital Editions of the Papers of the Winthrops and the Adamses*, ed. C. James Taylor (Boston: Massachusetts Historical Society, 2014), http://www.masshist.org/apde2/.

59. John Adams to Abigail Adams, August 13, 1776, *The Adams Papers, Digital*. On Adams's use of British maps during the war, see also John Adams to Joseph Gridley, August 28, 1780, *The Adams Papers, Digital*; Joseph Gridley to John Adams, [August 28, 1780], *The Adams Papers, Digital*. G. R. Crone, "John Green. Notes on a Neglected Eighteenth Century Geographer and Cartographer," *Imago Mundi* 6 (1949): 90–91; John Green, *Remarks, in Support of the New Chart of North and South-America in Six Sheets* (London, 1753), 8. Thomas Jefferys had published the first map on Adams's list, John Green's *A Chart of North and South America*, in 1753, and featured it in his *A General Topography* as well as his *American Atlas*.

60. "Declaration of Independence."

61. The other two Pennsylvania maps were Lewis Evans, *A general map of the middle British Colonies* (Philadelphia, 1755) and Nicholas Scull, *To the Honourable Thomas Penn and Richard Penn, Esqrs., true & absolute proprietaries & Governours of the Province of Pennsylvania & counties of New-Castle, Kent &*

Sussex on Delaware this map of the improved part of the Province of Pennsylvania (Philadelphia, 1759). P. J. Marshall, "A Nation Defined by Empire, 1755–1776," in *Uniting the Kingdom? The Making of British History*, ed. Alexander Grant and Keith J. Stringer (New York: Routledge, 1995), 220; Jack P. Greene, *Creating the British Atlantic: Essays on Transplantation, Adaptation, and Continuity* (Charlottesville: University of Virginia Press, 2013), chap. 2.

62. Stephen Hornsby, *Surveyors of Empire: Samuel Holland, J. F. W. Des Barres, and the Making of the Atlantic Neptune* (Montreal: McGill-Queen's University Press, 2011); "An estimate of the expence attending general surveys of H. M. Dominions in N. America for the year 1776," CO 324/18: 192–93, British Public Record Office Transcripts, Manuscripts Division, Library of Congress.
63. Ruddock F. Mackay, ed., *The Hawke Papers: A Selection 1743–1771* (London: Naval Records Society, 1991), 351.
64. Hornsby, *Surveyors of Empire*, 72. Several charts in the *Atlantic Neptune* did reveal signs of the war, including J. F. W. Des Barres, "Sketch of the Operations of His Majesty's Fleet & Army under the Command of Vice Admiral the Rt. Hbls. Lord Viscount Howe and Genl. Sr. Wm. Howe, K.B., in 1776," 1777, and "A Sketch of the Operations before Charlestown the capitol of South Carolina," 1780, in *The Atlantic Neptune* (London, 1777–[1781]). From Geography and Map Division, Library of Congress, Washington, DC, http://www.loc.gov/resource/g3801s.ar105700.
65. Jack P. Greene, "Transatlantic Colonization and the Redefinition of Empire in the Early Modern Era: The British-American Experience," in *Negotiated Empires: Centers and Peripheries in the Americas, 1500–1820*, ed. Christine Daniels and Michael Kennedy (New York: Routledge, 2002), xx.
66. On the relationship between emotion and political ideology in the American Revolution, see T. H. Breen, *American Insurgents, American Patriots: The Revolution of the People* (New York: Hill and Wang, 2010), especially 11–12.

CONCLUSION

1. Thomas Pownall, *The Administration of the Colonies* (London, 1764), 11–12; "List of maps, plans, &c. belonging to the right Hon. the Lords Commissioners for trade and plantations under the care of Francis Ægidius Assiotti, draughtsman," 1780, GA 195.L7 G7 folio ref., photocopy, Geography and Map Division, Library of Congress, Washington, DC.
2. James Boswell, *Boswell's London Journal 1762–1763* (London: Reprint Society, 1952), 64; William Mylne, *Travels in the Colonies in 1773–1775: Described in the Letters of William Mylne*, ed. Ted Ruddock (Athens: University of Georgia Press, 1993), 6.
3. David Hancock, *Citizens of the World: London Merchants and the Integration of the British Atlantic Community, 1735–1785* (New York: Cambridge University Press, 1995), chaps. 5–6; Bernard Bailyn, *Voyagers to the West: A Passage*

in the Peopling of America on the Eve of the Revolution* (New York: Vintage, 1986), 438; Daniel L. Schafer, "Plantation Development in British East Florida: A Case Study of the Earl of Egmont," *Florida Historical Quarterly* 63 (1984): 172–183; Harry Gordon to George Washington, May 17, 1783, *George Washington Papers at the Library of Congress, 1741–1799,* accessed May 2, 2016, http://memory.loc.gov; George C. Rogers Jr., "The East Florida Society of London, 1766–1767," *Florida Historical Quarterly* 54 (1976): 488; Biography of Thomas Thoronton (1723–1794), University of Nottingham Manuscripts and Special Collections, accessed April 29, 2016, https://www.nottingham.ac.uk/manuscriptsandspecialcollections/collectionsindepth/family/thorotonhildyard/biographies/biographyofthomasthoroton(1723–1794).aspx. Rutland was the father of John Manners, Lord Granby, the commander of the forces during the Seven Years' War and the army's commander in chief from 1766 to 1769.

4. Leland J. Bellot, *William Knox: The Life and Thought of an Eighteenth-Century Imperialist* (Austin: University of Texas Press, 1977), 92–93; Burke, quoted in Arthur H. Bayse, *The Lords Commissioners of Trade and Plantations Commonly Known as the Board of Trade, 1748–1782* (New Haven, CT: Yale University Press, 1925), 199.
5. Edmund Burke, *The Works of the Right Honorable Edmund Burke,* rev. ed., 9 vols. (Boston: Little, Brown, and Company, 1866), 2:340–349. See Dennis Stephen Klinge, "Edmund Burke, Economical Reform, and the Board of Trade, 1777–1780," *Journal of Modern History* on-demand supplement 51 (1979): D1185–D1200; David Bromwich, *The Intellectual Life of Edmund Burke: From the Sublime and Beautiful to American Independence* (Cambridge, MA: Belknap / Harvard University Press, 2014), 358–359.
6. "Mr. Pownal's Sketch of a Report concerning the Cessions in Africa and America at the Peace of 1763," in R. A. Humphreys, "Lord Shelburne and the Proclamation of 1763," *English Historical Review* 49 (1934): 258–264.
7. T. H. Breen, *The Marketplace of Revolution: How Consumer Politics Shaped American Independence* (New York: Oxford University Press, 2004), 37–38, 163–164, 325–331; Henry Laurens (1774), quoted in Breen, *Marketplace of Revolution*, 327; *The Substance of the Evidence on the Petition Presented by the West-India Planters and Merchants, to the Hon. House of Commons* ([New York], [1775]), 33.
8. Locke, quoted in Antonia LoLordo, *Locke's Moral Man* (Oxford: Oxford University Press, 2012), 25; LoLordo, *Locke's Moral Man*, 46; James C. Scott, *Seeing Like a State: How Certain Schemes to Improve the Human Condition Have Failed* (New Haven, CT: Yale University Press, 1998), 345.
9. John Cannon, "Petty, William, second earl of Shelburne and first marquess of Lansdowne (1737–1805)," *Oxford Dictionary of National Biography* (Oxford: Oxford University Press, 2004); Shelburne (1775, 1778), quoted in Cannon,

"Petty, William"; *Cobbett's Parliamentary History of England,* [vol. 20, *1778–1780*] (London, 1814), 39–40.

10. "Report on Instructions on Peace Negotiations, [7 January 1782]," *Founders Online,* http://founders.archives.gov/documents/Madison/01-04-02-0002; Benjamin Franklin to Robert R. Livingston, August 12, 1782, in *The Papers of Benjamin Franklin, Digital Edition,* Packard Humanities Institute, http://franklinpapers.org/franklin/framedVolumes.jsp?vol=37&page=730a; see also James D. Drake, *The Nation's Nature: How Continental Presumptions Gave Rise to the United States of America* (Charlottesville: University of Virginia Press, 2011), 271, 218–221, 225–227.
11. "Franklin: Journal of the Peace Negotiations, May 9, 1782," in *Franklin Papers,* http://franklinpapers.org/franklin/yale?vol=37&page=291a. The map was John Mitchell, *Amérique septentrionale avec les routes, distances en miles, limites et etablissements françois et anglois,* published in Paris in one of several possible editions between 1756 and 1776. "Boundary Discussions between Jay and Aranda, 3–30 August 1782," in *John Jay: The Winning of the Peace, Unpublished Papers 1780–1784,* ed. Richard B. Morris, 2 vols. (New York: Harper & Row, 1980), 2:270–283. For a map of various boundary proposals, see Richard B. Morris, *The Peacemakers: The Great Powers and American Independence* (New York: Harper & Row, 1965), following 350.
12. Matthew H. Edney, "The Mitchell Map: An Irony of Empire," 1997, Osher Map Library, University of Southern Maine, accessed September 1, 2016, http://www.oshermaps.org/special-map-exhibits/mitchell-map; Edney, "The Most Important Map in U.S. History," 2012, Osher Library, accessed September 1, 2016, http://www.oshermaps.org/exhibitions/map-commentaries/most-important-map-us-history; Andrew Hamilton, *Trade and Empire in the Eighteenth-Century Atlantic World* (Newcastle upon Tyne, UK: Cambridge Scholars, 2008), 33–48; James Ashley Morrison, "Before Hegemony: Adam Smith, American Independence, and the Origins of the First Era of Globalization," *International Organization* 66 (2012): 410–420. See also Esmond Wright, "The British Objectives, 1780–1783: 'If Not Dominion Then Trade,'" in *Peace and the Peacemakers: The Treaty of 1783,* ed. Ronald Hoffman and Peter J. Albert (Charlottesville: University of Virginia Press, 1986), 3–29.
13. "Benjamin Franklin to Richard Oswald," September 5, 1782, note 4, *Founders Online,* http://founders.archives.gov/documents/Franklin/01-38-02-0057; "Preliminary Articles of Peace: First Draft Treaty, [5–7 October 1782]," *Founders Online,* http://founders.archives.gov/documents/Franklin/01-38-02-0143; "Preliminary Articles of Peace: Second Draft Treaty, [4–7 November 1782]," in *Founders Online,* http://founders.archives.gov/documents/Franklin/01-38-02-0205. On these boundaries and their discontents after 1783, see J. P. D. Dunbabin, "Red Lines on Maps: The Impact of Cartographical Errors on the Border between the United States and British North

America, 1782–1842," *Imago Mundi* 50 (1999): 105–125; see also Francis M. Carroll, *A Good and Wise Measure: The Search for the Canadian–American Boundary, 1783–1842* (Toronto: University of Toronto Press, 2001), chap. 1; see also Lawrence B. A. Hatter, "Taking Exception to Exceptionalism: Geopolitics and the Founding of an American Empire," *Journal of the Early Republic* 34 (2014): 653–660.

14. "Representation to His Majesty, proposing publick Seals for the Provinces of Quebec, East Florida, & West Florida, and for the Government of Grenada," October 5, 1763, CO 324 / 21: 294–297, British Public Record Office Transcripts, Manuscript Division, Library of Congress, Washington, DC; Virgil, *The Aeneid,* trans. Robert Fitzgerald (New York: Random House, 1983), 190, 12. For an interlinear translation of these passages, see Levi Hart and V. R. Osborn, *The Works of P. Virginlius Maro* (Philadelphia: D. McKay, 1882), 186, 174–175. See also Phiroze Vasunia, "Virgil and the British Empire, 1760–1880," in *Lineages of Empire: The Historical Roots of British Imperial Thought,* ed. Duncan Kelly (Oxford: Oxford University Press, 2009), 83–116.
15. "Representation . . . proposing publick Seals"; Robert R. Rea, "The Deputed Great Seal of British West Florida," *Alabama Historical Quarterly* 40 (1978): 162–168; see also Peter Walne, "The Great Seals of British East Florida," *Florida Historical Quarterly* 61 (1982): 49–53; *Acts of the Privy Council of England, Colonial Series,* vol. 5, *A.D. 1766–1783,* ed. James Munro (London: H. M. Stationery Office, 1912), 15. My thanks to Jon E. Lendon and Elizabeth Meyer for assistance and advice with Latin translation.
16. Peter S. Onuf, *The Origins of the Federal Republic: Jurisdictional Controversies in the United States, 1775–1787* (Philadelphia: University of Pennsylvania Press, 1983); Martin Brückner, *The Geographic Revolution in Early America: Maps, Literacy, & National Identity* (Chapel Hill: University of North Carolina Press, 2006); Peter S. Onuf, *Statehood and Union: A History of the Northwest Ordinance* (Bloomington: Indiana University Press, 1987).
17. Eran Shalev, " 'A Republic Amidst the Stars': Political Astronomy and the Intellectual Origins of the Stars and Stripes," *Journal of the Early Republic* 31 (2011): 39–73; Jay quoted in Drake, *Nature's Nation,* 295.

Map Bibliography

The maps listed below are referenced in the text by number, in brackets.

Chapter 1. A Vision for American Empire

Representing American Spaces, 1660–1762

1. [Newfoundland], [before 1671?], Cabinet Blathwayt 7, Blathwayt Atlas, John Carter Brown Library, Brown University, Providence, RI, JCB Map Collection, http://jcb.lunaimaging.com/luna/servlet/detail/JCBMAPS~1~1~1571~101940004.
2. Abraham Peyrounin, *Carte de lisle de Sainct Chrisophle* (Paris, [ca. 1667]). From John Carter Brown Library, Brown University, Providence, RI, Cabinet Blathwayt 27, Blathwayt Atlas, JCB Map Collection, http://jcb.lunaimaging.com/luna/servlet/detail/JCBMAPS~1~1~1600~102110001.
3. John Seller, *A Chart of the Caribe Islands* (London [ca. 1675]). From John Carter Brown Library, Brown University, Providence, RI, Cabinet Blathwayt 25, Blathwayt Atlas, JCB Map Collection, http://jcb.lunaimaging.com/luna/servlet/detail/JCBMAPS~1~1~1597~102030003.
4. Robert Morden and William Berry, *A New Map of the English Plantations in America, both Continent and Islands* (London, [1673]). From John Carter Brown Library, Brown University, Providence, RI, Cabinet Blathwayt 3, Blathwayt Atlas, JCB Map Collection, http://jcb.lunaimaging.com/luna/servlet/detail/JCBMAPS~1~1~1450~100860002.
5. F[rancis] Lamb, *Jamaicae Descriptio* (n.p., [ca. 1675]). From John Carter Brown Library, Brown University, Providence, RI, Cabinet Blathwayt 33, Blathwayt Atlas, JCB Map Collection, http://jcb.lunaimaging.com/luna/servlet/detail/JCBMAPS~1~1~1606~102040003.
6. Richard Ford, *A New Map of the Island of Barbadoes* ([London], [1675]). From John Carter Brown Library, Brown University, Providence, RI, Cabinet Blathwayt 2, Blathwayt Atlas, JCB Map Collection, http://jcb.lunaimaging.com/luna/servlet/detail/JCBMAPS~1~1~1133~100880001.
7. James Lancaster, [Albemarle Sound, North Carolina], 1679, Cabinet Blathwayt 21, Blathwayt Atlas, John Carter Brown Library, Brown University, Providence, RI, JCB Map Collection, http://jcb.lunaimaging.com/luna/servlet/detail/JCBMAPS~1~1~1583~102030001.
8. Thomas Clarke, "Mapp or Description of Sommer Islands Sometime called Bermudas," 1678, Cabinet Blathwayt 24, Blathwayt Atlas, John Carter Brown Library, Brown University, Providence, RI, JCB Map Collection, http://jcb.lunaimaging.com/luna/servlet/detail/JCBMAPS~1~1~1587~102120002.
9. [New England, Showing Massachusetts's Boundaries], [1678], Cabinet Blathwayt 8, Blathwayt Atlas, John Carter Brown Library, Brown University,

Providence, RI, JCB Map Collection, http://jcb.lunaimaging.com/luna/servlet/detail/JCBMAPS~1~1~1572~101870004.

10. "Mountserrat Island," 1673, Cabinet Blathwayt 30, Blathwayt Atlas, John Carter Brown Library, Brown University, Providence, RI, JCB Map Collection, http://jcb.lunaimaging.com/luna/servlet/detail/JCBMAPS~1~1~1604~102040002.
11. Father Louis Hennepin, "A Map of a Large Country Newly Discovered in the Northern America," in *A New Discovery of a Vast Country in America . . .* (London, 1698). From Beinecke Rare Book and Manuscript Library, Yale University, New Haven, CT, http://brbl-dl.library.yale.edu/vufind/Record/3518265.
12. Guillaume Delisle, *Carte du Canada ou de la Nouvelle France et des decouvertes qui y ont été faites* (Paris, 1703). From Geography and Map Division, Library of Congress, Washington, DC, https://www.loc.gov/resource/g3400.ct003997.
13. "A new map of Louisiana and the River Mississippi," in *Some considerations on the consequences of the French settling colonies on the Mississippi* (London, 1720). From Tracy W. McGregor Library of American History, Small Special Collections, University of Virginia Libraries, Charlottesville, VA.
14. Herman Moll, *A new and exact map of the dominions of the King of Great Britain on ye continent of North America* ([London], 1715). From Geography and Map Division, Library of Congress, Washington, DC, https://www.loc.gov/resource/g3300.ct000232.
15. Henry Popple, (Composite of) *A map of the British Empire in America with the French and Spanish settlements adjacent thereto* (London, 1733). From David Rumsey Historical Map Collection, http://www.davidrumsey.com/luna/servlet/detail/RUMSEY~8~1~887~70081.
16. John Mitchell, *A map of the British and French dominions in North America, with the roads, distances, limits, and extent of the settlements* (London, 1755). From Geography and Map Division, Library of Congress, Washington, DC, https://www.loc.gov/resource/g3300.np000009.

The New Map of Empire

17. Emanuel Bowen, *An accurate map of North America. Describing and distinguishing the British, Spanish and French dominions on this great continent; exhibiting the present seat of war, and the French encroachments* (London, [1755?]). From Geography and Map Division, Library of Congress, Washington, DC, https://www.loc.gov/resource/g3300.ar002001.
18. Emanuel Bowen, *An Accurate Map of North America Describing and distinguishing the British, Spanish and French Dominions on this Great Continent According to the Definitive Treaty Concluded at Paris 10th Feby. 1763* (London,

[ca. 1763]). From The National Archives of the UK, Kew, Richmond, Surrey, MR/1 26, http://discovery.nationalarchives.gov.uk/details/r/C4559053.

Chapter 2. Commanding Space after the Seven Years' War

Wartime Mapmaking

19. George Washington, "Map of the river systems from Fort Presque Isle on Lake Erie to the Potomac River," 1754, MPG 1/118, The National Archives of the UK, Kew, Richmond, Surrey, http://discovery.nationalarchives.gov.uk/details/r/C3980807.
20. William Alexander, "A Sketch of the several Routes of the French from Quebec to the Mississippi," 1755, MPG 1/331, The National Archives of the UK, Kew, Richmond, Surrey, http://discovery.nationalarchives.gov.uk/details/r/C3981020.
21. [Map showing rivers and forts in North America], 1758, Add MS 57711.5, The British Library, from Norman B. Leventhal Map Center at the Boston Public Library, Boston, MA, http://maps.bpl.org/id/n51858.
22. G. Wright, "A Plan of Pitts Fort at Pittsburg Octr. 1759," 1759, MPG 1/588, The National Archives of the UK, Kew, Richmond, Surrey, http://discovery.nationalarchives.gov.uk/details/r/C3981277.
23. Mary Ann Rocque and John Rocque, "A Plan of the New Fort at Pitts-Burgh or Du Quesne," in *A Set of Plans and Forts in America* (London, 1765). From Small Special Collections, University of Virginia Libraries, Charlottesville, VA.
24. "Chart of the East Coast of Cape Breton," [175-?], G3422.C29P5 175- .C5 Howe 7, Richard Howe Collection, Geography and Map Division, Library of Congress, Washington, DC, https://lccn.loc.gov/gm72003563.
25. Francisco Mathias Celi, "Plano de la gran Bahi de Nipe in ya. de Cuba," [174-?], G4922.N53 174- .C4 Howe 44, Richard Howe Collection, Geography and Map Division, Library of Congress, Washington, DC, https://lccn.loc.gov/73691514.
26. Cyprian Southack, "Chart of the Gulf and River St. Lawrence," [1710?], CO 700/Canada 1, The National Archives of the UK, Kew, Richmond, Surrey, http://discovery.nationalarchives.gov.uk/details/r/C3476313.
27. "The River St. Laurence," [175-?], G3312.S5P5 175- .R5 Howe 2, Richard Howe Collection, Geography and Map Division, Library of Congress, Washington, DC, https://lccn.loc.gov/gm72003558.
28. "A Sketch of the River St. Laurence from La Gallette to the Island of Perrot with the Encampments of the Army, 1760," 1760, MPG 1/5, The National Archives of the UK, Kew, Richmond, Surrey, http://discovery.nationalarchives.gov.uk/details/r/C3980694.
29. [John Green], "Chart of North and South America, including Atlantic and Pacific Oceans, with coasts of Africa, Asia & Europe" (orig. 1753), in Thomas

Jefferys, *A General Topography of North America and the West Indies* (London, 1768). From Geography and Map Division, Library of Congress, Washington, DC, https://www.loc.gov/resource/g3300m.gar00003/?sp=8.

30. [Thomas Jefferys and John Green], "Plate 1. The Claims of the French in 1756," part of [John Green], "Chart of the Atlantic Ocean," in Thomas Jefferys, *A General Topography of North America and the West Indies* (London, 1768). From Geography and Map Division, Library of Congress, Washington, DC, https://www.loc.gov/resource/g3300m.gar00003/?sp=21.
31. [Thomas Jefferys and John Green], "Plate 2. The French Dominions and Neutral Lands, as proposed by M. de Buffy in 1761," part of [John Green], "Chart of the Atlantic Ocean," in Thomas Jefferys, *A General Topography of North America and the West Indies* (London, 1768). From Geography and Map Division, Library of Congress, Washington, DC, https://www.loc.gov/resource/g3300m.gar00003/?sp=22.
32. [Thomas Jefferys and John Green], "Plate 3. The Dominions ceded by France and Spain to Great Britain in 1762," part of [John Green], "Chart of the Atlantic Ocean," in Thomas Jefferys, *A General Topography of North America and the West Indies* (London, 1768). From Geography and Map Division, Library of Congress, Washington, DC, https://www.loc.gov/resource/g3300m.gar00003/?sp=23.
33. Thomas Jefferys, "A New Map of Nova Scotia and Cape Britain" (orig. 1755), in Thomas Jefferys, *A General Topography of North America and the West Indies* (London, 1768). From Geography and Map Division, Library of Congress, Washington, DC, https://www.loc.gov/resource/g3300m.gar00003/?sp=33.
34. John Green, "A Map of the Most Inhabited part of New England" (orig. 1755), in Thomas Jefferys, *A General Topography of North America and the West Indies* (London, 1768). From Geography and Map Division, Library of Congress, Washington, DC, https://www.loc.gov/resource/g3300m.gar00003/?sp=38.
35. Lewis Evans, "A General Map of the Middle British Colonies in America" (orig. 1758), in Thomas Jefferys, *A General Topography of North America and the West Indies* (London, 1768). From Geography and Map Division, Library of Congress, Washington, DC, https://www.loc.gov/resource/g3300m.gar00003/?sp=44.
36. Joshua Fry and Peter Jefferson, "A Map of the most Inhabited part of Virginia" (orig. 1755), in Thomas Jefferys, *A General Topography of North America and the West Indies* (London, 1768). From Geography and Map Division, Library of Congress, Washington, DC, https://www.loc.gov/resource/g3300m.gar00003/?sp=66.
37. William De Brahm, "A Map of South Carolina and a part of Georgia" (orig. 1757), in Thomas Jefferys, *A General Topography of North America and the West Indies* (London, 1768). From Geography and Map Division, Library of Congress, Washington, DC, https://www.loc.gov/resource/g3300m.gar00003/?sp=71.

38. Thomas Jefferys, "Florida, from the latest Authorities," in Thomas Jefferys, *A General Topography of North America and the West Indies* (London, 1768). From Geography and Map Division, Library of Congress, Washington, DC, https://www.loc.gov/resource/g3300m.gar00003/?sp=75.
39. Robert Stobo, [Plan of Fort Du Quesne], 1754, DAR.1925.05, Darlington Collection, Special Collections Department, University of Pittsburgh, PA, http://digital.library.pitt.edu/u/ulsmanuscripts/pdf/31735060225889.pdf.
40. [Robert Stobo], "A Plan of Fort De Quesne," 1754, Maps K.Top. 122.16, The British Library, from Norman B. Leventhal Map Center at the Boston Public Library, Boston, MA, http://maps.bpl.org/id/n51560.
41. [Robert Stobo], "Plan of Fort Le Quesne, Built by the French, At the Fork of the Ohio and Monongahela in 1754," in Thomas Jefferys, *A General Topography of North America and the West Indies* (London, 1768). From Geography and Map Division, Library of Congress, Washington, DC, https://www.loc.gov/resource/g3300m.gar00003/?sp=57.
42. Robert Orme, "Six Plans of the different Dispositions of the English Army under the Command of the late General Braddock in North America," in Thomas Jefferys, *A General Topography of North America and the West Indies* (London, 1768). From Geography and Map Division, Library of Congress, Washington, DC, https://www.loc.gov/resource/g3300m.gar00003/?sp=58, https://www.loc.gov/resource/g3300m.gar00003/?sp=59, https://www.loc.gov/resource/g3300m.gar00003/?sp=60, https://www.loc.gov/resource/g3300m.gar00003/?sp=61, https://www.loc.gov/resource/g3300m.gar00003/?sp=62, https://www.loc.gov/resource/g3300m.gar00003/?sp=63.
43. Thomas Jefferys, "An Exact Chart of the River St. Laurence, from Fort Frontenac to the Island of Anticosti, copied from the French," in Thomas Jefferys, *A General Topography of North America and the West Indies* (London, 1768). From Geography and Map Division, Library of Congress, Washington, DC, https://www.loc.gov/resource/g3300m.gar00003/?sp=25.
44. "A Map of the Several Dispositions of the English Fleet & Army on the River St. Laurence, to the taking of Quebec," in Thomas Jefferys, *A General Topography of North America and the West Indies* (London, 1768). From Geography and Map Division, Library of Congress, Washington, DC, https://www.loc.gov/resource/g3300m.gar00003/?sp=31.
45. "A Correct Plan of the Environs of Quebec, and of the Battle fought on the 13th September, 1759," in Thomas Jefferys, *A General Topography of North America and the West Indies* (London, 1768). From Geography and Map Division, Library of Congress, Washington, DC, https://www.loc.gov/resource/g3300m.gar00003/?sp=29; see also, "Second Plate," https://www.loc.gov/resource/g3300m.gar00003/?sp=30.
46. "A Plan of Quebec, The Capital of New France or Canada" (orig. 1758), in Thomas Jefferys, *A General Topography of North America and the West Indies*

(London, 1768). From Geography and Map Division, Library of Congress, Washington, DC, https://www.loc.gov/resource/g3300m.gar00003/?sp=28.

47. Thomas Jefferys, "A New Chart of the West Indies, drawn from the best Spanish Maps," in Thomas Jefferys, *A General Topography of North America and the West Indies* (London, 1768). From Geography and Map Division, Library of Congress, Washington, DC, https://www.loc.gov/resource/g3300m.gar00003/?sp=83.
48. "A Map of the Isle of Cuba, with the Bahama Islands, Gulf of Florida, and Windward Passage: Drawn from English and Spanish Surveys," in Thomas Jefferys, *A General Topography of North America and the West Indies* (London, 1768). From Geography and Map Division, Library of Congress, Washington, DC, https://www.loc.gov/resource/g3300m.gar00003/?sp=91.
49. [Eleven charts of Cuban port towns and harbors], in Thomas Jefferys, *A General Topography of North America and the West Indies* (London, 1768). From Geography and Map Division, Library of Congress, Washington, DC, https://www.loc.gov/resource/g3300m.gar00003/?sp=92, https://www.loc.gov/resource/g3300m.gar00003/?sp=93, https://www.loc.gov/resource/g3300m.gar00003/?sp=94, https://www.loc.gov/resource/g3300m.gar00003/?sp=95, https://www.loc.gov/resource/g3300m.gar00003/?sp=96, https://www.loc.gov/resource/g3300m.gar00003/?sp=97.

Mapping Conquered Lands

50. Archibald Campbell, "Sketch of the coast round the island of Dominique," [1761], G5100 1761 .C3 Vault, Geography and Map Division, Library of Congress, Washington, DC, https://lccn.loc.gov/74695907.
51. [Richard Tyrell], "A Plan of the Island of Dominica," [1764?], MFQ 1 / 1173 / 1, The National Archives of the UK, Kew, Richmond, Surrey, http://discovery.nationalarchives.gov.uk/details/r/C8955582.
52. [Area around Clunie, in Perthshire], [n.d.], Roy Map 17 / 3c, Roy Military Survey of Scotland, 1747–55, The British Library, London.
53. James Young, "A Survey of the Carenage and Lagoon with the Heights and Town of St. Georges Grenada," [n.d.], ADM 352 / 243, The National Archives of the UK, Kew, Richmond, Surrey, http://discovery.nationalarchives.gov.uk/details/r/C11490287.
54. "Parish of Long Point. Parish of Longeuil," 1761, sheet 10, Atl 1761 Mu, Murray Atlas of Canada, William L. Clements Library, University of Michigan, Ann Arbor, http://mirlyn.lib.umich.edu/Record/006960071.
55. "Plan of the Settled Part of the Province of Quebec Reduced from the large Surveys to Serve as an Index Plan To the Larger Sheets," [1761], sheet 1, NMC135035, local class no. A / 300 / [1761], Library and Archives of Canada, Ottawa, ON, http://collectionscanada.gc.ca/pam_archives/index.php?fuseaction=genitem.displayItem&lang=eng&rec_nbr=4134077.

56. Samuel Lewis, "A map of the river St. Lawrence, reduced from the actual surveys of Samuel Holland Esq., surveyor-general of the northern district of America, and assistants; surveyed agreeable to the orders and instructions of the Lords Commissioners for Trade and Plantations," 1773, Maps K.Top. 119.28, The British Library, London, http://primocat.bl.uk/F/?func=direct&local_base=PRIMO&doc_number=004987830.

Settling Inward in West Florida

57. Daniel Paterson, "Cantonment of His Majesty's forces in N. America according to the disposition now made & to be compleated as soon as practicable taken from the general distribution dated at New York 29th. March 1766," [1767], G3301.R2 1767 .P3, Geography and Map Division, Library of Congress, Washington, DC, https://www.loc.gov/resource/g3301r.ar011800.
58. "Plan of Fort Rosalia or the Natches, on the River Missisippi," [1765?], CO 700 / Florida 26, The National Archives of the UK, Kew, Richmond, Surrey, http://discovery.nationalarchives.gov.uk/details/r/C3477634.
59. Jean Baptiste D'Anville, *Carte de la Louisiane Par le Sr. d'Anville. Dressee en Mai 1732.* (Paris, 1752). From David Rumsey Historical Map Collection, http://www.davidrumsey.com/luna/servlet/detail/RUMSEY~8~1~3038~410037.
60. James Cook, "A Draught of West Florida from Cape Blaze to Ibberville," [1765], Vv3, UK Admiralty Library, Portsmouth.
61. [Philip Pittman], "Sketch Map of the River Iberville," [1765], CO 700 / Florida 16, The National Archives of the UK, Kew, Richmond, Surrey, http://discovery.nationalarchives.gov.uk/details/r/C3477625.
62. W[illiam] Brasier, "Plan of Point Ibberville with the situation of the Fort," 1767, MPG 1 / 17, The National Archives of the UK, Kew, Richmond, Surrey, http://discovery.nationalarchives.gov.uk/details/r/C3980706.
63. "Map of part of Florida West, from the Bay Pascagoula to the River Amit beyond Lake Maurepas, with reference to the granted Lands on the River Amit, etc.," [1770?], CO 700 / Florida 41, The National Archives of the UK, Kew, Richmond, Surrey, http://discovery.nationalarchives.gov.uk/details/r/C3477649.
64. Elias Durnford, "Sketch Map of part of the Rivers Iberville, Amit and Comit," [1771?], CO 700 / Florida 49, The National Archives of the UK, Kew, Richmond, Surrey, http://discovery.nationalarchives.gov.uk/details/r/C3477657.
65. Elias Durnford, "Map of the Mississip[p]i River, from the River Iberville to the North boundary line of the Province of West Florida on the River Yazous," [1770?], CO 700 / Florida 38, The National Archives of the UK, Kew, Richmond, Surrey, http://discovery.nationalarchives.gov.uk/details/r/C3477646.

66. William Brasier and Philip Pittman, "Plan of Mobile," [1763], Maps K.Top. 122.94.1, The British Library, from Norman B. Leventhal Map Center at the Boston Public Library, Boston, MA, http://maps.bpl.org/id/n51597.
67. Ph[ilip] Pittman, "Plan of the Fort at Mobile," [1770?], CO 700 / Florida 42 / 1, The National Archives of the UK, Kew, Richmond, Surrey, http://discovery .nationalarchives.gov.uk/details/r/C3477650.
68. Ph[ilip] Pittman, (Plate for) "Plan of the Fort at Mobile," [1770?], CO 700 / Florida 42 / 2, The National Archives of the UK, Kew, Richmond, Surrey, http://discovery.nationalarchives.gov.uk/details/r/C3477650.
69. "Plan of Fort Tombeckbee," [1765?], CO 700 / Florida 31, The National Archives of the UK, Kew, Richmond, Surrey, http://discovery.nationalarchives .gov.uk/details/r/C3477639.
70. "A Draught of the River Mobile," [1765?], CO 700 / Florida 28, The National Archives of the UK, Kew, Richmond, Surrey, http://discovery.nationalarchives .gov.uk/details/r/C3477636.
71. Elias Durnford, "Field Survey of the River Mobile and part of the Rivers Alabama and Tensa, with the different Settlements and Lands marked thereon. The Old Settlements made by the French are marked with black, those granted by the English are marked red," [1770?], CO 700 / Florida 40, The National Archives of the UK, Kew, Richmond, Surrey, http://discovery .nationalarchives.gov.uk/details/r/C3477648.
72. Joseph Purcell, "Survey of the Bay and River Mobile," [1775], CO 700 / Florida 51, The National Archives of the UK, Kew, Richmond, Surrey, http://discovery .nationalarchives.gov.uk/details/r/C3477659.
73. "Plan and Sections of the Fort at Pensacola," 1764, MPG 1 / 518, The National Archives of the UK, Kew, Richmond, Surrey, http://discovery.nationalarchives .gov.uk/details/r/C3981217.
74. Elias Durnford, "Plan of the New Town of Pensacola and country adjacent," [1765?], CO 700 / Florida 20 / 1, The National Archives of the UK, Kew, Richmond, Surrey, http://discovery.nationalarchives.gov.uk/details/r/C6744131.
75. Joseph Purcell, "A plan of Pensacola and its environs in its present state," [1778], G3934.P4 1778 .P8, Geography and Map Division, Library of Congress, Washington, DC, https://www.loc.gov/resource/g3934p.ar166100.
76. George Gauld, "A plan of the harbour of Pensacola in West-Florida," 1764, G3932.P45 1764 .G3, Geography and Map Division, Library of Congress, Washington, DC, https://www.loc.gov/resource/g3932p.ar165600.
77. George Gauld, "A Survey of the Bay of Espiritu Santo," 1765, x64 Jv, Archive of the UK Hydrographic Office, Taunton.
78. George Gauld, "A Survey of the Coast of West Florida from Pensacola to Cape Blaise: including the Bays of Pensacola, Santa Rosa, St. Andrew, and St. Joseph, with the Shoals lying off Cape Blaise," 1766, A9464 31c, Archive of the UK Hydrographic Office, Taunton.

79. W[illiam] Brasier, "Plan of the Fort at Appalache called Fort St. Mark, with Projects for its reparation and defence," [1765?], CO 700 / Florida 18, The National Archives of the UK, Kew, Richmond, Surrey, http://discovery.nationalarchives.gov.uk/details/r/C3477627.
80. George Gauld and Phillip Pittman, "A sketch of the entrance from the sea to Apalachy and part of the environs," [1767], Maps 5-J-1, William L. Clements Library, University of Michigan, Ann Arbor, http://mirlyn.lib.umich.edu/Record/004673946.
81. David Taitt, "A plan of part of the rivers Tombecbe, Alabama, Tensa, Perdido, & Scambia in the province of West Florida; with a sketch of the boundary between the nation of upper Creek Indians and that part of the province which is contiguous thereto, as settled at the congresses at Pensacola in the years 1765 & 1771," [1772?], G3971.P53 1771 .T3, Geography and Map Division, Library of Congress, Washington, DC, http://www.loc.gov/resource/g3971p.ar164900.
82. Samuel Lewis, "Plan of the Rivers Mississippi, Iberville, and Bay of Pensacola; in the Province of West Florida," 1772, Maps K.Top. 122.90, The British Library, London, http://primocat.bl.uk/F?func=direct&local_base=ITEMV&doc_number=004987553.
83. Samuel Lewis, "A New Map of West Florida, Georgia, and South Carolina, with part of Louisiana, laid down from different surveys by order of John Stuart, Esq.," [1774], Maps K.Top. 122.89, The British Library, London, http://primocat.bl.uk/F?func=direct&local_base=ITEMV&doc_number=004987550.
84. Joseph Purcell, "Map of the Boundary Line of the Lands ceded to His Majesty by the Chactaw Indians, from the north boundary of West Florida on the Yazo River to the River Pasca Ocoola," 1779, CO 700 / Florida 56, The National Archives of the UK, Kew, Richmond, Surrey, http://discovery.nationalarchives.gov.uk/details/r/C3477664.

Chapter 3. Securing the Maritime Northeast

85. Samuel Holland, "Map in two sheets of the inhabited parts of Canada, with Chart of the River St. Lawrence from the Isle of Orleans upwards," 1760, CO 700 / Canada 18, The National Archives of the UK, Kew, Richmond, Surrey, http://discovery.nationalarchives.gov.uk/details/r/C3476330.

Lessons from Canso

86. Cyprian Southack, *The Harbour and Islands of Canso, part of the Boundaries of Nova Scotia* (Boston, 1720). From The National Archives of the UK, Kew, Richmond, Surrey, CO 700 / Nova Scotia 6, http://discovery.nationalarchives.gov.uk/details/r/C3477867.

87. Cyprian Southack, [Charts of the New England Coast], in *The New England coasting pilot from Sandy Point of New York, unto Cape Canso in Nova Scotia and part of Island Breton* ([London, 1734?]). From Geography and Map Division, Library of Congress, Washington, DC, https://www.loc.gov/resource/g3301pm.gct00083/?sp=5, https://www.loc.gov/resource/g3301pm.gct00083/?sp=6, https://www.loc.gov/resource/g3301pm.gct00083/?sp=7, https://www.loc.gov/resource/g3301pm.gct00083/?sp=8, https://www.loc.gov/resource/g3301pm.gct00083/?sp=9, https://www.loc.gov/resource/g3301pm.gct00083/?sp=10, https://www.loc.gov/resource/g3301pm.gct00083/?sp=11.
88. "Nova Scotia. Draught of Pemaquid, with the Land cleared by Col. Dunbar: recd. with Captn. Well's Lr. of Septr 10th. 1730," 1730, MPG 1/181, The National Archives of the UK, Kew, Richmond, Surrey, http://discovery.nationalarchives.gov.uk/details/r/C3980870.
89. William Bollan, "A Map of North America as far as relates to the English Settlements. taken from the Sieur Bellin 1746," in William Bollan, *The Importance and Advantage of Cape Breton . . .* (London, [1746]). From Small Special Collections, University of Virginia Libraries, Charlottesville, VA.
90. William Bollan, "A Map of the Island of Cape Breton as laid down by the Sieur Bellin 1746," in William Bollan, *The Importance and Advantage of Cape Breton . . .* (London, [1746]). From Small Special Collections, University of Virginia Libraries, Charlottesville, VA.
91. John Henry Bastide and Philip Durell, [An outline plan of the harbour and fortifications of Louisbourg], 1745, Maps K.Top. 119.88.1, The British Library, from Norman B. Leventhal Map Center at the Boston Public Library, Boston, MA, http://maps.bpl.org/id/n51614.
92. John Henry Bastide and Phi[lip] Durell, "A Plan of the Harbour and Fortifications of Louisbourg," 1745, Maps K.Top. 119.90, The British Library, from Norman B. Leventhal Map Center at the Boston Public Library, Boston, MA, http://maps.bpl.org/id/n51616.
93. John Brewse, "Project for Fortifying the Town of Hallifax; in Nova Scotia," 1749, Maps K.Top. 119.77, The British Library, from Norman B. Leventhal Map Center at the Boston Public Library, Boston, MA, http://maps.bpl.org/id/n51610.
94. Thomas Jefferys, *A new map of Nova Scotia and Cape Britain, with the adjacent parts of New England and Canada* (London, 1755). From David Rumsey Historical Map Collection, http://www.davidrumsey.com/luna/servlet/detail/RUMSEY~8~1~29569~1141032.
95. T[homas] Pownall, "Chart [which] represents . . . the several English Colonies & the British Territories up to the River St. Laurence & the Great Lakes," 1755, LO 740, Loudoun Papers, Huntington Library, San Marino, CA.
96. Winckworth Tonge, "A Plan Of the River of Chibenaccadie from its Source To its Discharge into the Bay of Mines Surveyed in August 1754," 1754, K.Top.

119.61.a, The British Library, from Norman B. Leventhal Map Center at the Boston Public Library, Boston, MA, http://maps.bpl.org/id/n51608.

97. Charles Morris, "A Chart of the Sea Coasts of the Peninsula of Nova Scotia," 1755, Maps K.Top. 119.58, The British Library, from Norman B. Leventhal Map Center at the Boston Public Library, Boston, MA, http://maps.bpl.org/id/n51603.
98. Charles Morris, "Plan of Part of the Province of Nova Scotia or Accadie. 1765," 1765, CO 700 / Nova Scotia 40, The National Archives of the UK, Kew, Richmond, Surrey, http://discovery.nationalarchives.gov.uk/details/r/C3477901.

Discovering the Gulf of St. Lawrence

99. James Cook, "A Plan of the Islands of St. Peters, Langly and Miquelon / Survey'd by Order of His Excellency Thos. Graves Esqr. Governor of Newfoundland," [1763], Maps K.Top. 119.111, The British Library, London, http://primocat.bl.uk/F/?func=direct&local_base=PRIMO&doc_number=004987935.
100. James Cook, "A Sketch of the Island of Newfoundland: Done from the latest observations," 1763, MS 368, MS Charts and Maps of America, vol. 1, UK Admiralty Library, Portsmouth.
101. James Cook, "A Plan of York Harbour on the Coast of Labradore Surveyed by Order of His Excellency Thos. Graves Esqr. Governor of Newfoundland &c. &c. &c.," 1763, ADM 352 / 60, The National Archives of the UK, Kew, Richmond, Surrey, http://discovery.nationalarchives.gov.uk/details/r/C11446559.
102. James Cook, *A chart of part of the south coast of Newfoundland: including the islands St. Peters and Miquelon* (London, 1766). From Beinecke Rare Book & Manuscript Library, Yale University, New Haven, CT, http://brbl-dl.library.yale.edu/vufind/Record/3546546.
103. James Cook, *A chart of the Straights of Bellisle: with part of the coast of Newfoundland and Labradore from actual surveys* (London, 1766). From Beinecke Rare Book & Manuscript Library, Yale University, New Haven, CT, http://brbl-dl.library.yale.edu/vufind/Record/3442423.
104. James Cook, *A chart of the west coast of Newfoundland* (London, 1794; orig. 1768). From Memorial University Libraries, St. John's, Newfoundland, http://collections.mun.ca/cdm/ref/collection/maps/id/841.
105. [H. Coates], "A Sketch of the Island of St Johns in the Gulf of St Lawrence," 1764, CO 700 / Prince Edward Island 2, The National Archives of the UK, Kew, Richmond, Surrey, http://discovery.nationalarchives.gov.uk/details/r/C3478003.
106. [Henry] Mowatt, "Richmond Bay," [ca. 1765], MS 368, MS Charts and Maps of America, vol. 1, UK Admiralty Library, Portsmouth.
107. [Henry] Mowatt, "Bay of Hillsborough," [ca. 1765], MS 368, MS Charts and Maps of America, vol. 1, UK Admiralty Library, Portsmouth.

108. [Henry] Mowatt, "Cardigan Bay," [ca. 1765], MS 368, MS Charts and Maps of America, vol. 1, UK Admiralty Library, Portsmouth.
109. Thomas Wright, [St. John Island survey sketches], 1765, Thomas Wright Field Book, Public Archives and Record Office of Prince Edward Island [Canada], Charlottetown, PE.
110. [Samuel] Holland, "Plan of the Island of St John in the Province of Nova Scotia, as surveyed agreeable to the order and instructions of The Lords Commissioners of Trade and Plantations," 1765, CO 700 / Prince Edward Island 3, The National Archives of the UK, Kew, Richmond, Surrey, http://discovery.nationalarchives.gov.uk/details/r/C3478004.
111. [Samuel Holland] and John Lewis, "A Plan of the Island of St. John in the Province of Nova Scotia," 1765, MR 1 / 1785, The National Archives of the UK, Kew, Richmond, Surrey, http://discovery.nationalarchives.gov.uk/details/r/C4560812.
112. Samuel Holland, "A Map of the Island of St. John in the Gulph of S. Lawrence in North America from an actual Survey made in 1763 By Order of The Right Honourable The Lords Commissioners for Trade and Plantations," 1767, Maps K.Top. 119.96.2, The British Library, London, http://primocat.bl.uk/F?func=direct&local_base=ITEMV&doc_number=004987877.
113. [Egmont Bay], [ca. 1767?], B5300 Uk, Archive of the UK Hydrographic Office, Taunton.
114. Samuel Holland, *A plan of the island of St. John with the divisions of the counties, parishes, & the lots as granted by government, likewise the soundings round the coast and harbours* (London, [1775]). From Geography and Map Division, Library of Congress, Washington, DC, https://www.loc.gov/resource/g3426f.ct002331.

Envisioning Order on Remote Coasts

115. John Montrésor, *Map of Nova Scotia or Acadia, with the Islands of Cape Breton and St. John's* (London, 1768). From Geography and Map Division, Library of Congress, Washington, DC, https://www.loc.gov/resource/g3420.ar302300.
116. "Isle Royalle," 1740, Maps K.Top. 119.85, The British Library, from Norman B. Leventhal Map Center at the Boston Public Library, Boston, MA, http://maps.bpl.org/id/n51611.
117. Samuel Holland, "A Plan of the Island of Cape Britain reduced from the large Survey made according to the Orders and Instructions of the Right Honorable the Lords Commissioners for Trade and Plantations," 1767, Add MS 55701, The British Library, from Norman B. Leventhal Map Center at the Boston Public Library, Boston, MA, http://maps.bpl.org/id/n51791.
118. John Pringle, "A Plan of Chaleur or Sterling Bay, with the entrance of the River Rustigusche, Shipegan and Miscou Islands," [1765], CO 700 / Canada

25, The National Archives of the UK, Kew, Richmond, Surrey, http://discovery.nationalarchives.gov.uk/details/r/C3476337.

119. John Collins, "A Plan of Grand River in the Bay of Chaleur in the Province of Quebec," 1765, CO 700 / Canada 21, The National Archives of the UK, Kew, Richmond, Surrey, http://discovery.nationalarchives.gov.uk/details/r/C3476333.
120. John Collins, "A Plan of Port Daniel in the Bay of Chaleur in the Province of Quebec," 1765, CO 700 / Canada 22, The National Archives of the UK, Kew, Richmond, Surrey, http://discovery.nationalarchives.gov.uk/details/r/C3476334.
121. John Collins, "A Plan of Paspebiac in the Bay of Chaleurs in the Province of Quebec," 1765, CO 700 / Canada 23, The National Archives of the UK, Kew, Richmond, Surrey, http://discovery.nationalarchives.gov.uk/details/r/C3476335.
122. John Collins, "A Plan of Bonaventur in the Bay of Chaleurs in the Province of Quebec," 1765, CO 700 / Canada 24, The National Archives of the UK, Kew, Richmond, Surrey, http://discovery.nationalarchives.gov.uk/details/r/C3476336.
123. Thomas Wright, "A Plan of the Island of Anticosti in the Gulph of St. Lawrence Surveyed under the Directions of Samuel Holland Esqr. His Majesty's Surveyor General of Lands for the Northern District of North America," 1767, Add MS 57704.1, The British Library, from Norman B. Leventhal Map Center at the Boston Public Library, Boston, MA, http://maps.bpl.org/id/n51798.
124. [Peter] Frederick Haldimand, "A Plan of the Magdalen, Brion, Bird, Entry, and Deadmans Islands in the Gulph of St. Lawrence. Surveyed . . . under the directions of Captain Holland, Surveyor General of the Northern District of America," [ca. 1765], CO 700 / Canada 27, The National Archives of the UK, Kew, Richmond, Surrey, http://discovery.nationalarchives.gov.uk/details/r/C3476339.
125. Samuel Holland, "A Sketch of the Country Between New Hampshire and Nova Scotia," 1770, MPG 1 / 346, The National Archives of the UK, Kew, Richmond, Surrey, http://discovery.nationalarchives.gov.uk/details/r/C3981035.
126. James Grant and Thomas Wheeler, "A Plan of the Sea Coast from Little Rocks near Hampton to Normans Woe near Cape Ann, including Cape Ann, Ipswich, Newbury and Hampton Harbours," [1771 or 1772], CO 700 / Massachusetts 13, The National Archives of the UK, Kew, Richmond, Surrey, http://discovery.nationalarchives.gov.uk/details/r/C3477723.
127. James Grant and Thomas Wheeler, "A Plan of Piscataqua Harbour," [1771 or 1772], CO 700 / New Hampshire 6, The National Archives of the UK, Kew, Richmond, Surrey, http://discovery.nationalarchives.gov.uk/details/r/C3477739.
128. James Grant, "A Plan of the Sea Coast from Ogunkett River to Cape Elizabeth, including the Bays of Wells, Saco and Black Point," 1771, CO 700 / Maine

15, The National Archives of the UK, Kew, Richmond, Surrey, http://discovery.nationalarchives.gov.uk/details/r/C3477701.

129. George Sproule, "A Plan of the Sea Coast from Cape Elizabeth, to the entrance of Sagadahock, or Kennebeck River, including Casco Bay with all its islands, harbors, &c," 1772, Maps K.Top. 120.19, The British Library, from Norman B. Leventhal Center at the Boston Public Library, Boston, MA, http://maps.bpl.org/id/n51480.

130. George Sproule, "A Plan of the Coast from Kennebeck River to Round Pond on the West Side of Muscongus Bay," 1772, Maps K.Top. 120.20, The British Library, London, http://primocat.bl.uk/F?func=direct&local_base=ITEMV&doc_number=004987819.

131. Charles Blaskowitz and James Grant, "A drawn Plan of the Coast from Pleasant River to the west end of Penobscot Bay," 1772, Maps K.Top. 120.21, The British Library, London, http://primocat.bl.uk/F?func=direct&local_base=ITEMV&doc_number=004987816.

132. Thomas Wright, "A Plan of the Coast from the West Passage of Passamiquodi Bay To the River St. John in the Bay of Fundy," 1772, Maps K.Top. 119.50, The British Library, London, http://primocat.bl.uk/F?func=direct&local_base=ITEMV&doc_number=004987879.

133. Samuel Holland, "A Plan of the Sea Coast from Cape Elizabeth on the West Side of Casco Bay to the St. [J]ohns River in the Bay of Fundy," 1772, Maps K. Top. 120.18, The British Library, London, http://primocat.bl.uk/F?func=direct&local_base=ITEMV&doc_number=004987817.

Chapter 4. Marking the Indian Boundary

The Imagined Line, 1721–1763

134. (Copy of) [Map of the several nations of Indians to the Northwest of South Carolina], or [Catawba Deerskin Map] (1929; orig. [1724?]), G3860 1724 .M2 1929, Geography and Map Division, Library of Congress, Washington, DC, https://www.loc.gov/resource/g3860.ct000734.

135. Lewis Evans, *A general map of the middle British colonies, in America* ([Philadelphia], 1755). From Geography and Map Division, Library of Congress, Washington, DC, https://www.loc.gov/resource/g3710.ar070900.

136. [Charles Thompson], "A Map of the Province of Pen[n]sylvania," in *An enquiry into the causes of the alienation of the Delaware and Shawanese Indians from the British interest* (London, [1759]). From Tracy W. McGregor Library of American History, Small Special Collections, University of Virginia Libraries, Charlottesville, VA.

137. J[ohn] Gibson, "The British Governments in Nth. America Laid down agreeable to the Proclamation of Octr. 7. 1763," *Gentleman's Magazine* (October 1763). From Florida Map Collection, University of South Florida Libraries, Digital Collections, http://digital.lib.usf.edu/SFS0024542/00001.

138. "A new map of North America: shewing the advantages obtain'd therein to England, by the peace," *Royal Magazine* (May 1763). From Beinecke Rare Book and Manuscript Library, Yale University, New Haven, CT.
139. Thomas Kitchin, *A new map of the British Dominions in North America; with the limits of the governments annexed thereto by the late Treaty of Peace, and settled by Proclamation, October 7th 1763* ([London], [1763]). From Geography and Map Division, Library of Congress, Washington, DC, http://hdl.loc.gov/loc.gmd/g3300.ar010300.

The Negotiated Line, 1763–1768

140. John Stuart, "A Map of the Cherokee Country," [ca. 1761], Add MS 14,036.e, The British Library, London, http://primocat.bl.uk/F?func=direct&local_base=ITEMV&doc_number=004987621.
141. Samuel Wyly, "A Map of the Catawba Indians Land surveyed agreeable to an Agreement made with them by His Majesty's Governors of South Carolina, North Carolina, Georgia and Virginia, and Superintendent of Indian Affairs, at a Congress lately held at Augusta by His Majesty's special command," 1764, CO 700/Carolina 24, The National Archives of the UK, Kew, Richmond, Surrey, http://discovery.nationalarchives.gov.uk/details/r/C3477600.
142. John Stuart, "A Map of the Southern Indian District 1764," 1764, Add MS 14,036.d, The British Library, London, http://primocat.bl.uk/F?func=direct&local_base=ITEMV&doc_number=004987549.
143. William de Brahm, "A Map of the inhabited Part of Georgia laid down to Shew the Latitudes & Longitudes, of the Places, that are proposed to be Fortified, in order to Judge of there [sic] communications by William de Brahm," [1755], CO 700/Georgia 12, The National Archives of the UK, Kew, Richmond, Surrey, http://discovery.nationalarchives.gov.uk/details/r/C3477679.
144. Henry Yonge and William De Brahm, "A Map of the sea coast of Georgia & the inland parts thereof extending to the westward of that part of Savannah called broad River including the Several Inlets, Rivers, Islands, Sounds, Creeks, Rivulets, Towns, Roads, Forts & most remarcable places therein: performed at the request of His Exclly. James Wright," 1763, Maps 6-E-9, William L. Clements Library, University of Michigan, Ann Arbor, http://mirlyn.lib.umich.edu/Record/005217242.
145. Bernard Romans and Samuel Savory, "Sketch of the Boundary Line between Georgia and the Creek Indian Nation," 1769, MPG 1/337, The National Archives of the UK, Kew, Richmond, Surrey, http://discovery.nationalarchives.gov.uk/details/r/C3981026.
146. John Pickens, "Boundary Line Between the Province of South Carolina and the Cherokee Indian Country, Marked Out In Presence of the Head Men of the Upper, Middle and Lower Cherokee Towns, Whose Hands and Seals are Affixed," 1766, CO 700/Carolina 26, The National Archives of the UK,

Kew, Richmond, Surrey, http://discovery.nationalarchives.gov.uk/details/r/C3477602.

147. "Map showing boundary lines between the British colonies and the country of the Six Nations and the Southern Indians, as recommended in the Representation of 7 March 1768 from the Lords Commissioners for Trade and Plantations to the King," 1768, MPG 1/280, The National Archives of the UK, Kew, Richmond, Surrey, http://discovery.nationalarchives.gov.uk/details/r/C3980969.

The Eroding Line, 1768–1774

148. [Simon Metcalf], "A Sketch of part of the Province of New York, exhibiting the boundary-line established at the Treaty at Ft. Stanwix 5 Nov 1768," 1768, CO 700/New York 34, The National Archives of the UK, Kew, Richmond, Surrey, http://discovery.nationalarchives.gov.uk/details/r/C3477795.

149. Guy Johnson, "Map of the Frontiers of the Northern Colonies with the Boundary Line established between them and the Indians at the treaty held by Sr. Will. Johnson at Ft. Stanwix in Novr. 1768. Corrected and Improved from Evans Map," 1768, MPG 1/197, The National Archives of the UK, Kew, Richmond, Surrey, http://discovery.nationalarchives.gov.uk/details/r/C3980886.

150. [Philip Yonge], "A Map of the Lands Ceded to His Majesty by the Creek and Cherokee Indians at a Congress held in Augusta the 1st June 1773. . . . Containing 1616298 Acres," 1773, MPG 1/2, The National Archives of the UK, Kew, Richmond, Surrey, http://discovery.nationalarchives.gov.uk/details/r/C3980691.

151. Patrick Colhoun, [Map of lands on the Saluda River, showing two tracts of land reserved for two Cherokee 'half-breeds' and lands laid out for the King for his payment of Cherokee debts due to Edward Wilkinson], 1770, MPG 1/338, The National Archives of the UK, Kew, Richmond, Surrey, http://discovery.nationalarchives.gov.uk/details/r/C3981027.

152. [Map of the lands purchased at Fort Stanwix, and those which remain to be purchased of the Cherokees in order to secure peace], 1768, CO 700/Virginia 18, The National Archives of the UK, Kew, Richmond, Surrey, http://discovery.nationalarchives.gov.uk/details/r/C3477833.

153. John Stuart, "A Sketch of the Cherakee Boundaries with the Province of Virginia etc.," 1771, MPG 1/348, The National Archives of the UK, Kew, Richmond, Surrey, http://discovery.nationalarchives.gov.uk/details/r/C3981037.

154. John Donelson, [Plan of the Boundary Line Between the Colony of Virginia and the Cherokee Indians hunting grounds], 1771, CO 700/Virginia 19, The National Archives of the UK, Kew, Richmond, Surrey, http://discovery.nationalarchives.gov.uk/details/r/C3477834.

155. Joseph Purcell, "A Map of the Southern Indian District of North America: Compiled Under the Direction of John Stuart Esqr. His Majesty's Superintendent of Indian Affairs," [1775], CO 700 / North American Colonies General 12, The National Archives of the UK, Kew, Richmond, Surrey, http://discovery.nationalarchives.gov.uk/details/r/C3477551.
156. Joseph Purcell, "A Map of the Road from Pensacola in W. Florida to St. Augustine in East Florida. From a Survey made by order of the late Hon. Col. John Stuart, Esquire. H. M.'s Superintendent of Indian Affairs, Southern District, in 1778. By Joseph Purcell. With Itinerary along the roads from Pensacola towards St. Augustine, and General Remarks," 1778, CO 700 / Florida 54, The National Archives of the UK, Kew, Richmond, Surrey, http://discovery.nationalarchives.gov.uk/details/r/C3477662.
157. Joseph Purcell, "A New Map of the Southern District of North America from Surveys taken by the Compiler and Others, from Accounts of Travellers and from the Best Authorities etc. etc. Compiled in 1781 for Lieut. Colonel Thomas Brown, His Majesty's Superintendent of Indian Affairs etc.," 1781, CO 700 / North American Colonies General 15, The National Archives of the UK, Kew, Richmond, Surrey, http://discovery.nationalarchives.gov.uk/details/r/C3477554.
158. Harry Gordon, "River of Ohio," [1766], G3707.O5 1766.G6, Geography and Map Division, Library of Congress, Washington, DC, https://www.loc.gov/resource/g3707o.ar078200.
159. "British America, Bounded and Divided as proposed by the Author of American Independence," in John Cartwright, *American Independence: The Interest and Glory of Great-Britain* ([London], 1775). From John Carter Brown Library, Brown University, Providence, RI, JCB Map Collection, http://jcb.lunaimaging.com/luna/servlet/detail/JCBMAPS~1~1~1334~115900921.
160. David Hartley, [A map of the United States east of the Mississippi River in which the land ceded by the Treaty of Paris is divided by parallels of latitude and longitude into fourteen new states], 1784, Small Maps 1784, William L. Clements Library, University of Michigan, Ann Arbor, http://quod.lib.umich.edu/w/wcl1ic/x-813/wcl000907.
161. [Eight survey tracts along the Kanawha River, W.Va. showing land granted to George Washington and others], [1774?], G3892.K3G46 1774, Geography and Map Division, Library of Congress, Washington, DC, https://www.loc.gov/resource/g3892k.ct000363.

Chapter 5. Charting Contested Caribbean Space

162. Nathaniel Uring, "A New and Correct Draught of the Caribbee Islands," in *A History of the Voyages and Travels of Capt. Nathaniel Uring: With New Draughts of the Bay of Honduras and the Caribbee islands; and Particularly of St. Lucia, and the Harbour of Petite Carenage* (London, 1726). From Small Special Collections, University of Virginia Libraries, Charlottesville, VA.

163. [John Campbell], "Map of the Caribbee islands and the Guianas," in *Candid and Impartial Considerations on the Nature of the Sugar Trade* (London, 1763). From John Carter Brown Library, Brown University, Providence, RI, JCB Map Collection, http://jcb.lunaimaging.com/luna/servlet/detail/JCBMAPS~1~1~1245~115900908.

Inscribing a Plantation Landscape

164. John Byres, *Plan of the island of Dominica laid down by actual survey under the direction of the Honorable the Commissioners for the Sale of Lands in the Ceded Islands* (London, 1776). From Geography and Map Division, Library of Congress, Washington, DC, G5100 1776 .B91, https://www.loc.gov/resource/g5100.ar203201.
165. John Byres, *Plan of the island of St. Vincent laid down by actual survey under the direction of the Honorable the Commissioners for the Sale of Lands in the Ceded Islands* (London, 1776). From Geography and Map Division, Library of Congress, Washington, DC, G5121.G46 1776 .B9, https://www.loc.gov/resource/g5121g.ar208100.
166. John Byres, *Plan of the island of Tobago, laid down by actual survey under the direction of the Honorable the Commissioners for the Sale of Lands in the Ceded Islands* (London, 1776). From Geography and Map Division, Library of Congress, Washington, DC, G5160 1776 .B9, https://www.loc.gov/resource/g5160.ar211100.
167. "Plan of Prince Rupert's Bay," 1763, w9 Ag2, Archive of the UK Hydrographic Office, Taunton.
168. James Simpson, "Plan relating to the Town Lotts, Streets, Lanes and certain appropriations in the Town of Portsmouth in the Island of Dominica," 1765, CO 700 / Dominica 2, The National Archives of the UK, Kew, Richmond, Surrey, http://discovery.nationalarchives.gov.uk/details/r/C3476650.
169. James Simpson, "Plan relating to the appropriated and reserved lands about the Town of Portsmouth in the Island of Dominica," 1765, CO 700 / Dominica 1, The National Archives of the UK, Kew, Richmond, Surrey, http://discovery.nationalarchives.gov.uk/details/r/C3476649.
170. James Simpson, "Lands near the Town of Portsmouth in the Island of Dominica granted to poor settlers," 1765, CO 700 / Dominica 4, The National Archives of the UK, Kew, Richmond, Surrey, http://discovery.nationalarchives.gov.uk/details/r/C3476652.
171. Robert Brereton, "Plan of Prince Rupert's & Douglas bays in the Island of Dominica; and the adjacent country from Rollo's Head to Douglas Point including the town of Portsmouth as it was laid out, with the works and batteries proposed for their defence," [1785], G5102.P7 1785 Howe 53, Richard Howe Collection, Geography and Map Division, Library of Congress, Washington, DC, https://lccn.loc.gov/73691523.

172. Richard Tyrell, "A Draft of the Harbour of Fort Royal with the Bay and Town of Basse Terre in the Island of Grenada," [1764?], MFQ 1 / 1173 / 2, The National Archives of the UK, Kew, Richmond, Surrey, http://discovery.nationalarchives.gov.uk/details/r/C8955583.
173. Will Cockburn, "Grenada. Plan of the Town and Harbour of Fort Royal, with a perspective view of the Fort, Queen's Battery at the Hospital, and Monckton's Redoute," 1763, CO 700 / Grenada 1, The National Archives of the UK, Kew, Richmond, Surrey, http://discovery.nationalarchives.gov.uk/details/r/C3476923.
174. D. Imbert, "A plan of Georgetown in the Island of Grenada. Surveyed by the direction of the Commissioners for the sale and disposal of His Majesty's Lands," 1765, CO 700 / Grenada 5, The National Archives of the UK, Kew, Richmond, Surrey, http://discovery.nationalarchives.gov.uk/details/r/C3476927.
175. Isaac Werden, "A Plan of the Rosalij Compy. Estates[,] The Property of His Excelly. Charles O'Harra, the Honble. L[ie]ut. Gov. Will. Stuart, James Clarke & Rob. & Phill. Browne[,] Esqrs., Situate at Rosalÿ in the Parish of St. David[,] DOMINICA," [1776], G5104.R6A1 1776.W4, Geography and Map Division, Library of Congress, Washington, DC, https://www.loc.gov/resource/g5104r.ct000290.
176. Isaac Werden, "A Plan of That Part of the Rosalÿ Estate called THE GRAND FONDS[,] The Property of His Excellency Charles O'Harra[,] The Honble. L[ie]ut. Gov. William Stuart[,] James Clarke and Robt. & Phill. Browne[,] Esqrs., Situate at Rosalÿ in the Parish of St. David[,] DOMINICA," [1776], G5104.R6A1 1776, Geography and Map Division, Library of Congress, Washington, DC, https://www.loc.gov/resource/g5104r.ar203400.
177. Isaac Werden, "A Plan of that part of the Rosalÿ Estate call'd NEW-FOUNDLAND[,] The Property of His Excellcy. Chas. O'Harra, the Hon. Lt. Gov. Wm. Stuart[,] James Clarke & Rob. & Phill Browne[,] Esqrs., Situate at Rosalij in the Parish of St. David[,] DOMINICA," [1776], G5104.R6A1 1776 .W43, Geography and Map Division, Library of Congress, Washington, DC, https://www.loc.gov/resource/g5104r.ar203600.
178. Isaac Werden, "A Plan of that Part of the Rosalÿ Estate Called ROSALIJ VALLEY[,] The Property of His Excellcy. Charles O'Harra, the Honorable L[ie]ut. Governor William Stuart[,] James Clarke and Robert & Phillip Browne[,] Esqrs., Situate at Rosalÿ in the Parish of St. David[,] DOMINICA," [1776], G5104.R6A1 1776 .W44, Geography and Map Division, Library of Congress, Washington, DC, https://www.loc.gov/resource/g5104r.ar203700.
179. Isaac Werden, "A Plan of that Part of the ROSALIJ Estate call'd THE RETREAT[,] The Property of His Excellency Charles O'Harra[,] the Honorable Lieutenant Governor William Stuart[,] James Clarke and Robert & Phill. Browne[,] Esqrs., Situate at Rosalÿ in the Parish of St[.] David[,] DOMINICA," [1776], G5104.R6A1 1776 .W45, Geography and Map Division, Library of Congress, Washington, DC, https://www.loc.gov/resource/g5104r.ar203800.

180. Isaac Werden, [A Plan of the Rosalij Compy. Estates showing the impracticable lands], [1776], G5104.R6A1 1776 .W46, Geography and Map Division, Library of Congress, Washington, DC, https://www.loc.gov/resource/g5104r.ar203900.
181. M. Pinel, *Plan de l'Isle de la Grenade . . . fait par ordre de son Excellence Mr. George Scott, Gouverneur pour sa Majesté de la dite Isle et de ses dependances* ([London], 1763). From Special Collections Centre, Toronto Public Library, Toronto, ON, http://www.torontopubliclibrary.ca/detail.jsp?Entt=RDMDC-912-72984J24&R=DC-912-72984J24.
182. James Casey, "A plan of the Town of St. George in the Island of Grenada with the Harbour and environs," 1778, CO 700 / Grenada 5, The National Archives of the UK, Kew, Richmond, Surrey, http://discovery.nationalarchives.gov.uk/details/r/C3476928.
183. Jacques-Nicolas Bellin, "Port et fort Royal de la Grenade," in *Le Petit Atlas Maritime Recueil De Cartes et Plans Des Quatre Parties Du Monde* (Paris, 1764). From David Rumsey Historical Map Collection, http://www.davidrumsey.com/luna/servlet/detail/RUMSEY~8~1~232807~5509428.
184. Daniel Paterson, *A new plan of the island of Grenada, from the original French survey of Monsieur Pinel; taken in 1763 by order of government, and now published with the addition of English names, alterations of property, and other improvements to the present year 1780* (London, 1780). From Geography and Map Division, Library of Congress, Washington, DC, G5130 1780 .P3, https://www.loc.gov/resource/g5130.ar210100.
185. John Byres, *Plan of the island of Bequia laid down by actual survey under the direction of the Honorable the Commissioners for the Sale of Land in the Ceded Islands* (London, 1776). From John Carter Brown Library, Brown University, Providence, RI, Cabinet Es776 ByJ(1), JCB Map Collection, http://jcb.lunaimaging.com/luna/servlet/detail/JCBMAPS~1~1~3306~101585.
186. Walter Fenner, *A New and Accurate Map of the Island of Carriacou in the West Indies* (London, 1784). From The British Library, London, Maps 79457.(3), http://primocat.bl.uk/F?func=direct&local_base=ITEMV&doc_number=004816234.

Geographies of the Carib War

187. Jacques-Nicolas Bellin, "Carte de I'isle Saint Vincent," in *Le Petit Atlas Maritime Recueil De Cartes et Plans Des Quatre Parties Du Monde* (Paris, 1764). From David Rumsey Historical Map Collection, http://www.davidrumsey.com/luna/servlet/detail/RUMSEY~8~1~232801~5509425.

Colonizing Tobago

188. Thomas Spencer and Nicholas Benoist, "The Island of Tobago," [n.d.], L6544 Ba2M, Archive of the UK Hydrographic Office, Taunton.

189. Johannes Vingboons, "De bay en't fort Nieuw Vlissingen op't eylandt Tabago," 1665, 4.VELH, Nationaal Archief, Den Haag, Netherlands, http://www.gahetna.nl/collectie/afbeeldingen/kartencollectie/zocken/weergave/detail/q/id/af879ae4-d0b4-102d-bcf8-003048976d84.
190. [John Seller], *The Island of Tobago* ([London], [ca. 1682]). From John Carter Brown Library, Brown University, Providence, RI, Cabinet Blathwayt 29, Blathwayt Atlas, JCB Map Collection, http://jcb.lunaimaging.com/luna/servlet/detail/JCBMAPS~1~1~1603~102040001.
191. Robert Egerton, "To Sr. Charles Wager this chart of Tobago is dedicated," 1722, Cabinet Eu722 1 Ms, John Carter Brown Library, Brown University, Providence, RI, JCB Map Collection, http://jcb.lunaimaging.com/luna/servlet/detail/JCBMAPS~1~1~3345~101610.
192. "The Plan of the Isl.d Tobago, scituated in the Latitudeof 11d:10m North, and 59d:24m West Longitude from the Meridian of London," 1749, q56 Ag2, Archive of the UK Hydrographic Office, Taunton.
193. John Stott, "Plan of the Island of Tobago," 1764, ADM 352 / 82, The National Archives of the UK, Kew, Richmond, Surrey, http://discovery.nationalarchives.gov.uk/details/r/C11446605.
194. James Simpson, "A map of part of the Island of Tobago, with the several Lotts of Land as . . . laid out for sale or other purposes. Laid down from actual traverses and surveys in 1764 and 1765," [1765], CO 700 / Tobago 3, The National Archives of the UK, Kew, Richmond, Surrey, http://discovery.nationalarchives.gov.uk/details/r/C3478456.
195. James Simpson, "Plan of Georgetown, Barbados Bay," 1765, CO 700 / Tobago 2, The National Archives of the UK, Kew, Richmond, Surrey, http://discovery.nationalarchives.gov.uk/details/r/C3478455.
196. [Edward] Columbine, "Trinidad as far as it has been survey'd by Capt. Columbine R. N.," 1762, s84 1, Archive of the UK Hydrographic Office, Taunton.

Chapter 6. Defining East Florida

Imagining Florida

197. Abraham Ortelius, "La Florida," in *Theatrum Orbis Terrarum* (Antwerp, 1584). From Outer Banks History Center, North Carolina State Archives, North Carolina Maps, http://dc.lib.unc.edu/cdm/ref/collection/ncmaps/id/1097.
198. Ed[ward] Crisp, *A compleat description of the province of Carolina in 3 parts* ([London, 1711?]). From Geography and Map Division, Library of Congress, Washington, DC, https://www.loc.gov/resource/g3870.ct001123.
199. Guillaume Delisle, *Carte de la Louisiane et du cours du Mississip[p]i: dressée sur un grand nombre de mémoires entrautres sur ceux de Mr. le Maire* (Paris, 1718). From Geography and Map Division, Library of Congress, Washington, DC, https://www.loc.gov/resource/g3700.ct000666.

200. Antonio de Arredondo, (Copy of) "Descripcion geographica de la parte que los españoles poseen actualmente en el continente de la Florida . . . ," (1914; orig. 1742), G3860 1742 .A7 1914, Geography and Map Division, Library of Congress, Washington, DC, http://www.loc.gov/resource/g3860.np000145.
201. Jacques-Nicolas Bellin, *Carte réduite des costes de la Louisiane et de la Floride* ([Paris], 1764). From Geography and Map Division, Library of Congress, Washington, DC, https://www.loc.gov/resource/g3862c.ar160200.
202. Thomas Wright, "A Map of Georgia and Florida, Taken from the latest and most Accurate Surveys," 1763, CO 700 / Georgia 13, The National Archives of the UK, Kew, Richmond, Surrey, http://discovery.nationalarchives.gov.uk/details/r/C3477680.
203. Thomas Jefferys, *Florida from the Latest Authorities* (London, 1763). From Norman B. Leventhal Map Center at the Boston Public Library, Boston, MA, http://maps.bpl.org/id/n50826.
204. J[ohn] Gibson, *A Map of the New Governments, of East & West Florida* (London, 1763). From Norman B. Leventhal Map Center at the Boston Public Library, Boston, MA, http://maps.bpl.org/id/n50823.
205. William Stork, *A new Map of East Florida* ([London], [1767]). From The National Archives of the UK, Kew, Richmond, Surrey, CO 700 / Florida 34/1, http://discovery.nationalarchives.gov.uk/details/r/C3477642.
206. Thomas Jefferys, "East Florida from surveys made since the last peace, adapted to Dr. Stork's history of that country," in William Stork, *A description of East-Florida,* 3rd ed. (London, 1769). From Tracy W. McGregor Library of American History, Small Special Collections, University of Virginia Libraries, Charlottesville, VA.
207. Emanuel Bowen, *An accurate map of North America* (London, 1772). From Norman B. Leventhal Map Center at the Boston Public Library, Boston, MA, http://maps.bpl.org/id/m8634.
208. Thomas Jefferys, *Plan of the town and harbour of St. Augustine* (London, 1763). From The British Library, from Norman B. Leventhal Map Center at the Boston Public Library, Boston, MA, http://maps.bpl.org/id/n51643.
209. John de Solis, *A new & accurate plan, of the town of St. Augustine* ([n.p.], [1764?]). From Geography and Map Division, Library of Congress, Washington, DC, https://www.loc.gov/resource/g3934s.ar164200.
210. Philip Pittman, "Plan of the Fort St. Augustine," 1763, MPG 1 / 287, The National Archives of the UK, Kew, Richmond, Surrey, http://discovery.nationalarchives.gov.uk/details/r/C3980976.
211. James Moncrief, "Map of part of East Florida from St. John's River to Bay of Mosquitos, showing names of proprietors of estates. Drawn from the originall plan of John Gordons, Esquire, given to Governor Grant," [1764], CO 700 / Florida 7, The National Archives of the UK, Kew, Richmond, Surrey, http://discovery.nationalarchives.gov.uk/details/r/C3477616.

212. Sam[uel] Roworth, "Plan of the inlet, strait, & town of St. Augustine," [176_?], G3934.S2A1 176- .R61, Geography and Map Division, Library of Congress, Washington, DC, https://www.loc.gov/resource/g3934s.ct003704.
213. Sam[uel] Roworth, "A plan of the land between Fort Mossy and St. Augustine in the province of East Florida," [176_?], G3934.S2A1 176- .R6 Vault, Geography and Map Division, Library of Congress, Washington, DC, https://www.loc.gov /resource/g3934s.ar163500/.
214. [James Moncrief], "Plan of the Town of St. Augustine and its environs," [1765], ADM 352/160, The National Archives of the UK, Kew, Richmond, Surrey, http://discovery.nationalarchives.gov.uk/details/r/C11490146.
215. [James Moncrief], "Scetch of the city and environs of St. Augustine," [1765], CO 700/Florida 23, The National Archives of the UK, Kew, Richmond, Surrey, http://discovery.nationalarchives.gov.uk/details/r/C3477631.
216. James Moncrief, "A Plan of the Fort and Harbour of Matanzas, distant from St. Augustine Five Leagues," [1765], CO 700/Florida 10, The National Archives of the UK, Kew, Richmond, Surrey, http://discovery.nationalarchives.gov.uk /details/r/C3477619.
217. James Moncrief, "Plan of Fort Picalata on St. John's River, distant from St. Augustine Seven Leagues," [1765], The National Archives of the UK, Kew, Richmond, Surrey, http://discovery.nationalarchives.gov.uk/details/r/C3477618.
218. James Moncrief, "Plan of the Harbour of Musquitos distant from St. Augustine 72 miles," [1765], CO 700/Florida 11, The National Archives of the UK, Kew, Richmond, Surrey, http://discovery.nationalarchives.gov.uk/details/r /C3477620.
219. James Moncrief, "Sketch of part of the Coast of East Florida from St. Augustine to the Bay of Musquitos," 1765, CO 700/Florida 14, The National Archives of the UK, Kew, Richmond, Surrey, http://discovery.nationalarchives.gov.uk /details/r/C3477623.

The Search for Atlantic Inlets

220. William De Brahm, *A map of South Carolina and a part of Georgia* (London, 1757). From Geography and Map Division, Library of Congress, Washington, DC, http://hdl.loc.gov/loc.gmd/g3910.ar151700.
221. William De Brahm, [Charts of the coast of Florida], [1765], G3932.C6 svar .D4 Vault, Geography and Map Division, Library of Congress, Washington, DC, https://lccn.loc.gov/75693274.
222. William De Brahm, "Special Chart of Cape Florida," in [Charts of the coast of Florida], [1765], G3932.C6 svar .D4 Vault, Geography and Map Division, Library of Congress, Washington, DC, https://www.loc.gov/resource/g3932c.ct000730.
223. William De Brahm, "Special Chart of Muskito Inlet," in [Charts of the coast of Florida], [1765], G3932.C6 svar .D4 Vault, Geography and Map Division, Library of Congress, Washington, DC, https://lccn.loc.gov/75693274.

224. William De Brahm, [Charts of the East Florida coast], in "Report of the general survey in the southern district of North America," 1772, King's MS 211, The British Library, London, http://primocat.bl.uk/F?func=direct&local_base=ITEMV&doc_number=004987548.
225. W[illiam] G[erard] De Brahm, "New Bermuda in East Florida," 1766, MPG 1/187/2, The National Archives of the UK, Kew, Richmond, Surrey, http://discovery.nationalarchives.gov.uk/details/r/C8925498.

Planting East Florida

226. John and Samuel Lewis, "A Plan of Part of the Coast of East Florida, including St. John's River, from an actual survey by Wm. Gerard de Brahm, Esqr. Surveyor General of the Southern District of North America," 1769, K.Top. 122.81, The British Library, London, http://primocat.bl.uk/F?func=direct&local_base=ITEMV&doc_number=004987573.

Mapping East Florida's Edges

227. William Gerard De Brahm, "Map of the General Surveys of East Florida [from St. Mary's River to Row's Hammock]," in "Report of the general survey in the southern district of North America," 1772, King's MS 211.1, The British Library, London, http://primocat.bl.uk/F?func=direct&local_base=ITEMV&doc_number=004987566.
228. William Gerard De Brahm, "Map of the General Surveys of East Florida [from Row's Hammock to Cape Florida]," in "Report of the general survey in the southern district of North America," 1772, King's MS 211.6, The British Library, London, http://primocat.bl.uk/F?func=direct&local_base=ITEMV&doc_number=004987575.
229. William Gerard De Brahm, "Chart of the South-end of East Florida, and Martiers," in "Report of the general survey in the southern district of North America," 1772, King's MS 211.12, The British Library, London, http://primocat.bl.uk/F?func=direct&local_base=ITEMV&doc_number=004987567.
230. Manuel de Rueda, "Plano y, Descripcion D[e] L[os] Cayos del Norte Des de la Costa de la Habana," in *Atlas americano, desde la isla de Puerto-Rico, hasta el puerto de Vera-Cruz para los navios de el Ray, y el comercio* ([n.p.], [1766?]). From Geography and Map Division, Library of Congress, Washington, DC, http://lccn.loc.gov/74650093.
231. William De Brahm, "A Sketch of the Antient figure of the Southermost part of the Promontory formerly called Tegeste," 1771, MPG 1/341, The National Archives of the UK, Kew, Richmond, Surrey, http://discovery.nationalarchives.gov.uk/details/r/C3981036.
232. William De Brahm, "Hydrographical map of the southernmost part of the promontory of East Florida," 1771, MPG 1/347, The National Archives

of the UK, Kew, Richmond, Surrey, http://discovery.nationalarchives.gov.uk/details/r/C3981036.

233. William De Brahm, "Hydrographical map of the Atlantic Ocean, extending from the southernmost part of North America to Europe," in *The Atlantic Pilot* (London, 1772). From Norman B. Leventhal Map Center at the Boston Public Library, Boston, MA, http://maps.bpl.org/id/12058.

Chapter 7. Atlases of Empire

Completing the Map of America

234. Ezra Stiles, "The Bloody Church," 1774, Stiles MP 627, Beinecke Rare Book and Manuscript Library, Yale University, New Haven, CT.

235. William Faden, *The British Colonies in North America* ([London], 1777). From Lionel Pincus and Princess Firyal Map Division, New York Public Library Digital Collections, http://digitalcollections.nypl.org/items/510d47db-c691-a3d9-e040-e00a18064a99.

Atlases of Empire

236. Thomas Jefferys, "A Chart of the Gulf of St. Laurence," in *The North-American pilot for Newfoundland, Labradore, the Gulf and River St. Laurence* (London, 1775). From Tracy W. McGregor Library of American History, Small Special Collections, University of Virginia Libraries, Charlottesville, VA.

237. Samuel Holland, "The Provinces of New York, and New Jersey; with part of Pensilvania, and the Province of Quebec," in *The American Atlas* (London, 1776). From David Rumsey Historical Map Collection, http://www.davidrumsey.com/luna/servlet/detail/RUMSEY~8~1~1913~120019.

238. Thomas Jefferys and William Scull, "A Map Of Pennsylvania Exhibiting not only The Improved Parts of that Province, but also Its Extensive Frontiers," in *The American Atlas* (London, 1776). From David Rumsey Historical Map Collection, http://www.davidrumsey.com/luna/servlet/detail/RUMSEY~8~1~1915~120021.

239. Henry Mouzon, (Composite of) "An Accurate Map Of North And South Carolina With Their Indian Frontiers," in *The American Atlas* (London, 1776). From David Rumsey Historical Map Collection, http://www.davidrumsey.com/luna/servlet/detail/RUMSEY~8~1~1921~120027.

240. Emanuel Bowen, (Composite of) "An Accurate Map of North America. Describing and distinguishing the British and Spanish Dominions on the great Continent; According to the Definitive Treaty Concluded at Paris 10th Feby. 1763," in *The American Atlas* (London, 1776). From David Rumsey Historical Map Collection, http://www.davidrumsey.com/luna/servlet/detail/RUMSEY~8~1~1901~120008.

241. Charles Blaskowitz, "A Topographic Chart of the Bay of Narraganset in the Province of New England," in *The North American Atlas* (London, 1777). From

Geography and Map Division, Library of Congress, Washington, DC, https://www.loc.gov/resource/g3300m.gar00002/?sp=12.

242. Thomas Jefferys, "An Index Map to the following Sixteen Sheets, being A Compleat Chart of the West Indies" and "A Compleat Chart of the West Indies," in *The West-India atlas* (London, 1775). From Geography and Map Division, Library of Congress, Washington, DC, https://www.loc.gov/resource/g4900m.gar00006/?sp=25, https://www.loc.gov/resource/g4900m.gar00006/?sp=26, https://www.loc.gov/resource/g4900m.gar00006/?sp=27, https://www.loc.gov/resource/g4900m.gar00006/?sp=28, https://www.loc.gov/resource/g4900m.gar00006/?sp=29, https://www.loc.gov/resource/g4900m.gar00006/?sp=30, https://www.loc.gov/resource/g4900m.gar00006/?sp=31, https://www.loc.gov/resource/g4900m.gar00006/?sp=32, https://www.loc.gov/resource/g4900m.gar00006/?sp=33, https://www.loc.gov/resource/g4900m.gar00006/?sp=34, https://www.loc.gov/resource/g4900m.gar00006/?sp=35, https://www.loc.gov/resource/g4900m.gar00006/?sp=36, https://www.loc.gov/resource/g4900m.gar00006/?sp=37, https://www.loc.gov/resource/g4900m.gar00006/?sp=38, https://www.loc.gov/resource/g4900m.gar00006/?sp=39, https://www.loc.gov/resource/g4900m.gar00006/?sp=40, https://www.loc.gov/resource/g4900m.gar00006/?sp=41.

243. "Map of the Bearings of Grenada & Tobago to the Spanish main & Trinidada," 1777, in Lord Macartney to Lord George Germain, October 27, 1777, CO 101/21: 86, The National Archives of the UK, Kew, Richmond, Surrey, http://discovery.nationalarchives.gov.uk/details/r/C2993322.

244. Samuel Dunn, "A compleat map of the West Indies," in Robert Sayer and John Bennett, *The American military pocket atlas; being an approved collection of correct maps, both general and particular; of the British colonies* (London, 1776). From Geography and Map Division, Library of Congress, Washington, DC, https://www.loc.gov/resource/g3300m.gar00005/?sp=6.

245. B[ernard] Romans, "A general map of the Northern British Colonies in America," in Robert Sayer and John Bennett, *The American military pocket atlas; being an approved collection of correct maps, both general and particular; of the British colonies* (London, 1776). From Geography and Map Division, Library of Congress, Washington, DC, https://www.loc.gov/resource/g3300m.gar00005/?sp=7.

246. B[ernard] Romans, "A general map of the Middle British Colonies in America," in Robert Sayer and John Bennett, *The American military pocket atlas; being an approved collection of correct maps, both general and particular; of the British colonies* (London, 1776). From Geography and Map Division, Library of Congress, Washington, DC, https://www.loc.gov/resource/g3300m.gar00005/?sp=8.

247. B[ernard] Romans, "A general map of the Southern British Colonies in America," in Robert Sayer and John Bennett, *The American military pocket atlas;*

being an approved collection of correct maps, both general and particular; of the British colonies (London, 1776). From Geography and Map Division, Library of Congress, Washington, DC, https://www.loc.gov/resource/g3300m.gar00005/?sp=9.

248. Peter Bell, *A new and accurate map of North America, drawn from the famous Mr. d'Anville with improvements from the best English maps* (London, 1771). From Geography and Map Division, Library of Congress, Washington, DC, https://www.loc.gov/resource/g3300.ar012000.
249. Thomas Jefferys and [John Green], *A map of the most Inhabited part of New England; containing the provinces of Massachusets Bay and New Hampshire, with the colonies of Konektikut and Rhode Island, divided into counties and townships* ([London], 1755). From Geography and Map Division, Library of Congress, Washington, DC, https://www.loc.gov/resource/g3720.ar079700.
250. William Scull, *To the Honorable Thomas Penn and Richard Penn, Esquires, true and absolute proprietaries and Governors of the Province of Pennsylvania and the territories thereunto belonging and to the Honorable John Penn, Esquire, Lieutenant-Governor of the same, this map. Of the Province of Pennsylvania* (Philadelphia, 1770). From Geography and Map Division, Library of Congress, Washington, D.C., https://www.loc.gov/resource/g3820.ar129500.
251. J. F. W. Des Barres, [Chart of Mecklenburgh Bay], 1776, in *The Atlantic Neptune* (London, 1777–[1781]). From Norman B. Leventhal Map Center at the Boston Public Library, Boston, MA, http://maps.bpl.org/id/15132.
252. J. F. W. Des Barres, [Chart of Hell Gate, Oyster Bay and Huntington Bay], 1778, in *The Atlantic Neptune* (London, 1777–[1781]). From Norman B. Leventhal Map Center at the Boston Public Library, Boston, MA, http://maps.bpl.org/id/rb15755.
253. J. F. W. Des Barres, "Cape Blowmedown," 1777, in *The Atlantic Neptune* (London, 1777–[1781]). From Norman B. Leventhal Map Center at the Boston Public Library, Boston, MA, http://maps.bpl.org/id/15117.
254. J. F. W. Des Barres, "The Entrance to Barrington Bay N by E," 1776, in *The Atlantic Neptune* (London, 1777–1781). From Norman B. Leventhal Map Center at the Boston Public Library, Boston, MA, http://maps.bpl.org/id/n42233.
255. J. F. W. Des Barres, [Chart of Falmouth Harbor], 1776, in *The Atlantic Neptune* (London, 1777–[1781]). From Norman B. Leventhal Map Center at the Boston Public Library, Boston, MA, http://maps.bpl.org/id/12630.

Conclusion

256. John Mitchell, *A Map of the British Colonies in North America,: With the Roads, Distances, Limits, and Extent of the Settlements, Humbly Inscribed to the Right Honourable The Early of Halifax, and the other Right Honourable The Lords Commissioners for Trade & Plantations* (London, 1775). From The British

Library, Maps K.Top. 118.49.b, from Norman B. Leventhal Map Center at the Boston Public Library, Boston, MA, http://maps.bpl.org/id/n52532.

257. Abel Buell, *A New and Correct Map of the United States of North America: Layd down from the Latest Observations and Best Authorities Agreeable to the Peace of 1783* (New Haven, CT, 1784).

Acknowledgments

The idea for this book came out of my participation in discussions on the broad theme of the South when I was a faculty fellow at the Illinois Program for Research in the Humanities in 2002–2003. I approached the diffuse, interdisciplinary discussions led by Suvir Kaul and Matti Bunzl with some skepticism at first, but they soon prompted me to ask fundamental questions about space and empire, which proved highly productive. My research began in earnest in 2007 in the Geography and Map Reading Room at the Library of Congress, where John Hébert, Ed Redmond, Mike Klein, and Jim Flatness were exceptionally generous with their advice and assistance. I spent a year studying maps and delving deeply into the documentary record of colonial British America at the Library of Congress's Manuscript Division. My thanks to the staff members who alerted me to the Special Index, an early-twentieth-century card catalog that enabled me to look up the names of surveyors whose maps I had examined in the morning and read troves of transcribed correspondence relating to them in the afternoon. During a sabbatical supported by the University of Illinois at Urbana-Champaign (UIUC) and a Kislak Fellowship in American Studies, the John W. Kluge Center at the Library of Congress offered a congenial home base during my year in Washington, DC.

A J. B. Harley Research Fellowship in the History of Cartography from the J. B. Harley Research Trust, as well as funds from the dean of the College of Arts and Sciences and vice president for research and graduate studies at the University of Virginia, supported an intensive research trip to the United Kingdom in 2010. I am grateful to the remarkable staff at the National Archives at Kew and especially to Rose Mitchell for her assistance. Peter Barber at the British Library, Mary Pedley and Brian Dunnigan at the William L. Clements Library at the University of Michigan, Jenny Wraight at the Admiralty Library in Portsmouth, and Guy Hannaford at the Archive of the UK Hydrographic Office in Taunton all helped me a great deal as I examined maps in their collections. I am grateful for Anna Brown's hospitality during my two weeks at King's

Bench Road in Southwark, and to Anna and Dan MacCannell for their hilarious companionship, especially during our Oxford nostalgia tour.

One track of research for this book involved working with original maps and manuscripts; another focused on designing and developing a digital platform for displaying them online. A National Endowment for the Humanities (NEH) Digital Start-up Grant, supplemented by an Arnold O. Beckman Research Award from the UIUC Research Board, gave me the time to experiment with tools and strategies for map visualization for a project I then called the Cartography of American Colonization Database. Alan Craig and Robert McGrath at UIUC's National Center for Supercomputing Applications helped me think through the early goals for the project.

When I moved to the University of Virginia (UVA) in 2009, I continued this work as part of Research Professor Bill Ferster's "Visualization Cohort" at the Sciences, Humanities, and Arts Network of Technological Initiatives (SHANTI). The American Council of Learned Societies provided me with a year of teaching release and research support through a Digital Innovation Fellowship as Bill and I developed a prototype platform for historical map visualization built on VisualEyes, his web-based authoring tool. A three-year NEH Digital Humanities Implementation Grant enabled us to create MapScholar at SHANTI, where Raf Alvarado and others provided valuable assistance. My collaboration with Bill Ferster has been one of the most productive and enjoyable of my career—and not only because we achieved our goal of using distributed resources on the web to present and interact with old maps online. Our work designing and developing MapScholar went beyond technical means to shed light on how these digital tools might, and might not, achieve the ultimate end of humanities and social science scholarship: generating new knowledge and sharing it with others in compelling ways. Martin Öhman, Doug MacGregor, Rebecca Green, Lee Wilson, Andrew McGee, Jim Ambuske, Mary Draper, Ryan Bibler, Shane Lin, and Adele McInerney made significant contributions to this project as MapScholar research assistants. Additional support from Jama Coartney at the UVA Library's Digital Media Lab; Worthy Martin at the Institute for Advanced Technology in the Humanities (IATH); Bethany Nowviskie, Wayne Graham, and Eric Rochester at the Library's Scholars' Lab; and especially the Lab's

GIS team, Kelly Johnston and Chris Gist, was indispensable. The Library's Digitization Services produced beautiful, high-resolution scans of maps held by the Small Special Collections Library; its Interlibrary Services department helped secure images of maps in rare books held elsewhere. Madelyn Wessel, formerly UVA's associate general counsel, patiently fielded many obtuse questions regarding copyright law, image reproduction, and the internet.

MapScholar grew out of an effort to organize a large number of images and to reunite maps that had been scattered among far-flung repositories so that they could be analyzed together. I began assembling my digital map collection by taking photographs in the archives and entering metadata in a FileMaker Pro database. Creating MapScholar visualizations was more than a way to illustrate this book; working with map images on the screen and thinking about how to design a computer interface to display them was an essential part of its research and sparked what I think are its best insights. None of this work would be possible without the increasing digital preservation of important map collections and the efforts of those who curate them to make them widely available. My work draws on the remarkable online collections of the Geography and Map Division of the Library of Congress, the David Rumsey Historical Map Collection, the Map Division of the New York Public Library, the Norman B. Leventhal Map Center at the Boston Public Library, the John Carter Brown Library's JCB Map Collection, the University of South Florida Libraries' Digital Collections, the collaborative North Carolina Maps project, and those of many other libraries and archives that provide high-resolution map images without charge. By embracing the ethos and practices of open access, these far-sighted institutions make possible an unprecedented engagement with maps as sources that promise new windfalls from the study of spatial humanities. When it was not possible to simply download images of the maps I have included in this study, I worked with many helpful librarians to obtain high-resolution photographs and scans and the permission to display them. I am thankful to Paul Johnson at the UK National Archives and to image specialists and librarians at the British Library, the John Carter Brown Library, the William L. Clements Library, the Huntington Library, the Public Archives and Record Office of Prince Edward Island, the Toronto Public Library, and the Archive of the UK Hydrographic Office. Kathleen Miller, office manager for the Corcoran Department of History at UVA, completed complicated transatlantic purchases with good grace.

The research and writing of this book has occupied me for nearly a decade, and I have accumulated a large number of debts to those who have contributed their advice and suggestions and shared materials. These include C. J. Anderson, Jennifer Anderson, Andrew Beaumont, Amy Turner Bushnell, Mary Draper, Matthew Edney, Simon Finger, David Flaherty, Alec Haskell, Stephen Hornsby, Gillian Hutchinson, Michael Jarvis, Alex Johnson, Joel Kovarsky, Charles Lesser, Timo Lindman, Dan MacCannell, Elizabeth Mancke, the late Pauline Meier, James Merrell, Phil Morgan, Anthony Mullan, Michele Navakas, the late Steve Ritchie, Benjamin Sacks, Steve Sarson, Philip Schwartzberg, Kelvin Smith, Simon Smith, Owen Stanwood, Gary Thompson, Lee Wilson, Natalie Zacek, and Nuala Zahedieh. University of Virginia history research librarians Gary Treadway and Keith Weimer helped obtain important books and access to essential primary source databases.

My colleagues in the Corcoran Department of History have not only tolerated my cartographic obsessions but also lent their special expertise when I've pressed them for advice on translating Latin mottos, mathematical terminology, early modern English legal sources, battle plan iconography, and other matters. I owe special thanks to Brian Balogh, Ted Lendon, Elizabeth Meyer, Karen Parshall, Gary Gallagher, Paul Halliday, John Stagg, and Robert Stolz. Claire Weiss, as our department's technology support specialist, helped process map images.

I presented drafts of these chapters to the Rocky Mountain Seminar in Early American History, the USC Early Modern Studies Institute's American Origins Seminar at the Huntington Library, the History Seminar at the Johns Hopkins University, the Omohundro Institute for Early American History and Culture Colloquium, the Garrett Lectures in the History of Cartography at the University of Texas at Arlington, and the University of Edinburgh-University of Virginia Transatlantic Videoconference Seminar. The generous feedback I received from scholars who saw this work at its rough early stages helped me improve it. Comments and questions from David Buisseret, Jessica Cattelino, Jonathan Eacott, Mike Johnson, Eric Hinderaker, Daniel Walker Howe, Peter Mancall,

Paul Mapp, Tobie Meyer-Fong, Phil Morgan, Jenny Pulsipher, Mary Ryan, Carole Shammas, Neil York, and others at these forums helped guide my research and writing, probably more than they might have suspected. Christa Dierksheide, Peter Onuf, Andrew O'Shaughnessy, Alan Taylor, and the other scholars who make up my own such community—the University of Virginia Early American Seminar at Monticello—have provided the best possible intellectual home in Charlottesville. My participation in two Liberty Fund conferences, "Safeguarding British Liberty: The British Debate over Colonial Resistance, 1764–1776" in 2012 and "The Stamp Act Crisis and the Debate over Liberty and Imperial Authority, 1764–1766," in 2015 gave me the opportunity to read and discuss a broad selection of important political writings.

I am especially grateful to those who read drafts of the manuscript. I thank Douglas Fordham, Jeffers Lennox, Louis Nelson, Andrew O'Shaughnessy, and Daniel Schafer for their comments on particular chapters. Linda Colley, Eliga Gould, Jack Greene, Peter Onuf, and Alan Taylor read the manuscript in its entirety. Where I have found ways to respond to their critiques and suggestions, the book has been made much the better for it. Jamie Thaman and Louise Robbins reviewed the final manuscript with thoughtfulness and care. My editor at Harvard University Press, Joyce Seltzer, helped me conceptualize this book and bring it to completion after many long delays. Along the way, she has provided just the right advice to sharpen its arguments and tighten its structure.

Although I researched this book largely within the confines of map reading rooms and in front of a computer screen, it seemed important to spend at least some time on the deck of a ship, viewing the spaces I was writing about as an eighteenth-century surveyor would have. I was glad to have the opportunity to gain a ten-day berth aboard the NOAA ship *Thomas Jefferson* as its officers and crew surveyed the waters off the Florida Keys during the summer of 2010. Rear-Admiral Shepard Smith and Captain Richard Brennan helped me conduct plane-table and lead-line surveys of the Coast Guard's Trumbo Point Station at Key West and gave me a crash course in advanced hydrographic methods. Louis Maltais, manager of hydrographic operations for the Canadian Hydrographic Service, was kind enough to calculate the extent of the Gulf of St. Lawrence's coastline for me.

My family has helped me in countless ways over the many years I have spent researching and writing this book. I owe special thanks to my mother, Fran Edelson, and all the Minnesota Edelsons. My late father, Bob Edelson, would have been delighted to hold *The New Map of Empire* in his hands. I thank my East Coast family—Eric Dawson, Daniela Gilbert, Elizabeth Pugh, Bill Seedyke, Gerry Seedyke, and Jessica Seedyke—for their company and support. For making it possible for me to do this work, and for distracting me from it, I dedicate this book to Jen Edelson and our sons, Benny, Leo, and Will.

Index